D1534727

Ring-Opening Polymerization

ACS SYMPOSIUM SERIES 286

Ring-Opening Polymerization
Kinetics, Mechanisms, and Synthesis

James E. McGrath, EDITOR

Developed from a symposium sponsored by
the Division of Polymer Chemistry
at the 187th Meeting
of the American Chemical Society,
St. Louis, Missouri,
April 8–13, 1984

American Chemical Society, Washington, D.C. 1985

Library of Congress Cataloging in Publication Data
Ring-opening polymerization.
(ACS symposium series, ISSN 0097-6156; 286)

"Developed from a symposium sponsored by the
Division of Polymer Chemistry at the 187th Meeting of
the American Chemical Society, St. Louis, Missouri,
April 8-13, 1984."

Bibliography: p.
Includes indexes.

I. Polymers and polymerization—Congresses.

I. McGrath, James E. II. American Chemical
Society. Division of Polymer Chemistry. III. American
Chemical Society. Meeting (187th: 1984: St. Louis,
Mo.) IV. Series.

QD380.R56   1985   547.7   85-13352
ISBN 0-8412-0926-X

ACS Symposium Series

M. Joan Comstock, *Series Editor*

Advisory Board

FOREWORD

The ACS SYMPOSIUM SERIES was founded in 1974 to provide a
medium for publishing symposia quickly in book form. The
format of the Series parallels that of the continuing ADVANCES
IN CHEMISTRY SERIES except that, in order to save time, the
papers are not typeset but are reproduced as they are submitted
by the authors in camera-ready form. Papers are reviewed under
the supervision of the Editors with the assistance of the Series
Advisory Board and are selected to maintain the integrity of the
symposia; however, verbatim reproductions of previously published papers are not accepted. Both reviews and reports of
research are acceptable, because symposia may embrace both
types of presentation.

CONTENTS

vii

PREFACE

POLYMERIZATION REACTIONS CAN BE EASILY CLASSIFIED into either step-growth (polycondensation) or chain-reaction (addition) polymerizations. In the latter type, the reaction mechanisms that can be considered depend upon the polymerization conditions and the structure of the monomer. Although researchers have investigated many types of polymerizations, ring-opening polymerizations, important for many years, have been relatively neglected. These polymerizations can be classified as either step growth or chain reaction; the distinction is often a function of how molecular weight varies with conversion. Polymers obtained by ring-opening polymerizations have already found many important applications in industry. These applications range from water-soluble materials that are useful as oil additives or in cosmetics to reactive intermediates for segmented urethane or urea foams and high-performance elastomers. Moreover, carpet fiber textile materials and significant new biomaterials are also often based on ring-opening polymerization.

The symposium upon which this book is based provided a comprehensive discussion of ring-opening polymerizations; presentations by 12 international speakers were included in the 4-day sessions. Thus, the chapters contained within this book should represent an up-to-date view of the entire field of ring-opening polymerizations. The ideas discussed and results tabulated herein should provide the basis for many additional academic experiments as well as industrial research and product development in this important area.

JAMES E. MCGRATH
Blacksburg, Virginia
January, 1985

Ring-Opening Polymerization: Introduction

JAMES E. McGRATH

Department of Chemistry and Polymer Materials and Interfaces Laboratory,
Virginia Polytechnic Institute and State University, Blacksburg, VA 24061

The principal purpose of this paper is to provide a suitable introduction to the remaining 25 papers that comprise this book. As
mentioned earlier in the Preface, this book is a result of a
symposium presented in 1984 under the auspices of the Polymer
Division of the American Chemical Society. Most of the speakers at
the symposium have chosen to prepare manuscripts to further
illustrate the research conducted in their own laboratories. A
minority of these speakers chose to publish their work elsewhere. We
have in fact been able to publish 26 of the 32 lectures which
represents a substantial majority of those presented at the meeting.
The range of topics discussed is quite broad and therefore should be
of considerable interest to many scientists and engineers in both
academic and industrial institutions around the world.

The subjects include fundamental and applied research on the
polymerization of cyclic ethers, siloxanes, N-carboxy anhydrides,
lactones, heterocyclics, aziridines, phosphorous containing
monomers, cycloalkenes, and acetals. Block copolymers are also
discussed where one of the constituents is a ring opening monomer.
Important new discussions of catalysis via not only the traditional
anionic, cationic and coordination methods, but related UV initiated
reactions and novel free radical mechanisms for ring opening
polymerization are also included.

It will be appropriate here to cite pioneering work in
establishing a number of the fundamental thermodynamic and kinetic
features associated with polymerization in general and ring opening
reactions in particular (1-35). Fortunately, Professor Ivin along
with Professor Saegusa have recently edited an excellent three-
volume treatise on ring opening polymerization (1). In this book,
various experts in their own areas have contributed quite defini-
tive chapters on a number of the important polymerization aspects
related to ring opening reactions. The book thus provides an excellent survey of the state-of-the-art and many references up
through 1980 or perhaps later are included. Their important review
should thus be very complementary to the current book which should
be considered as more of a effort directed toward new research in
ring opening polymerization reactions. However, the availability of

0097–6156/85/0286–0001$06.50/0

the Ivin/Saegusa book will be utilized in this introductory article
to illustrate a number of key points.

Ring opening polymerizations have led to a number of commercially important materials, some of which are illustrated in
Scheme I. For example, polyoxymethylene is an important article of
commerce which can be prepared by the cationic ring opening of
trioxane 1 to yield the oxymethylene unit. In practice, these
materials are copolymers with minor amounts of ethylene oxide
included to stabilize the macromolecule end groups. The polymerization of ethylene oxide 3 to yield polyethylene oxide is an
example of the synthesis of an important water soluble macromolecule. Structure 4 currently produced at both low molecular
weights of a few hundred to a few thousand, as well as in very high
molecular weight, e.g. up into the millions (25). The closely
related polypropylene oxide is an important intermediate for polyurethane foams and the production of nearly all of the automobile
seating in the United States is based upon this technology. Cyclic
ethers, based on tetrahydrofuran 5 can also be polymerized by
oxonium ion methods to yield polytetramethylene oxide 6 (PTMO).
Ordinarily this material is generated with dihydroxy end-groups and
is an important intermediate, not only for thermoplastic segmented
polyurethanes and ureas but also for the thermoplastic polyester
ether elastomers based upon polybutylene terephthalate and PTMO.

Lactones, such as ε—caprolactone, were one of the earliest ring
opening polymerizations studied by Carothers some 50 years ago (1).
At that point, it was demonstrated that the low melting behavior (Tm
~50°C) observed with this polymer meant that 8 would not be of
interest as a fiber. Relatively little further work was conducted
until the 1970's, when ε—caprolactone monomer became an article of
commerce. Since that time, many interesting features such as the
biodegradability of 8 have been demonstrated, along with the fact
that it shows interesting compatibility behavior in blends with a
variety of other polymers. In addition, the ring opening polymerization of 7 can be initiated with diols or triols and hence important intermediates for polyurethanes are also made from this
material.

The polymerization of ε—caprolactam is one of the older and yet
extremely important examples of ring opening polymerizations. The
polyamide, commonly referred to as nylon-6, has many important
applications; perhaps the most important would be its application as
a textile fiber. It especially finds utility in textile carpeting
materials. From a structure property point of view, it is
interesting to note that the polyamide 10 has a crystalline melting
point of ~220°C relative to the low value of 50°C for its polyester
analog 8. This again was one of the major reasons for emphasizing
polyamides in early pioneering polymerization studies.

As a last example, we illustrate the ring opening polymerization of octamethyltetrasiloxane (D$_4$) to yield polydimethylsiloxane. The linear polymer 12 is the basis of both silicone oils
at low molecular weights and silicone rubber at very high molecular
weights (M$_w$ ≅ 500,000). Of course, in order to provide for
vulcanization and enhance low temperature behavior, other units such
as methyl-vinyl and possibly diphenyl are also incorporated along
the chain, as will be discussed later in this brief review and in

SCHEME I

Some Commercially Important Ring Opening Polymerizations

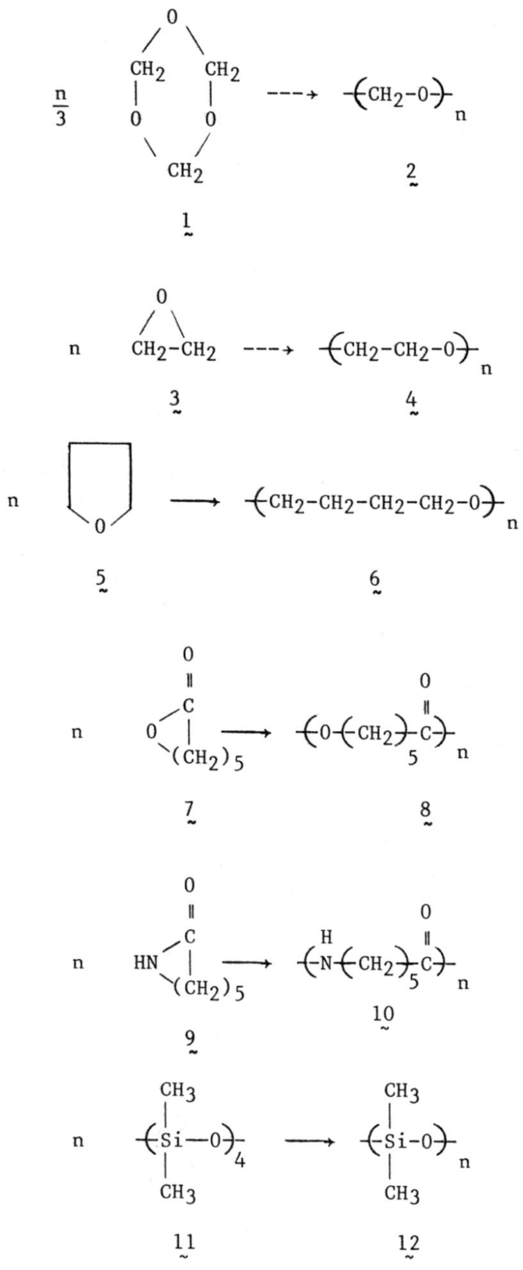

$$\frac{n}{3} \quad \underset{\underset{\displaystyle 1}{}}{\underset{\text{CH}_2}{\overset{\displaystyle \overset{O}{\diagup\quad\diagdown}}{\underset{\text{O}\diagdown\quad\diagup\text{O}}{\text{CH}_2\qquad\text{CH}_2}}}} \quad \dashrightarrow \quad \underset{2}{\left(\!\!\text{CH}_2-\text{O}\!\!\right)_n}$$

$$n \quad \underset{3}{\overset{O}{\overset{\diagup\quad\diagdown}{\text{CH}_2-\text{CH}_2}}} \quad \dashrightarrow \quad \underset{4}{\left(\!\!\text{CH}_2-\text{CH}_2-\text{O}\!\!\right)_n}$$

$$n \quad \underset{5}{\boxed{}_O} \quad \longrightarrow \quad \underset{6}{\left(\!\!\text{CH}_2-\text{CH}_2-\text{CH}_2-\text{CH}_2-\text{O}\!\!\right)_n}$$

$$n \quad \underset{7}{\overset{\displaystyle \overset{O}{\parallel}}{\underset{\diagdown(\text{CH}_2)_5}{\text{O}\diagdown^{\text{C}}}}} \quad \longrightarrow \quad \underset{8}{\left(\!\!\text{O}\!\!\left(\text{CH}_2\right)_5\!\!\overset{\overset{O}{\parallel}}{\text{C}}\!\!\right)_n}$$

$$n \quad \underset{9}{\overset{\displaystyle \overset{O}{\parallel}}{\underset{\diagdown(\text{CH}_2)_5}{\text{HN}\diagdown^{\text{C}}}}} \quad \longrightarrow \quad \underset{10}{\left(\!\!\overset{H}{\text{N}}\!\!\left(\text{CH}_2\right)_5\!\!\overset{\overset{O}{\parallel}}{\text{C}}\!\!\right)_n}$$

$$n \quad \underset{11}{\left(\!\!\underset{\text{CH}_3}{\overset{\text{CH}_3}{\text{Si}-\text{O}}}\!\!\right)_4} \quad \longrightarrow \quad \underset{12}{\left(\!\!\underset{\text{CH}_3}{\overset{\text{CH}_3}{\text{Si}-\text{O}}}\!\!\right)_n}$$

some of the research papers published in this symposium.
     The classes of structures that may be appropriate for ring
opening polymerizations can be at least simplistically generalized
as illustrated in Scheme II. One could, for example, have a cyclic

SCHEME II

Classes of Ring Opening Polymerizations

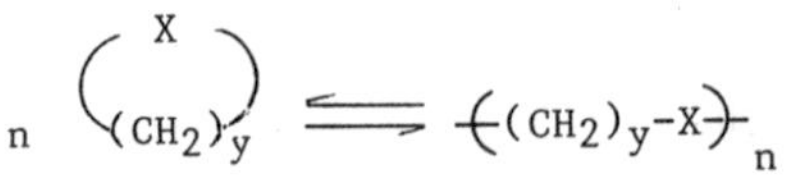

$$n \left(\begin{array}{c} X \\ (CH_2)_y \end{array} \right) \rightleftharpoons \left((CH_2)_y - X \right)_n$$

$$\underset{\sim}{13} \qquad\qquad \underset{\sim}{14}$$

$$X = O,\ S,\ NH,\ O-\overset{\overset{\textstyle O}{\|}}{C},\ -\overset{\overset{\textstyle H}{N}}{N}-\overset{\overset{\textstyle O}{\|}}{C}-,\ -CH=CH-$$

$$y \overset{>}{=} 2$$

alkylene oxide, sulfide, amine, lactone, lactam, or cycloalkene.
Other groups could of course could be cited but this list should
suffice for the present purposes. The opening of the cyclic 13 to
yield the linear chain 14 can in principle proceed with all of the
mentioned groups. However, it is usually necessary that the nature
of X be such that the heteroatom(s) provides a site for coordination
with an appropriate anionic, cationic, or coordination type
initiator. Thus one could imagine that if, for example, X was a
relatively inert methylene group, that a ring potentially able to
polymerize might not have an available mechanism to do so since no
available "site" for attack could be defined.

Thermodynamic Considerations

The general transformation of a cyclic ring compound to a linear
macromolecular chain-like molecule was illustrated already in
Scheme II. One often here may deal with a possible equilibrium,
wherein certain concentrations of the cyclic monomer remain after
the polymerization reaction has reached equilibrium. Thus,
polymerizations of this type are analogous in several ways to the
usual chemical reactions that can only proceed to high yield if the
equilibrium between reactants and product(s) favor the product(s).
Moreover, a suitable mechanism must be available which will permit
the reaction to proceed. In the case of a ring opening polymerization a large number of monomer units must be involved in a
propagation step to generate a macromolecule. This reminds us that
the Gibbs equation shown in Scheme III must yield a negative free
energy change for the propagation reaction if high molecular weight

Scheme III

$$\Delta G = \Delta H - T\Delta S$$

is to be achieved. It is appropriate to define also the criteria of a ceiling temperature, Tc, which requires one to consider the standard free energy change associated with any ring opening polymerization. This change is made up of enthalpy and entropy contributions which together with reaction temperatures define the magnitude and sign of the free energy. The chemical structure of the cyclic ring thus affects the free energy of the polymerization in a number of ways. These include the size and ring strain associated with the monomer, the presence of substituents on the cyclic ring, the geometrical or stereochemical chain isomerism and solid state morphology (e.g. crystallinity) that might be observed in the resulting macromolecule. As an illustration, in Schemes IV and V, data from the literature on the enthalpy of polymerization

SCHEME IV

Representative Enthalpy Values for Cyclic Ether
Polymerization (31)

Monomer	Ring Size	$-\Delta H$ (kJ/Mole)
Ethylene oxide (OXIRANE)	3	94.5
Trimethylene oxide (OXETANE)	4	81
Tetrahydrofuran (OXOLANE)	5	15
DIOXANE	6	~0
1,3 DIOXOCANE	8	53.5

SCHEME V

Representative Enthalpy and Entropy Values for
Lactam Polymerization (1c, 38)

Monomer	Ring Size	$-\Delta H$ (kJ/mole)	ΔS J/°Kmole
Butanolactam	5	4.6	−30.5
Pentanolactam	6	7.1	−25.1
Hexanolactam (Caprolactam)	7	13.8	4.6
Heptanolactam	8	22.6	16.7
Octanolactam	9	35.1	
Nonanolactam	10	23.4	
Decanolactam	11	11.7	
Undecanolactam	12	−2.1	
Dodecanolactam	13	2.9	

for some representative cyclic ethers and cyclic lactams are provided. One sees for ethylene oxide that the three-membered ring is quite highly strained and, accordingly, has a typical exo-thermic enthalpy which would also be characteristic of many vinyl monomers, such as ethylene or propylene. By contrast, with the

larger cyclic ethers such as THF there is a clear reduction in the
exothermic nature of the process. Moreover, one reaches essentially
zero enthalpy for the case of m-dioxane. In the case of the larger
rings, the enthalpy again appears to increase; no doubt related to
the different conformations that the larger rings can assume. In a
similar way one observes analogous data for the lactams, including
the most important case of ε-caprolactam (hexanolactam).

A summary of the most essential appropriate equations is
provided in Scheme VI.

SCHEME VI

Ceiling Temperature and Propagation Depropagation Behavior

$$\sim\sim\sim\sim\sim P_N^* \ + \ M \ \underset{k_d}{\overset{k_p}{\rightleftharpoons}} \ \sim\sim\sim\sim\sim P_{N+1}^*$$

● Equilibrium Constant is, $K = \dfrac{k_p}{k_d}$

● If $P_N \cong P_{N+1}$, $K = \dfrac{1}{[M]_e}$, where M_e is the equilibrium monomer concentration

● $\Delta G = -RT\ln K = RT\ln[M]_e$

● $\Delta G = \Delta H - T\Delta S = 0$ (at the Ceiling Temperature)

● $T_c = \dfrac{\Delta H}{\Delta S + R\ln[M]_e}$

Kinetic and Mechanistic Overview

The thermodynamic considerations discussed above are, of course,
central to the question of whether or not a particular ring opening
polymerization might proceed. However, in order to address the
question of how fast such an event might proceed and by what route,
one needs to discuss some aspects of the kinetics and mechanisms
that are possible for these ring opening polymerizations. The
reactions here are almost always ionic and thus proceed in
general via anionic, cationic or coordination catalysis. Histori-
cally, free radical processes have not been important. However,
Professor Bailey has pioneered the novel free radical routes to
ring opening polymerization as evidenced in (28). His manuscript in
this book further discusses the possibilities available with his
interesting systems.

It is also sometimes considered that ring opening reactions may
proceed via either step-growth, chain-growth or living type
polymerizations. Indeed, in some cases the differentiation is
primarily made on the basis of how molecular weight varies with
conversion (4). In this brief review we will attempt merely to
illustrate some of the breadth that is possible in ring opening

polymerizations as well as some of the accomplishments, and
challenges for the future.
   One of the oldest and yet most important ring opening
polymerizations as eluded to earlier is the hydrolytic
polymerization of ε-caprolactam (HEXANOLACTAM). The overall
processes are outlined in Scheme VII. For detailed discussions the

SCHEME VII

Hydrolytic Polymerization of ε-Caprolactam

$$n \quad \underset{\underset{9}{\overset{}{}}}{HN\overset{\overset{O}{\parallel}}{\underset{(CH_2)_5}{\overset{C}{\diagdown}}}} \quad + \quad \underset{\underset{13}{}}{H_2O} \quad \overset{250°C}{\rightleftharpoons} \quad \underset{\underset{14}{}}{HO-\!\!\left(\!\!-C-(CH_2)_5-N-\!\!\right)\!\!-H}\,\underset{n}{\overset{\overset{O}{\parallel}\qquad\quad H}{}}$$

STEP 1: $\underset{9}{9} + \underset{13}{13} \rightarrow H_2N(CH_2)_5COOH$

$$\underset{15}{15} \quad \text{and}$$

$$\overset{\oplus}{\underset{16}{H_3N-(CH_2)_5COO}}\overset{}{\overset{\ominus}{}}$$

STEP 2: $\underset{15}{15}$ or $\underset{16}{16}$ + $\underset{9}{9}$

$$\downarrow$$

$$\underset{\underset{17}{}}{HO(\overset{\overset{O}{\parallel}}{C}-(CH_2)_5-\overset{H}{N})-H}\underset{x}{} \quad + \quad \underset{\underset{18}{}}{HO-(\overset{\overset{O}{\parallel}}{C}-(CH_2)_5-\overset{H}{N})-H}\underset{y}{}$$

$$\downarrow \text{ HEAT}$$

$$\underset{\underset{19}{}}{HO-(\overset{\overset{O}{\parallel}}{C}-(CH_2)_5-\overset{H}{N})-H}\underset{x+y}{} \quad + \quad H_2O$$

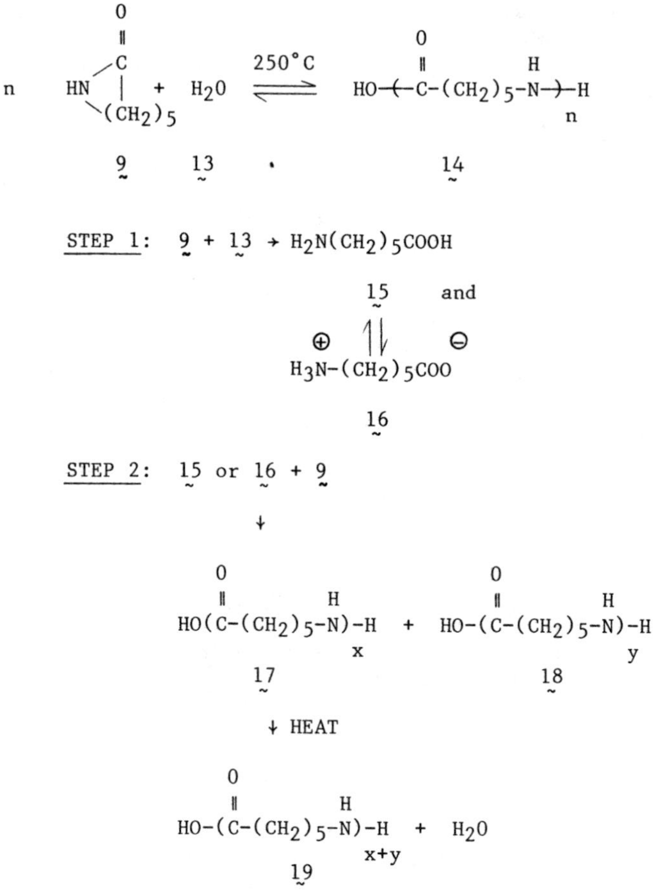

reader is referred to the reviews of Reimschuessel (35) Hedrick et
al (43) and Sekiguchi (1c). The hydrolytic polymerization is known
to involve several stages and a number of equilibria. Initially the
lactam is reacted with a small quantity of water which is capable of
opening at least a portion of the monomer to the open chain analog
15 or possible its zwitterion 16. The aminocaproic acid is capable
of further attacking the lactam to produce a series of short chain
oligomers such as 17 and 18. Eventually in the second stage of the

process the oligomers are believed to react with each other via basically a step-growth or polycondensation process to produce the high molecular weight polyamide 19. The removal of the water, in this case, plays one of the critical roles in determining the molecular weight of 19. Under the conditions utilized (approximately 250°C) there is a significant monomer polymer equilibrium which may be as high as 8-10 percent of the monomer, plus some very short chain oligomers. Much of the cyclic impurities can be removed by water, affording (after drying) a fiber grade nylon-6 by this process. Alternate methods of polymerization of caprolactam are known and discussed in some depth in the earlier cited reviews. The principal second method that has had much attention in recent years is outlined in Scheme VIII. Here we refer to the initiated anionic polymerization of ε-caprolactam. Indeed, for many years it has been known that at high temperatures strong bases such as 20 are capable of polymerizing 9 to a high molecular weight polymer very rapidly. At more moderate temperatures (150°C), the lactam anion 21 is incapable of polymerizing 9 in any reasonable period of time. This is no doubt related to the possible resonance forms that amides such as 9 could display, for example, the carbonyl group could be considered to be relatively negative due to the resonance of the -NH electrons. Thus amides, such as 9, are not readily attacked by the lactam anion such as 21. However, it was discovered many years ago (1c), that acylated lactams such as structure 22 are capable of inducing polymerization in the anionic system shown at much lower temperatures. For example, as indicated, it is possible to utilize some fraction of a mole percent of 21 and 23 in the presence of monomer 9 at temperatures at least as low as 140°C. The reaction is quite rapid due to the fact that the lactam anion 21 can readily attack the acylated lactam 23 to produce the open chain structure 24. Structure 24 is considered to be able to undergo exchange reactions with the larger amount of monomer present to produce the initiated species structure 25. In this process proton exchange takes place which regenerates the catalyst 21. Thus the the whole process can proceed very rapidly (minutes). The significance as outlined in Scheme VIII, is that the polymerization can be conducted below the melting point of the crystalline nylon-6 (approximately 220°C). As a result of the polymerization temperature and possibly also due to the fact that the polymer crystallizes, the anionically synthesized nylon-6 can have as little as one or two percent residual monomer, as compared to the eight to ten percent typically observed in the hydrolytic polymerization. Such a property is important since it allows, in some cases, the utilization of the cast nylon directly in the form of solid gears, fenders, etc. In recent years this process has been often termed reaction injection molding of nylon-6 or RIM (43).

Moving on to the area of epoxide polymerization, we would like to point out some of the basic chemistry known for the anionic polymerization of propylene and ethylene oxide. Scheme IX illustrates some basic ideas which have been established for many years due to the pioneering work of Price and others (20). Potassium hydroxide can attack propylene oxide at the primary carbon thus generating the alkoxide 28. For simplicity purposes we show the intermediate here as an alkoxide (anion and potassium cation).

SCHEME VIII

Initiated Anionic Polymerization of ε-Caprolactam

(A)

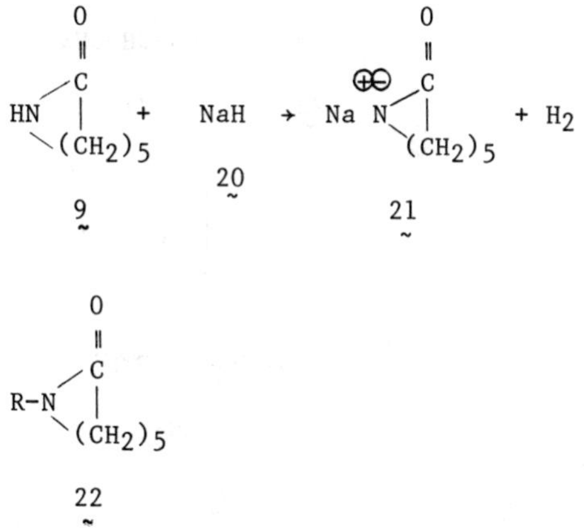

$$HN\overset{\overset{\displaystyle O}{\parallel}}{\underset{(CH_2)_5}{\diagup}}\overset{\displaystyle C}{\diagdown} \quad + \quad NaH \quad \rightarrow \quad Na\,\overset{\oplus\ominus}{N}\overset{\overset{\displaystyle O}{\parallel}}{\underset{(CH_2)_5}{\diagup}}\overset{\displaystyle C}{\diagdown} \quad + \quad H_2$$

9   20   21

$$R-N\overset{\overset{\displaystyle O}{\parallel}}{\underset{(CH_2)_5}{\diagup}}\overset{\displaystyle C}{\diagdown}$$

22

$$R = CH_3-\overset{\overset{\displaystyle O}{\parallel}}{C}-, \quad R'-N-\overset{\overset{\displaystyle O}{\parallel}}{\underset{H}{C}}-,$$ or in general, an electron withdrawing acyl lactam.

(B)   In a typical polymerization:

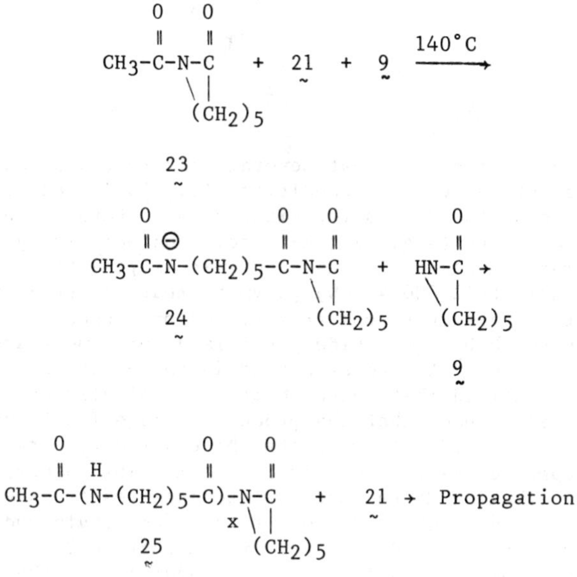

$$CH_3-\overset{\overset{\displaystyle O}{\parallel}}{C}-N-\overset{\overset{\displaystyle O}{\parallel}}{\underset{\underset{(CH_2)_5}{|}}{C}} \quad + \quad 21 \quad + \quad 9 \quad \xrightarrow{140°C}$$

23

$$CH_3-\overset{\overset{\displaystyle O}{\parallel}}{C}-\overset{\ominus}{N}-(CH_2)_5-\overset{\overset{\displaystyle O}{\parallel}}{C}-N-\overset{\overset{\displaystyle O}{\parallel}}{\underset{\underset{(CH_2)_5}{|}}{C}} \quad + \quad HN-\overset{\overset{\displaystyle O}{\parallel}}{\underset{\underset{(CH_2)_5}{|}}{C}} \rightarrow$$

24   9

$$CH_3-\overset{\overset{\displaystyle O}{\parallel}}{C}-(N-(CH_2)_5-\overset{\overset{\displaystyle O}{\parallel}}{C})_x-N-\overset{\overset{\displaystyle O}{\parallel}}{\underset{\underset{(CH_2)_5}{|}}{C}} \quad + \quad 21 \rightarrow \text{Propagation}$$

25

SCHEME IX

Anionic Polymerization of Propylene Oxide

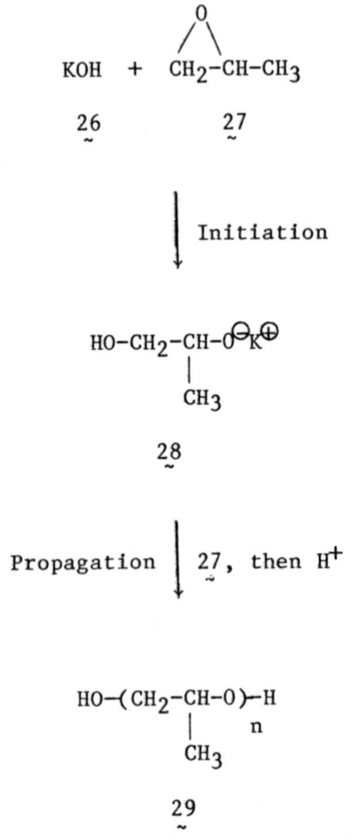

$$\text{KOH} \quad + \quad \overset{O}{\overset{\diagup\diagdown}{CH_2-CH-CH_3}}$$

$$\underset{\sim}{26} \qquad\qquad \underset{\sim}{27}$$

Initiation

$$HO-CH_2-CH-O^{\ominus}K^{\oplus}$$
$$\underset{|}{}$$
$$CH_3$$

$$\underset{\sim}{28}$$

Propagation $\quad 27$, then H^+

$$HO-(CH_2-CH-O)-H$$
$$\underset{|}{} n$$
$$CH_3$$

$$\underset{\sim}{29}$$

It is of course realized that several different ions and ion pairs may be present under these conditions (8,10). Indeed, in precise work, one needs to address the kinetic parameters in terms of their exact ionic nature (e.g., k±, k-, etc.) and not simply in terms of a "global" rate constant. In any event, it is possible to prepare low molecular weight (2,000-4,000) polymers that are predominately hydroxyl terminated with one primary and one secondary group. This general approach has been widely utilized for some years as a route for the important intermediates used in polyurethane foams. Since it is often desired that these foams be chemically crosslinked, one must consider methods that can produce average functionalities higher than 2 in order to generate the desired network behavior. One such approach is outlined in Scheme X. The intermediate oligomeric alkoxide 28 can undergo exchange reactions with poly-hydroxy compounds such as 30 to terminate one chain and initiate another from the alkoxide derived from structure 30. Assuming that most of the growth would take place by addition of the monomer 27 with the new alkoxide, one could see where an average functionality

SCHEME X

Average Molecular Weight and Functionality Control in
Poly(propylene oxide) Polyols

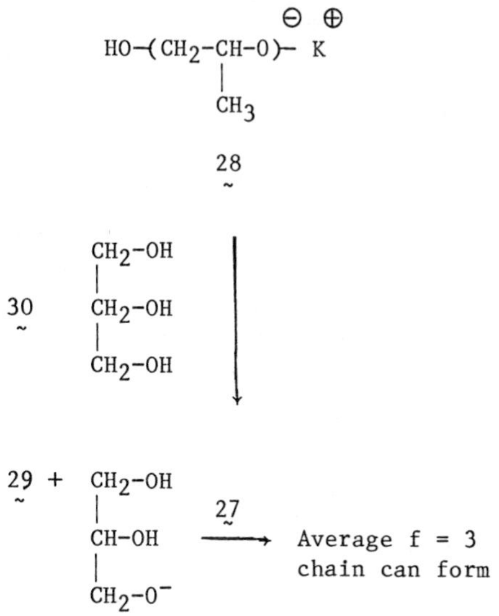

$$\text{HO} \overline{} (CH_2\text{-}CH\text{-}O) \overline{} \overset{\ominus}{} \overset{\oplus}{K}$$

$$| \atop CH_3$$

28

$$\begin{array}{l} CH_2\text{-}OH \\ | \\ 30 \quad CH_2\text{-}OH \\ | \\ CH_2\text{-}OH \end{array} \Big\downarrow$$

$$29 + \begin{array}{l} CH_2\text{-}OH \\ | \\ CH\text{-}OH \\ | \\ CH_2\text{-}O^- \end{array} \xrightarrow{27} \begin{array}{l} \text{Average f = 3} \\ \text{chain can form} \end{array}$$

● Polyhydroxy compound (e.g. 30)
  functions as a chain transfer
  agent also.

approaching 3 might be achieved. In addition, the compound 30 also
clearly would function as a chain transfer agent to regulate
molecular weight as well as functionality. Although one can obtain
reactive functionalities higher than 2 by this process, it should be
also obvious that there will be an average distribution of hydroxyl
groups including some monofunctional material. Other polyhydroxy
compounds, such as sucrose, are also utilized to produce higher
degrees of functionality. Despite the many commercial successes of
this approach it is recognized that additional fundamental characterization of molecular weights and functionalities need to be
achieved if the properties of the resulting foams are to be further
approved. One approach, often utilized, is to attempt to cap the
terminal secondary alkoxide chain ends with ethylene oxide in order
to produce a intermediate oligomer with a more reactive primary
hydroxyl terminal. Unfortunately, the initiation of ethylene oxide
by 28 is slower than the subsequent propagation. Essentially this
means that a significant amount of ethylene oxide must be incorporated if one wishes to have predominately primary alcohol
reactivity. Moreover, several side reactions are known for the
anionic polymerization of propylene oxide which are both molecular
weight limiting as well as detrimental to the development of
hydroxyl functionality. As outlined in Scheme XI, the alkoxide 28

SCHEME XI

Side Reactions in the Anionic Polymerization of Propylene Oxide

$$HO-(CH_2-CH-O)_x-\overset{\ominus}{K}\overset{\oplus}{} \quad + \quad CH_2-CH-CH_3$$
$$\underset{CH_3}{|} \qquad\qquad \diagdown\diagup$$
$$O$$

28   27

Hydrogen Abstraction

$$29 \quad + \quad \overset{\oplus}{K}\overset{\ominus}{CH_2}-CH-CH_2$$

31

$$CH_2=CH-CH_2-O\ \overset{\ominus}{}\overset{\oplus}{K}$$

32

27, then $H^{\oplus}$

$$CH_2=CH-CH_2-O-(CH_2-CH-O)_n-H$$
$$\underset{CH_3}{|}$$

33

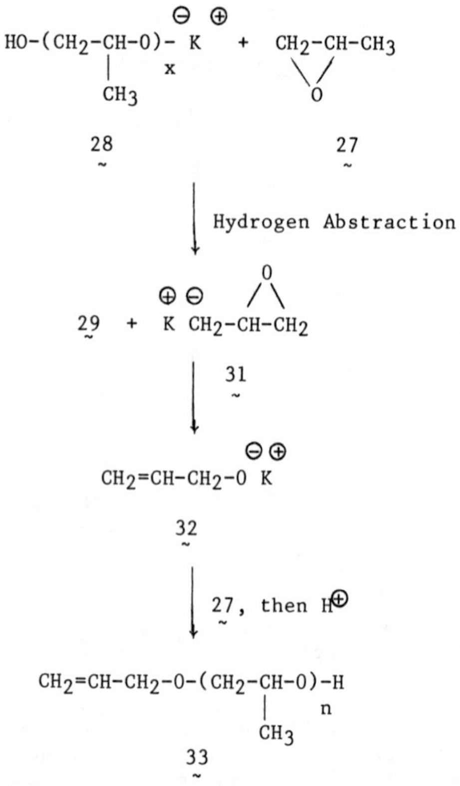

may occasionally attack 27 by abstracting a hydrogen atom and generating the strong base 31. This quickly isomerizes to the allyl alkoxide 32. This of course is capable of further reaction with monomer to produce a propylene oxide oligomer which can only bear, at best, one hydroxyl per molecule. This reaction can be a serious problem, particularly if one wishes to prepare linear propylene oxide polyols and/or to prepare high molecular weights species. By contrast, ethylene oxide of course does not have this problem and can be prepared in much higher molecular weight by simple anionic polymerization. In recent years important advances have been made in the coordination polymerization of propylene oxide as discussed in the papers and books edited by Vandenberg and Price (20,23) as well as in the review of Inoue and Aida (1). Moreover, Professor Inoue has contributed a very interesting paper to this book which further discusses his studies in coordination polymerization of epoxides and related compounds.

As discussed earlier, when one increases the ring size of the cyclic ethers the energetics become less favorable and the required

mechanisms are also changed. Whereas the three membered cyclic
ethers can be polymerized under either basic or cationic conditions,
the important five membered ring, tetrahydrofuran, is a Lewis base
and can only be polymerized by cationic or onium ion methods. Such
a possible route for polymerization is reviewed in Scheme XII. The
monomer 36 can be attacked either by a stable oxonium ion such as 35
or a proton source to generate the initiated species 37. The
oxonium ion thus produced can be interact with monomer 36 to
generate, essentially, a living polymerization. The terminal unit
of 38 should still bear an oxonium ion unit. Propagation would
involve nucleophilic attack of 36 at the carbon adjacent to the
oxonium ion. Termination with water, for example, would produce a
hydroxyl group at one end of the chain. If R, in structure 37 is a
hydrogen, the material should be difunctional. Additional methods
of producing difunctional materials have been identified and are
discussed both in this text by Franta as well as in the book by
Dreyfus (3) and (1). The kinetics and mechanisms of such processes
are extremely interesting and have been discussed by a number of
workers in Reference 1 as well as by Franta, Penczek and others in
the current book. The polymerization indicated in Scheme XII again

SCHEME XII

Cationic or Onium Ion Initiated Polymerization
of Tetrahydrofuran

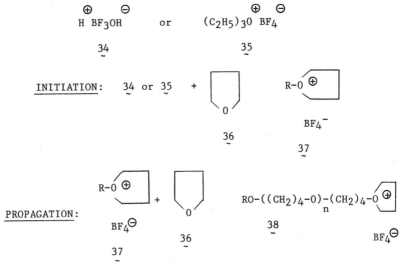

- Propagation involves nucleophilic attack of monomer
  at the carbon next to the oxonium ion.

- Termination (e.g. HOH) can produce desired functionality.

- Halide transfer can occur ($PF_6^- < BF_4^- < AlCl_4^-$).

involves ring-chain equilibria and in order to produce at least 90
percent of the chain composition the polymerization temperature is
typically about 0°C or lower. The polymerization is often analogous
to a living system, at least in the sense that the molecular weight
increases linearly with conversion and that relatively narrow
molecular weight distributions can be achieved. The nature of the
counter ion can be important and side reactions such as halogen
chain transfer have been identified. Such a process would destroy
the functionality and appears to increase as the counter ion is
varied in the order shown:

$$PF_6^- < BF_4^- < AlCl_4^-$$

Organosiloxane polymerizations have been of interest to our own
group (36,37) and others for sometime. In this book papers
featuring synthesis and kinetics as well more fundamental studies on
siloxanes are reported. Professor Yamashita and Dr. Kawakami have
prepared interesting polysiloxane macromonomers which can be used to
produce novel graft polymers. Their work in this area is discussed
in a later paper in this book. The field has also been recently
reviewed by Wright (1d). For purposes of our brief review here we
discuss (in Scheme XIII) three of the more important possible
situations. As has been recognized for some time (1d,30,34) the
cyclic trimer can be polymerized in solvents such as THF with
organolithiums. The initiation reaction could be represented by the
simple structure shown, although it is recognized to be possibly
more complicated. For example, residual carbanion might cleave
initially formed -Si-O-Si bonds before the crossover is completed.
The simple siloxanolate can further react with 40 to produce, by a
relatively straight forward propagation step, the linear polymer 41.
The progress of such a reaction is dependent upon several important
ideas. The cyclic trimer is known to be somewhat strained and to
polymerize with a significant exothermic enthalpy. If initiation is
comparable to propagation structure 41 behaves like a living
polymerization and produces fairly narrow molecular weight
distributions of predictable molecular weights. The ring strain in
the trimer is important as is the utilization of the organolithium
species. Lithium siloxanolates do not readily undergo interchange
(equilibration) processes, which are quite common for other alkali
metals with siloxane systems. The polar solvent THF is also
necessary to allow the reaction to proceed. No doubt this involves,
again, generation of various ionic intermediates and this question
is addressed in the paper of Boileau. The common route to
polysiloxanes utilizes the more readily available cyclic dimethyl
tetramer structure 42. In the presence of potassium hydroxide, the
ring is attacked to generate intermediate 43 which is certainly
capable of dimerizing to structure 44. Structures 43 or 44 can
interact with more of the cyclic tetramer 42 to produce a long chain
macromolecule 46. Indeed this route is often identified with the
preparation of silicone rubber or silicone oils. An important
variation in this type of chemistry is to utilize a disiloxane such
as structure 45 as both a chain transfer agent and a so called
end-blocker. This is discussed more fully in the paper of Sormani,
as well as in the reviews of Wright (1d) and others. Briefly, the

SCHEME XIII

Anionic Ring Opening Polymerization of Cyclic Dimethylsiloxanes

(A)   Promoted Organolithium/Trimer Reactions

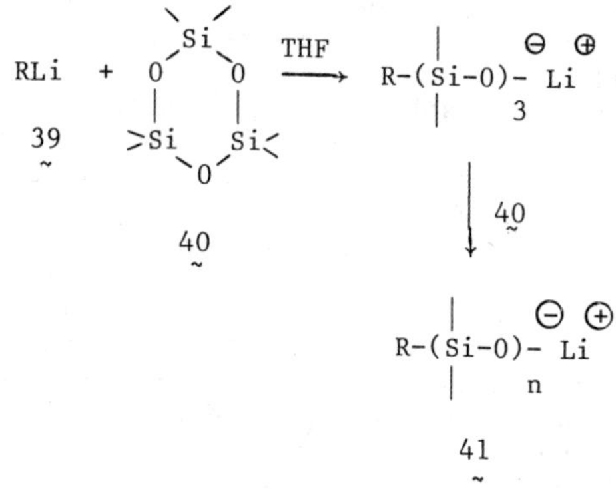

$$RLi \; + \; \underset{\underset{\underset{40}{\sim}}{\overset{39}{\sim}}}{\begin{array}{c} \diagup Si \diagdown \\ O \quad\quad O \\ >Si \quad Si< \\ \diagdown O \diagup \end{array}} \quad \xrightarrow{THF} \quad R-(Si-O)- \; \overset{\ominus}{} \; \overset{\oplus}{Li}_3$$

$$\downarrow \underset{\sim}{40}$$

$$R-(Si-O)- \; \overset{\ominus}{} \; \overset{\oplus}{Li}_n$$

$$\underset{\sim}{41}$$

(B)   Potassium Hydroxide Ring Opening and Equilibration of the
      Cyclic Tetramer

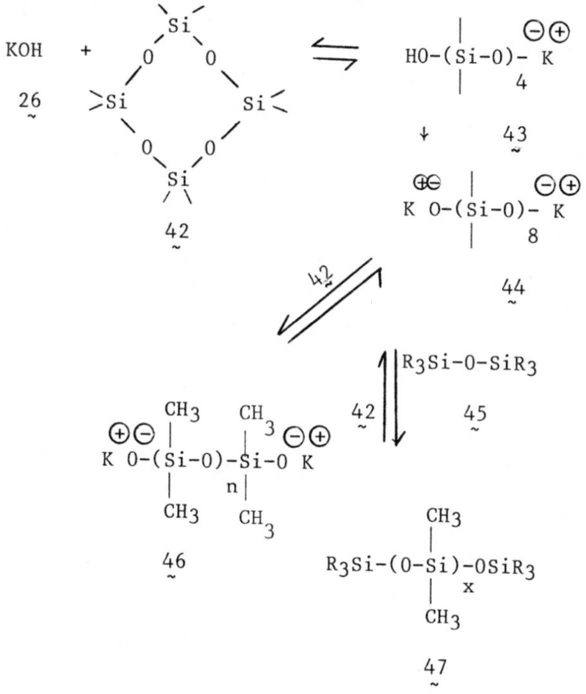

$$KOH \; + \; \underset{\underset{42}{\sim}}{\begin{array}{c} \diagup Si \diagdown \\ O \quad\quad O \\ >Si \quad\quad Si< \\ O \diagdown \quad \diagup O \\ Si \\ \diagup \diagdown \end{array}} \quad \underset{\longrightarrow}{\overset{\longleftarrow}{\rightleftharpoons}} \quad HO-(Si-O)- \; \overset{\ominus}{} \overset{\oplus}{K}_4$$

$$\downarrow \quad \underset{\sim}{43}$$

$$\overset{\oplus\ominus}{K} \; O-(Si-O)- \; \overset{\ominus}{} \overset{\oplus}{K}_8$$

$$\underset{\sim}{44}$$

$$\underset{\overset{42}{\sim}}{\diagup\diagup}$$

$$\uparrow \Big| R_3Si-O-SiR_3$$

$$\overset{\oplus\ominus}{K} \; O-(\underset{\underset{CH_3}{|}}{\overset{\overset{CH_3}{|}}{Si}}-O)_n\overset{\overset{CH_3}{|}}{-\underset{\underset{CH_3}{|}}{Si}}-O \; \overset{\ominus\oplus}{K} \qquad \underset{\sim}{42} \Big\Updownarrow \quad \underset{\sim}{45}$$

$$\underset{\sim}{46}$$

$$R_3Si-(O-\underset{\underset{CH_3}{|}}{\overset{\overset{CH_3}{|}}{Si}})_x-OSiR_3$$

$$\underset{\sim}{47}$$

growing chain can undergo various exchange reactions. These are
quite rapid in the case of the potassium cation but rather slow in
the case of lithium. One of the important features of compound 45
is that although it contains contains a silicon-oxygen-silicon bond
it also contains, usually, silicon carbon bonds which cannot undergo
the interchange reaction. As a result, one can generate oligomers
such as 47 which are effectively capped at both ends by the groups
present on the disiloxane 45. In the most common case, 45 is
hexamethyldisiloxane and therefore the end groups are all the
relatively inert and stable methyl groups. However, if one of the R
groups contains an effective functionality, for example aminopropyl,
it is possible to generate difunctional oligomers wherein the
molecular weight is controlled by the molar ratio of 42 to 45. Such
an approach has already been used effectively in our laboratory (37)
to produce valuable oligomeric intermediates for the ultimate
synthesis of block and segmented copolymers. In addition to the
anionic processes discussed in this book and briefly in Scheme XIII,
it is well known that the cyclic siloxanes can also be opened with
cationic reagents. Although there are practical materials prepared
this way the mechanism is still under intensive investigation
(1d,2a,42) and further comments would not be justified in this brief
review.

The important area of lactone polymerizations is illustrated in
Schemes XIV and XV. Perhaps the most important monomer currently
under investigation is structure 48, ε-caprolactone. This area has
been reviewed by Lundberg and Cox (15), Brode and Koleske (18), and
more recently by Lenz and Johns (1). Bases such as carbanions,
alkoxides and even hydroxyl groups can attack the monomer to
generate species such as 49, which could be considered to be the
initiation step. Attack of the alkoxide 49 on to more monomer can
quickly propagate the ring opening reaction to produce an aliphatic
polyester. Under certain conditions the monomer can yield a
predictable living chain end as well as a narrow distribution.
However, as pointed out by Yamashita (11) and Morton and Wu in this
book, side reactions such as ester interchange cyclization, etc. can
occur which will produce a series of low molecular weight oligomers
which are derived from the initial higher molecular weight narrow
distribution polymer. Interestingly, Hseih and Wang, in this book,
report that modification of the initial alkoxide with diethyl
aluminum chloride to produce carbon-oxygen-aluminum bonds can
markedly reduce the undesirable side reactions and permit one to
prepare a much better defined homopolymer and even block polymers.
The transformation to aluminum from, for example, a lithium cation,
thus is demonstrated to be very important. This clever transformation essentially alters the reaction from a simple anionic
process to what could perhaps be termed an anionic coordination
process such as those due to Teyssie (8,16), Inoue (1b) and others.
Thus the caprolactone represents an interesting system that can be
polymerized not only with the anionic and coordination catalysts but
also with cationic systems as well. As is often the case, the
cationic reactions appear to be more complicated and one is referred
to the literature references (15) for additional insight. One
should point out, however, that it is possible to prepare important
polyurethane intermediates from caprolactone by initiating the

SCHEME XIV

Ring Opening Polymerization of ε-Caprolactone

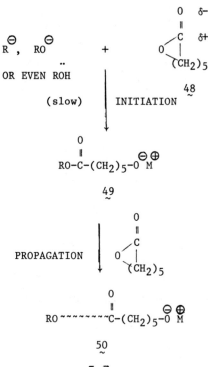

- ⟨Mn⟩ MAY BE GIVEN BY $\left[\dfrac{M}{I}\right]$

- ⟨Mw⟩ / ⟨Mn⟩ CAN BE NARROW

- SIDE REACTIONS SUCH AS ESTER INTERCHANGE CYCLIZATION VIA BACK-BITING CAN OCCUR

- ION-PAIR STRUCTURE, REACTION CONDITIONS, ETC. ARE IMPORTANT

- CATIONIC AND COORDINATION POLYMERIZATION ALSO KNOWN

- POLYOLS CAN BE GENERATED WITH e.g. DIOL OR TRIOL INITIATORS

SCHEME XV

Ring Opening Polymerization of β-Propiolactone and Derivatives

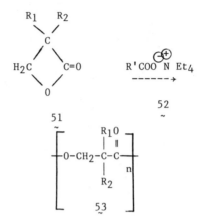

- CRYSTALLINE, HIGH T_m

- $R_1 = R_2 = CH_3$, $T_m \simeq 250°C$

- IMPROVED HYDROLYTIC STABILITY

- β,β-DIMETHYL, $T_m \sim 80°C$

- R_3N: ALSO AN EFFECTIVE CATALYST

- PROPAGATION VIA CARBOXYLATE ANION

polymerization with difunctional diols such as, for example, diethylene glycol or trifunctional hydroxyl bearing initiators. Indeed, these polyester polyols complement the polyether systems as intermediates for polyurethanes and related copolymers. Scheme XV draws our attention to the ring opening polymerization of β-propiolactone and its derivatives. Unlike the caprolactone systems, the β-lactones often undergo alkyl-oxygen scission to produce a growing carboxylate end group. As such they can be initiated with weaker bases, such as carboxylates, as opposed to the alkoxides previously discussed. The α,α'-dimethyl substituted system has been of considerable interest as a possible material for high melting fibers. Thus, one can produce polymers of pivalolactone which have melting points perhaps as high as 250°C. They are also believed to have improved hydrolytic stability due to steric factors. The utilization of such materials has been investigated in some detail and in this book the paper of Sharkey discusses additional aspects that have been not previously published on these interesting macromolecules. At this time the pivalolactone polymers are not commercially available, but certainly are of great interest from a structure property point of view.

The last system to be discussed in this introductory article is the interesting ring opening polymerization of 2-oxazolines as outlined briefly in Scheme XVI. This area has been pioneered by a number of people, including Professor Litt who contributes an additional article in this book. The polymerizations were also

SCHEME XVI

Ring Opening Polymerization of 2-Oxazolines

INITIATION:

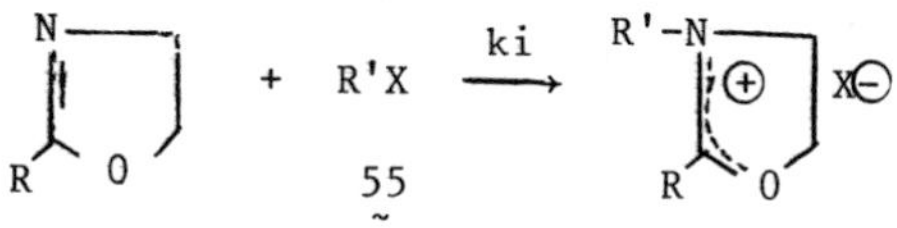

54   56

PROPAGATION:

56   54

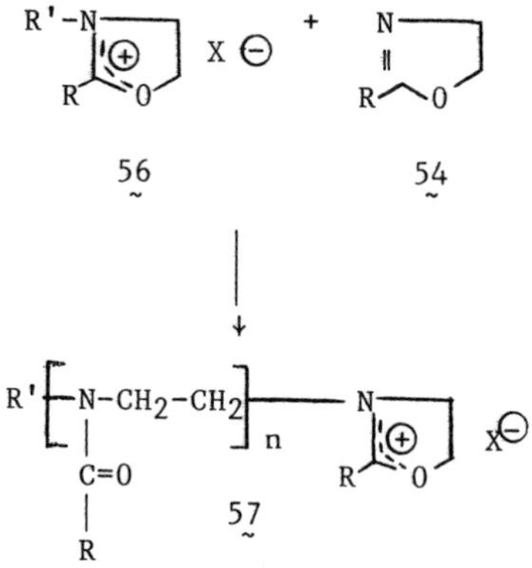

57

discussed recently by Kobayashi and Saegusa in (1). (Dr. Kobayashi has also contributed a paper on his interesting phosphorous containing monomers in this book.) The 2-ethyloxazoline, structure 54, where R as an ethyl group, has recently been commercialized and thus the polymerization will perhaps be studied in additional detail in the future. In general it is believed the polymerization involves the interaction of the monomer with a weak nucleophile, typified by R'X, to generate cationic species 56 in an initiation step. Propagation is then believed to involve nucleophilic attack by the nitrogen of monomer 54 on the carbon next to the oxygen, thus opening it up to produce the polymer 57. The polymers, such as 57, can also be hydrolyzed to polyalkylene amines which are very interesting materials for antistatic and other purposes. No doubt additional work on oxazoline polymerizations will be reported in the future.

Concluding Remarks

It is hoped that the brief discussions provided in this introductory
review article will encourage the reader to investigate the
remaining 25 research papers published in this book in great detail.
It is further hoped that the interesting research produced here will
stimulate both academic and industrial interest in kinetics,
mechanisms and synthesis problems and that new polymeric materials
based upon ring opening polymerizations will be discovered and
developed.

Acknowledgements

Support of the author's research in recent years by the ARO, ONR,
AFOSR, NASA, TACOM, the Petroleum Research Fund (PRF) and the Exxon
Foundation is gratefully acknowledged.

References

1. Ivin, K. J., and Saegusa, T., Editors, Ring Opening
    Polymerization (in three volumes), Elsevier, 1984.
1a. Ivin, K. J., and Saegusa, T., Chapter 1 in Reference 1.
1b. Inoue, S., and Aida, T., Chapter 4 in Reference 1.
1c. Sekiguchi, H, Chapter 12 in Reference 1.
1d. Wright, P. V., Chapter 14, in Reference 1.
1e. Kobayashi, S., and Saegusa, T., Chapter 11 in Reference 1.
2. Goethals, E. J., Editor, "Cationic Polymerization and Related
    Processes", Academic Press, 1984.
2a. Sauvet, G., Lebrun, J. J., and Sigwalt, P., pages 237-252 in
    Reference 2.
3. Dreyfus, P., "Polytetrahydrofuran", Gordon and Breach, 1982.
4. Odian, G., "Principles of Polymerization", 2nd Edition, Wiley,
    1981, Chapter 7.
5. Saegusa, T., Goethals, E., Editors, "Ring Opening
    Polymerization", ACS Symp. Series 59, 1977.
6. Furukawa, J., Vogl, O., Editors, "Ionic Polymerization:
    Unsolved Problems", Dekker, 1976.
7. Frisch, K. C., Reegen, S., Editors, "Ring Opening
    Polymerization", Dekker, 1969.
8. McGrath, J. E., Editor, "Anionic Polymerization: Kinetics,
    Mechanisms and Synthesis", ACS Symp. Series 166, 1981.
9. Ledwith, A., and Sherrington, D. C., in "Reactivity Structure
    and Mechanism in Polymer Chemistry", Wiley, 1974.
10. Penczek, S., in "Anionic Polymerization", J. E. McGrath,
    Editor, ACS Symp. 166, 1981, pp. 271-83.
11. Yamashita, Y., in "Anionic Polymerization", J. E. McGrath,
    Editor, ACS Symp. 166, 1981, pp. 199-211.
12. Young, R. H., Matzner, M., and Pilato, L. A., in "Ring Opening
    Polymerization", E. Goethals and T. Saegusa, Editors, ACS Symp.
    Series 59, 1977, pp. 152-165.
13. Lenz, R. W., et al., in "Ring Opening Polymerization",
    E. Goethals and T. Saegusa, Editors, ACS Symp. Series 59, 1977,
    pp. 210-216.
14. Hall, H. K., Macromolecules, 1969, 2, 488.

15. Lundberg, R. D., and Cox, E. F., in "Ring Opening Polymerization", K. C. Frisch and S. L. Reegen, Editors, Dekker, 1969, p. 266.
16. Teyssie, P., et al., J. Poly. Sci., 1977, 15, 685.
17. Brash, J. L., and Lyman, D. J., in "Cyclic Monomers", K. C. Frisch, Editor, Vol. 26, Wiley, 1972, pp. 147-178.
18. Brode, G. L., Koleske, J. V., J. Macro. Sci. Chem. A6, 1972, pp. 1109-1144.
19. Frisch, K. C., editor, "Cyclic Monomers", Vol. XXVI, High Polymer Series, Wiley-Interscience, 1972.
20. Price, C. C., and Vandenberg, E., editors, "Coordination Polymerization", Vol. 19, Polymer Science and Technology, Plenum, 1984.
21. Herold, R. J., and Livigni, R. A., N. Platzer, Ed., Adv. in Chemistry Series 128, 1973, p. 208.
22. Pruitt, M. E., and Baggett, J. M. (to Dow Chemical), U.S. Patent 2,706,181, April 12, 1955.
23. Vandenberg, E. J., editor, Polyethers: ACS Symposium Series 6 1975.
24. Livigni, R. A., Herold, R. J., Elmer, O. C., and Aggarwal, S. L., in Polyethers, E. J. Vandenberg, Editor, ACS Symp. Series 6, 1975, p. 20.
25. Bailey, F. E., and Koleske, J. V., "Polyethylene Oxide", Academic Press, 1976.
26. Hill, F. N., Bailey, F. E., and Fitzpatrick, J. T., Ind. Eng. Chem., (1958), 50, 5.
27. Allcock, H. R., J. Macromol. Sci., Revs., Macromol. Chem., 1970, C4, 149.
28. Endo, T., and Bailey, W. J., J. Polymer Sci., Polymer Letters, 1980, 18, 25; See also Bailey, W. J., this book.
29. Penczek, S., Kubisa, P., and K. Matyjazewski, "Cationic Ring Opening Polymerization", Adv. Polym. Sci., 1980, 37.
30. Bostick, E. E., Chapter 8, in "Ring Opening Polymerization", Frisch, K. C., and Reegen, S. C., Editors, Dekker, (1969).
31. Ivin, K. J., p. II-421-450 in "Polymer Handbook", Brandup, J., and Immergut, E. H., Editors, Wiley, (1975).
32. Ivin, K. J., Encyclopedia of Polymer Science and Technology, Bikales, N. M., Editor, Supplement Vol. 2, 1977, p. 700-45.
33. Ivin, K. J., "Olefin Metathesis", Academic, London, (1983).
33a. Patton, P. A., and McCarthy, T. J., Macromolecules, (1984), 17, 2940-2942.
34. McGrath, J. E., et al., "Ring Opening Polymerization of Siloxanes", in "Initiation of Polymerization", Bailey, F. E., Editor, ACS Symposium Series, (1983), No. 212.
35. Reimschuessel, H. K., J. Poly. Sci., Macromolecular Reviews, (1977), 12, 65.
36. Yilgor, I., Riffle, J. S., and McGrath, J. E., in "Reactive Oligomers", Harris, F. W., Editor, ACS Symp. Series, 1985.
37. Yilgor, I., Tyagi, D., Sha'ban, A., Steckle, W. S., Jr., Wilkes, G. L., and McGrath, J. E., Polymer (London), (1984), 25, 1800-1806.
38. Korshak, V. V., et al., Russian Chem. Rev., (1976), 45, 853.

39. Riffle, J. S., Yilgor, I., Banthia, A. K., Tran, C.,
     Wilkes, G. L., and McGrath, J. E., in "Epoxy Resins II", R. S.
     Bauer, Editor, ACS Symposium Series No. 221, 1983, Chapter 2.
40. Johnson, B. C., Ph.D Thesis, Virginia Polytechnic Institute
     and State University, June 1984.
41. Tran, C., Ph.D Thesis, Virginia Polytechnic Institute and State
     University, November 1984.
42. Yilgor, I., and McGrath, J. E., Polymer Preprints 26(1), 57,
     1985.
43. Hedrick. R. M., et al, ACS Symposium Series 270, p. 142,
     (1985).

RECEIVED June 21, 1985

Anionic Polymerization of Cyclosiloxanes with Cryptates as Counterions

SYLVIE BOILEAU

Laboratoire de Chimie Macromoléculaire associé au CNRS, Collège de France, 11 place Marcelin Berthelot, 75231 Paris Cédex 05, France

The anionic polymerization of cyclosiloxanes was examined in benzene and toluene with lithium cryptates as counterions. Only one type of active species is observed in the case of Li^+ + [211]; thus, the kinetics of the propagation and of the by-product cyclosiloxanes formation can be studied in detail for the first time. The reactivity of cryptated silanolate ion pairs toward the ring opening of D_3 is greatly enhanced compared to that of other systems. The amount of cyclic by-products is very low when the polymer yield reaches the maximum value. PDMS of narrow molecular weight distributions were obtained using a cryptand larger than the [211].

The anionic polymerization of cyclosiloxanes has been studied for a long time. However the knowledge of the mechanism of the process is still unfortunately rather limited (1,2). This is partially due to the fact that the polymerization is of the equilibrium type and besides linear polymers, cyclic structures exist in solution. It has been shown that under certain conditions, hexamethylcyclotri-siloxane (D_3) polymerizes to give linear polymers of negligible cyclosiloxane content and of narrow molecular weight distribution at a much more rapid rate than octamethylcyclotetra siloxane (D_4) (3). This can be explained by the ring strain of D_3 which enhances the reactivity of the Si-O bonds. A suitable initiator such as butyllithium can be mixed with D_3 in hydrocarbon solvents to form $Bu(-Si---(CH_3)_2-O)_3Li$ (4). No polymerization occurs even in the presence of excess D_3 until a donor solvent like THF (3,5), diglyme (6), triglyme (5), DME (7), HMPA (7) or DMSO (3) is added; then a reasonably rapid polymerization starts to give near monodisperse polymers.

Only few kinetic investigations have been performed on these systems though the side reactions which involve intra and inter-molecular attacks of the chains by living centers can be neglected in the first stage of polymerization (5,7). The nature of the active propagating species as well as their relative proportion is not well known in the investigated systems (5,8,9). For instance, the dependence of the rate of polymerization of D_3 on the concentration of active centers is complex in the case of a benzene/ THF (50/50) medium, with Li^+ as counterion (5). This behavior is obviously connected with the presence of associations of the lithium silanolates. Thus we thought that use of cryptands for the anionic polymerization of cyclosiloxanes might suppress the aggregates and simplify the mechanism (10,11,12).

These macrobicyclic ligands (I) discovered by Lehn (13) form extremely stable cation inclusion complexes, called cryptates. In the nomenclature system used herein, Structure I would be designated [211] when m=2 and n=p=1.

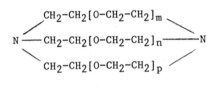

I

These complexes lead to a considerable increase of the interionic distance in the ion pairs and it has been shown that such ligands have a marked activating effect on anionic polymerizations (14,15, 16). Moreover, the aggregates are destroyed and simple kinetic results have been obtained in the case of propylene sulfide and ethylene oxide polymerizations (12).

Thus, in the present paper we present our data concerning the anionic polymerization of cyclosiloxanes, namely D_3 and D_4, with cryptates as counterions, in benzene or toluene solution.

Anionic Polymerization of D_3

It is known that catalysts involving bases generally do not polymerize D_3 at moderate temperatures, except in the presence of donor solvents (3). Preliminary experiments have shown however that polydimethylsiloxane (PDMS) of very high molecular weight was obtained in 60% yield after five minutes at room temperature using KOH complexed with [222] in benzene (14,16). However our kinetic investigations were performed using homogeneous conditions with lithium cryptates as counterions.

Living PDMS can be prepared by adding the [211] cryptand to benzene solutions of D_3 after reaction with n-butyllithium, at 25°C, as can be seen from the results of Table I. The propagation reaction is much faster with [211] than with THF.

Kinetics of the propagation reactions were followed by dilato-metry under high vacuum, at 20°C, at various living ends concentrations, with nearly the same value of $[M]_o$ (0.5 mole. 1^{-1}) (10,11,17). Some experiments were performed in toluene instead of benzene and in some cases, tertiary butyllithium was used as initiator instead of n-butyllithium. The results were nearly the same within the experimental errors. The values of the apparent propagation rate constant $k_p = R_p/[M] \times [C]$ were determined for each

Table I.   Polymerization of D_3 Initiated by n-BuLi, in Benzene at Room Temperature (Initiation time: 16 h. at 25°C. Termination reagent: trimethylchlorosilane)

[D_3] (mole.l^{-1})	[C] x 10^4 (mole.l^{-1})	[211]/[Li$^+$]	Propagation Time (hr.)	Yield (%)	M_n theor. x 10^{-5}	M_n osm. x 10^{-5}	M_n GPC x 10^{-5}
0.83	13.1	2.7	1	~100	~1.40	1.10	-
0.80	10.8	1.9	2	91	1.50	-	1.10
0.35	8.9	1.5	0.4	91	0.80	-	1.20
0.59[a]	8.9	0	4	59	0.86	-	0.91
0.39[a]	6.75	0	7	98	1.26	1.14	-

a) Benzene/THF ≈ 50/50 (v/v).

experiment (R_p = rate of polymerization, [M] and [C]: monomer and living ends concentrations). On plotting R_p/[M] versus [C] (Figure 1), a straight line passing throught the origin is obtained, showing that the reaction order in active centers is equal to 1 for a concentration range between 3.7×10^{-3} and 7.5×10^{-5} mole.1^{-1}. This result can be interpreted either by assuming that only one type of active species is present - presumably cryptated ion pairs - or by assuming that if there are different types of active centers, they might have the same reactivity.

In order to elucidate this point, viscosity measurements of living and deactivated PDMA solutions were performed in toluene, with Li^+ + [211] as counterion. As no significant change was observed, it can be deduced that the fraction of aggregates is negligible (<1% for [C] = 10^{-3} mole.1^{-1}). Moreover, conductance measurements made on model silanolates in THF indicate that the fraction of free ions is very low. In benzene or toluene which are less polar than THF, the contribution to the reactivity from free ions can be neglected. Thus it seems reasonable to assume that the main ionic species are cryptated ion pairs and

$$k_p = \frac{R_p}{[M][C]} = k\pm = 1.3 \ 1.mole^{-1}.sec^{-1}$$

in benzene or toluene at 20°C.

The amount of cryptated ion pairs corresponds to the amount of [211]. One experiment performed with a ratio [211]/[Li^+] lower than 1 led to the same value of k_p as those found in other experiments made with an excess of [211] providing that the concentration of [211] was used instead of that of n-BuLi.

Several kinetic experiments were made in toluene at 0° and -20°C. Thus it was possible to determine the value of the activation energy which is equal to 10 Kcal.$mole^{-1}$ in a temperature range of 40°C (17).

On the other hand, kinetic measurements were performed in toluene at 20°C on adding an excess of [221] (18). The rates of propagation are much lower than in the case of Li^+ + [211]. Moreover, the apparent rate constant of propagation k_p = R_p/[M][C] increases on decreasing the concentration of living centers. The reaction order in active centers is nearly equal to 0.25 for a concentration range between 8×10^{-3} and 9×10^{-4} mole.1^{-1}. This could be explained by the presence of associations of the lithium silanolates which are not destroyed by the ligand. Viscosity measurements of living and deactivated PDMS solutions performed in toluene with Li^+ + [221] show a tremendous change. However, it is not possible to make a quantitative study of the phenomena.

It is well known that cyclic oligomers (D_x) are formed as by-products through intramolecular attacks of the chains by active centers, during the propagation. The amount of consumed D_3 and formed D_4, D_5 and D_6 were determined by GLC analysis (11) of the polymerization mixture during the propagation. The results are shown in Figure 2 for [C] = 3.08×10^{-4} mole.1^{-1}. The maximum yield of polymer (~90%) is rapidly obtained whereas the proportion of cyclosiloxanes is very low (weight percentage of D_4 + D_5 + D_6 < 3%)

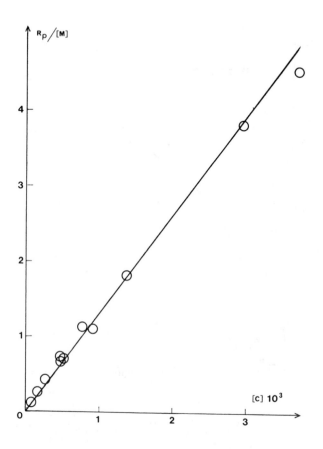

Figure 1.   Plot of Rp/[M] vs. [C] for the anionic polymerization of D3 with Li+ + [211] as counterion, in benzene and in toluene at 20°C.

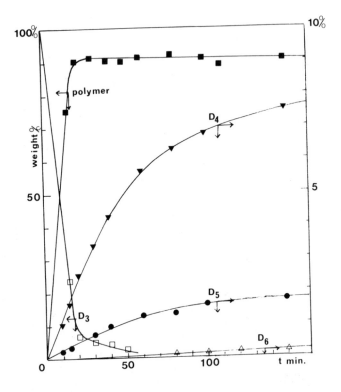

Figure 2. Conversion of D_3 and formation of D_4, D_5, and D_6 as a function of time: $[C] = 3.08 \times 10^{-4}$ mole.1^{-1}; $[M]_0 = 0.49$ mole.1^{-1}.

in the case of $Li^+ + [211]$ in benzene at 20°C. The amount of cyclic by-products is much lower (weight percentage of $D_4 + D_5 + D_6 < 0.5\%$) in the case of $Li^+ + [211]$ for a yield in polymer equal to 82% ($[C] = 9.7 \times 10^{-4}$ mole.l^{-1}).

Anionic Polymerization of D_4

It has been shown that addition of cryptands markedly increases the rate of the bulk polymerization of D_4 initiated by KOH at 160°C [19]. The anionic polymerization of D_4, in toluene, at room temperature, with $Li^+ + [211]$ as counterion appears to follow a different course than in the case of D_3. The rate of polymerization is much lower and cyclic oligomers formation is important as can be seen from the results of Figure 3 for $[C] = 2.76 \times 10^{-3}$ mole.l^{-1}.
Polymerization of D_4 is represented as follows:

$$\sim\sim Me_2SiO^- \ Li^+ \ + \ D_4 \ \xrightarrow{\ K_{p4}\ } \ \sim\sim Me_2SiO(Me_2SiO)_3Me_2SiO- \ Li^+$$

since it has been shown that the main ionic species are cryptated ion pairs in the case of the anionic polymerization of D_3 with $Li^+ + [211]$ as counterion, in benzene or toluene at 20°C; thus $k_p4 = k^+4$.
By-product cyclosiloxanes like D_5, D_6 and D_7 are formed during the polymerization. Equilibrium between linear and cyclic siloxanes are established as shown below [20]:

$$\sim\sim Me_2SiO(Me_2SiO)_{x-1} Me_2SiO^- \ Li^+ \ \underset{k_{px}}{\overset{k_{dx}}{\rightleftarrows}} \ \sim\sim Me_2SiO^- \ Li+ \ + \ D_x$$

with $K_x = k_{dx}/k_{px} = [D_x]$ eq. $(x > 3)$ $\qquad$ (1)

where k_{dx} and k_{px} are the rate constants of formation and of propagation of D_x, respectively, K_x is the molar cyclization equilibrium constant and $[D_x]$ eq. is the concentration of D_x in equilibrium.
The rates of consumption of D_4 and formation of cyclic oligomers D_x are given by kinetic equations (2) and (3):

$$-\frac{d[D_4]}{dt} = (k_p4[D_4] - k_d4) \ [C] \qquad (2)$$

$$\frac{d[D_x]}{dt} = (k_{dx} - k_{px}[D_x]) \ [C] \qquad (3)$$

This relation has been proposed by Shinohara [21] and applied by Yamashita et al. [22,23] to the anionic polymerization of ε-caprolactone. If the rate of formation of D_4 is assumed to be negligible compared to its rate of polymerization at the beginning of the reaction, Equation (2) becomes:

$$-\frac{d[D_4]}{dt} \simeq k_p4[D_4][C] \qquad (4)$$

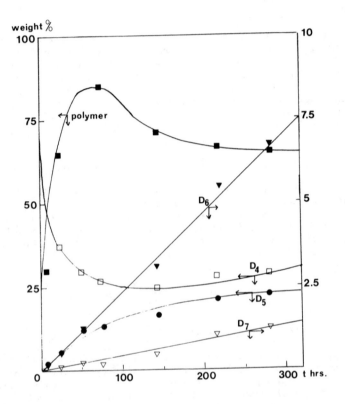

Figure 3.   Conversion of D_4 and formation of D_5, D_6, and D_7 as a function of time: $[C] = 2.76 \times 10^{-3}$ mole.1^{-1}; $[M]_0 = 1.06$ mole.1^{-1}.

Thus knowing k_{p4}, it is possible to evaluate k_{d4} from Equation (1) by determining [D_4] equation for each experiment.

Concerning the formation of by-product cyclosiloxanes, two types of approximations can be made. The first one consists in assuming that the rate of polymerization of D_x is negligible compared to its rate of formation at the beginning of the reaction. Thus Equation (3) becomes:

$$\frac{d[D_x]}{dt} \simeq k_{dx} [C] \qquad (5)$$

and k_{px} can be evaluated from Equation (1) by determining [D_x]eq. On the other hand, integration of Equation (3) gives:

$$[D_x] = \frac{k_{dx}}{k_{px}} (1 - e^{-kpx[C]t}) \qquad (6)$$

if [D_x]$_0$ = 0

If the product k_{px}[C]t is smaller than 1 (<0.7), the first terms of the series expansion of $1/e^{k_{px}[C]t}$ can be taken and Equation (6) becomes:

$$\frac{1}{[D_x]} \simeq \frac{k_{px}}{k_{dx}} + \frac{1}{k_{dx}[C]t} \qquad (7)$$

It is thus possible to determine k_{px} and k_{dx} by plotting $1/[D_x]$ versus $1/t$.

Both methods have been applied in the case of the anionic polymerization of D_4, in toluene, at room temperature, with Li$^+$ + [211] as counterion, which give nearly the same results. Moreover, k_{px} and k_{dx} do not vary significantly with [C] as expected since only one type of ionic species is present in the medium. The results are shown in Table II.

A comparison of reactivity of different cyclosiloxanes towards silanolates is shown in Table III. The results found in the base-catalysed solvolysis of siloxanes in alcohols (22) as well as those observed for the ring opening of cyclosiloxanes by potassium phenyldimethylsilanolate in heptane-dioxane (95/5, v/v) at 30°C (9) are also reported. Our results are in good agreement with those found by Lasocki et al. (22) whereas they do not fit well with the results of Chojnowski et al. (9). It is difficult to comment in detail on the results of these authors because the nature of the involved ionic species is not well known. However, while in an alcohol the free anions constitute the active form, in the nonpolar medium used by Chojnowski the attacking species have the structure of aggregates. It is reasonable to conceive that our results would better agree with those found in methanol than with those found in heptane since, though we are working in a non-polar medium, the addition of the [211] ligand suppresses the aggregates and leads to reactive ion pairs.

Table II. Rate Constants of Formation and of Propagation of D_x in
Toluene at Room Temperature, with Li^+ + [211] as
Counterion

D_x	D_4	D_5	D_6	D_7
k_{px}				
$1.mole^{-1}.hr.^{-1}$	17	6.5	1.2	–
k_{dx}				
$1.mole^{-1}.hr.^{-1}$	4	0.9	0.05	0.006

Table III. Comparison of the Reactivity of D_x Towards Silanolates

D_x	D_3	D_4	D_5	D_6
k_{px} a)				
$1.mole^{-1}.h^{-1}$	4700	17	6.5	1.2
k_{px}/k_{p4}	280	1	0.4	0.07
k_{dx}/k_{d4} b)	357	1	0.1	0.06
k_{dx}/k_{d4} c)	88	1	1.1	11.8

a) Found in toluene at 20°C with Li^+ + [211] (11,17).

b) Found in the KOH-catalyzed solvolysis of siloxanes in methanol
(22).

c) Found in the ring opening of D_x by potassium phenyldimethylsilanolate in heptane-dioxane, v/v (95:5) at 30°C (9).

Molecular Weight Distributions of PDMS

Gel permeation chromatography of PDMS prepared from D_4 with Li^+ + [211] as counterion, in toluene, shows a bimodal distribution. The ratio of the percentage of low molecular weight polymers over the percentage of high molecular weight PDMS is nearly constant during the course of the polymerization. Its value is rather important: $1/4 < R < 1/3$. Moreover M_w/M_n values are always equal to 2 for the main peak. This means that inter and intra molecular attacks of the chains by silanolate ends are important and occur at the beginning of the propagation.

GPC diagrams of the PDMS obtained from D_3 with Li^+ + [211] as counterion in benzene and in toluene show that the proportion of oligomers is very low. However, M_w/M_n values corresponding to high yields of conversion are nearly equal to 2.

Polymers of narrow molecular weight distribution can be obtained using a cryptand the cavity of which is larger than that of the [211] like the [221] or the [222]. The results are shown in Table IV. It can be noticed that the rates of propagation markedly decrease on increasing the size of the ligand. We have shown that a large proportion of aggregates remain on adding the [221] cryptand to a lithium silanolate solution in toluene. Moreover, the cyclic by-product formation is completely inhibited with Li^+ + [221] or Li^+ + [222] as counterions.

Table IV. Molecular Weight Distribution of PDMS Obtained from D_3 using Lithium Cryptates as Counterions

Counterion	Li^+ + [211]	Li^+ + [221]	Li^+ + [222]
$[D_3]_0$ mole.l^{-1}	0.48	1.07	2.10
$[C]$ x 10^3 mole.l^{-1}	0.49	1.8	6.3
Propagation Time	0.5 hr.	24 hr.	9 days
Temperature (°C)	20	20	35
Yield (%)	95	67	80
M_n theor.	213 000	88 000	54 000
M_n GPC	210 000	117 000	86 000
$I_p = M_w/M_n$	2.09	1.16	1.24

Conclusion

In conclusion, the advantages opf lithium cryptates for the anionic
polymerization of cyclosiloxanes are numerous. With Li$^+$ + [211],
only one type of active species is observed in benzene or toluene.
Thus, the kinetics of the propagation as well as that of the by-
prdoucts cyclosiloxanes formation can be studied in detail for the
first time. The reactivity of cryptated silanolate ion pairs
towards the ring opening of D$_3$ is tremendously enhanced compared to
that of the other systems used up to now. Moreover, the amount of
by-products cyclic compounds is very low when the polymer yield
reaches its maximum value. On the other hand, polymers of narrow
molecular weight distributions containing no cyclic-by-products
could be obtained using a cryptand the cavity of which is larger
than that of the [211] like the [221] or the [222] ligand. More
detailed studies in this field are in progress.

Acknowledgments

I would like to thank Drs. P. Hemery, H. Dang Ngoc, H. Porte, C.
Momtaz, S. Hubert and Mrs. N. D'Haeyer, who participated in the
investigations of the laboratory in this field.

Literature Cited

1. Bostick, E. E. In "Ring Opening Polymerization"; Frisch, K.
    C.; Reegen, S. L., Eds; Marcel Dekker, New York and London,
    1969; Chap. 8.
2. Bywater, S. Prog. Polym. Sci., 1975, 4, 27.
3. Lee, C. L.; Frye, C. L.; Johansson, O. K. Polym. Prep., 10,
    1971.
4. Frye, C. L.; Salinger, R. M.; Fearon, F. W. G.; Klosowski,
    J. M.; de Young, T. J. Org. Chem., 1970, 35, 1308.
5. Zilliox, J. G.; Roovers, J. E. L.; Bywater, S. Macromolecules,
    1975, 8, 573.
6. Saam, J. C.; Gordon, D. J.; Lindsey, S. Macromolecules, 1970,
    3, 1.
7. Holle, H. J.; Lehnen, B. R. Eur. Polym. J., 1975, 11, 663.
8. Chojnowski, J.; Mazurek, M. Makromolek. Chem., 1975, 176,
    2999.
9. Mazurek, M.; Chojnowski, J. Makromolek. Chem., 1977, 178,
    1005.
10. Dan Ngoc, H.; Porte, H.; Hemery, P.; Boileau, S. Internat.
    Symp. on Macromolecules, Mainz, Germany, 1979, Prep. Vol. I p.
    137.
11. Dang Ngoc, H. These Doctorat d'Etat, Paris, 1979.
12. Boileau, S. in "Anionic Polymerization, Kinetics, Mechanisms
    and Synthesis", McGrath, J. E. Ed.; ACS SYMPOSIUM SERIES No.
    166, American Chemical Society: Washington, D.C., 1981; pp.
    283-306.
13. Lehn, J. M. Structure and Bonding, 1973, 16, 1.
14. Lehn, J. M.; Schue, F.; Boileau, S.; Kaempf, B.; Cau, A.;
    Moinard, J.; Raynal, S. Fr. Patent, 1973, 2 201 304 and 1979,
    2 398 079.

15. Boileau, S.; Kaempf, B.; Lehn, J. J.; Schue, F. J. Polym.
    Sci., (B), 1974, 12, 203.
16. Boileau, S.; Hemery, P.; Kaempf, B.; Schue, F.; Viguier, M.
    J. Polym. Sci., (B), 1974, 12, 217.
17. Hubert, S. unpublished results.
18. Momtaz, C. These de Doctorat de 3° cycle, Paris, 1982.
19. Bargain, M.; Millet, C. Fr. Patent, 1975, 2 353 589.
20. Semlyen , J. A. Adv. Polym. Sci., 1976, 21, 41 and references
    therein.
21. Shinohara, M. Polym. Prep., 1973, 14, 1209.
22. Lasocki, Z.; Kulpinski, J.; Gador, W. Polimery, 1970, 15, 508.

RECEIVED February 1, 1985

Anionic Polymerization of Ethylene Oxide with Lithium Catalysts

Solution Properties of Styrene–Ethylene Oxide Block Polymers

RODERIC P. QUIRK[1] and NORMAN S. SEUNG[2]

[1]Institute of Polymer Science, The University of Akron, Akron, OH 44325
[2]Michigan Molecular Institute, Midland, MI 48640

The anionic polymerization of ethylene oxide has been investigated using poly(styryl)lithium, α-lithium poly(methyl methacrylate), cumylpotassium, and sodium tributyl magnesate as initiators. No ethylene oxide polymerization was detected in tetrahydrofuran using initiators with the lithium counterion. In a mixture of benzene and dimethylsulfoxide (2/1), ethylene oxide polymerization was observed using the ethylene oxide adduct of poly(styryl)lithium. Size exclusion chromatography and viscosity measurements indicate that the hydrodynamic volume of the triblock polymer, poly-(styrene-b-ethylene oxide-b-styrene) is very similar to that of the corresponding diblock polymer, poly(stryene-b-ethylene oxide) with half the molecular weight.

Ethylene oxide is an inherently reactive monomer from a thermodynamic point of view. Because of the ring strain in the three-membered ring the enthalpy of polymerization of ethylene oxide is comparable to that of cyclopropane, -27 kcal/mole($\underline{1}$). A variety of simple and complex catalysts and initiators can be used to effect the polymerization of ethylene oxide and homologous compounds ($\underline{2}$-$\underline{4}$). Therefore, it is somewhat surprising that lithium hydroxide and lithium alkoxides have been reported to be ineffective initiators for the polymerization of ethylene oxide and propylene oxide($\underline{5}$-$\underline{9}$). This apparent lack of reactivity of these lithium salts stands in sharp contrast with kinetic studies of the reactions of ethylene oxide with alkali metal derivatives of fluoradenyl($\underline{10}$) and polystyryl($\underline{11}$) carbanions where the lithium derivatives are the most reactive species by several powers of ten. This lack of polymerization activity has been used to advantage for the hydroxyethylation of simple($\underline{12}$) and polymeric ($\underline{2}$,$\underline{9}$,$\underline{13}$) organolithium compounds in high yields (Equation 1).

$$PLi \xrightarrow[\text{2) H}_2\text{O}]{\text{1) ethylene oxide}} PCH_2CH_2OH \qquad (1)$$

However, there are various, fragmentary reports that lithium-
based initiators may, under certain conditions, effect the polymerization of ethylene oxide (11,14,15). Since anionic polymerizations
of non-polar monomers involving lithium offer considerable advantages
in terms of control of the major variables affecting polymer proper-
ties(13), it is important to determine if conditions can be found to
use well-characterized, polymeric organolithium compounds to polymerize ethylene oxide. Herein are reported results of investigations of
reactions of polymeric organolithium compounds with ethylene oxide.
Also, preliminary results of the use of a new, hydrocarbon-soluble
initiator (sodium tributyl magnesate, NaMgBu₃) has been examined. In
addition, evidence for unusual hydrodynamic effects of poly(styrene-
b-ethylene oxide) vs. poly(styrene-b-ethylene oxide-b-styrene) will
be described.

Experimental

Styrene (99%, Aldrich) was purified by initial stirring and degassing
over freshly-crushed CaH_2 on a high vacuum line followed by distillation onto dibutylmagnesium (Lithium Corporation). Final purification involved distillation from this solution directly into calibrated ampoules. Ethylene oxide (99.7% min. purity, Matheson) was
condensed directly from the storage cylinder into a flask with
freshly-crushed CaH_2 on the vacuum line. After stirring and degassing, further purification involved distillation onto sodium dispersion (Alpha), stirring and degassing, distillation onto dibutylmagnesium and final distillation into a calibrated ampoule. Tetrahydrofuran (Fisher Scientific, certified spectralyzed, no preservatives)
was stirred and degassed over $LiAlH_4$ followed by distillation and
storage over sodium benzophenone ketyl. Dimethylsulfoxide (Fisher
Scientific) was stored over and distilled under vacuum from 4Å molecular sieves. Cumylpotassium in tetrahydrofuran was prepared from
methyl cumyl ether and sodium-potassium alloy(16). Sodium tributyl
nagnesate [1.4N(0.49 molar)] was used as received from Lithium Corporation. sec-Butyllithium (Lithium Corporation of America, 12.0 wt %
in cyclohexane) was used as received.
      Polymerizations were carried out in all-glass, sealed reactors
using breakseals and standard high vacuum techniques(17). Polymerizations using sodium tributyl magnesate were carried out in flasks
equipped with teflon Rotoflo stopcocks.
      Number-average molecular weights were determined in toluene
solutions using a membrane osmometer (Mechrolab, Hewlett-Packard 503
with S & S-08 membranes). Size exclusion chromatographic analyses in
chloroform were performed by HPLC (Perkin-Elmer 601 HPLC) using two
μ-Styragel columns (10^4, 10^3 Å) after calibration with standard polystyrene samples.
      Size exclusion chromatographic analyses in tetrahydrofuran were
performed with a Waters 150C GPC with six μ-Styragel columns having a
continuous porosity range of 10^6- 10^2 Å and also with the Perkin-
Elmer Model 601 HPLC with three μ-Styragel columns (10^5, 10^4, 10^3 Å)
after calibration with standard polystyrene samples. Intrinsic viscosities were measured in chloroform at 30.0°C and in tetrahydrofuran
at 40°C using an Ubbelohde type viscometer.

The ultracentrifugation experiment was performed using a Beckman Model E instrument with a capillary synthetic boundary cell at 30°C. Toluene was used as the solvent and solution concentrations were 1% (w/v).

Results and Discussion

Ethylene Oxide Polymerization. The apparent inability of lithium bases to effect the anionic polymerization of ethylene oxide and its homologues is unique among the alkali metals. At least part of the unreactivity of lithium alkoxides can be ascribed to their strong association in solution as shown in Table I(18). However, it can be seen that the corresponding sodium and potassium alkoxides are also highly associated in solution and yet they are active initiators for the polymerization of ethylene oxide.

Indeed, it has been reported that polymeric sodium and potassium alkoxides are associated into dimers even in hexamethylphosporictriamide at 40°C(19). It can be concluded that the lithium alkoxide unreactivity is not due to the phenomenon of association per se, but

Table I.   Degree of Association of Alkali Metal Alkoxides in Various Solvents(18).

Alkoxide	Degree of Association				
	Cyclohexane	Benzene	Diethyl ether	Tetra-hydrofuran	Pyridine
Lithium t-butoxide	5.8	6.2	5.9	4.1	4.0
Sodium t-butoxide	8.2	8.3	4.3	3.9	3.9
Potassium t-butoxide	–	–	3.9	4.0	–

to the strength of the association, i.e., the lack of dissociation. Two approaches have been taken to promote dissociation (and presumably reactivity) of lithium alkoxides: (a) addition of specific lithium-complexing agents (12-Crown-4 ether and N,N,N',N'-tetramethylethylenediamine); and (b) addition of dipolar aprotic solvents.

Poly(styryl)lithium ($\overline{M}_n$=15,000) in benzene solution was reacted with excess ethylene oxide in the presence of N,N,N',N'-tetramethylethylenediamine (TMEDA, [TMEDA]/[Li]=3.2). After 12 days at 25–30°C, size exclusion chromatographic analyses indicated no significant ethylene oxide polymerization. Hydroxyethylated polystyrene was recovered in essentially quantitative yield.

α-Lithium poly(methyl methacrylate) ($\overline{M}_n$=28,000) was prepared by polymerization of methyl methacrylate with 1,1-diphenyl(hexyl)lithium as initiator(20) at -78°C in tetrahydrofuran in the presence of 12-Crown-4 ether (Aldrich, [Crown]/[Li]=2). Once again after reaction with excess ethylene oxide for 12 days at 25°C, the homopolymer, poly(methyl methacrylate), presumably hydroxyethylated(20), was isolated essentially quantitatively and no significant ethylene oxide polymerization was evident by size exclusion chromatographic analyses. In view of the results described herein with dimethylsulfoxide, however, it would be imprudent to conclude that ethylene oxide polymerization would not occur with the lithium counterion at elevated temperatures in the presence of 12-Crown-4 ether or TMEDA.

Poly(styryl)lithium ($\overline{M}_n$=2400) was terminated with ethylene oxide in benzene solution at 25°C as shown in Equation 2. The GPC reten-

$$\text{PSLi} \xrightarrow{\text{ethylene oxide}} \text{PSCH}_2\text{CH}_2\text{OLi} \qquad (2)$$

$$\underline{A} \qquad\qquad\qquad\qquad \underline{B}$$

tion of this derivative (B) after hydrolysis had essentially the same retention volume as an aliquot of the original poly(styryl)lithium (A) which had been quenched with t-butanol (see Figure 1). The molecular weight distribution of A as calculated from the GPC data was 1.04 ($\overline{M}_w/\overline{M}_n$). An aliquot of the polymeric lithium alkoxide (B) (8 mmoles) was dissolved in ca. 300 mL of a 2/1 (v/v) mixture of benzene and dimethylsulfoxide using high vacuum techniques(17) after removal of the original benzene and excess ethylene oxide from the initial alkoxyethylation reaction (Equation 2). Ethylene oxide (0.52 moles) was then condensed into the high-vacuum reactor which was then placed in a 40°C bath for 4 days followed by 4 days at 60°C. The reaction was terminated by addition of a few mL of degassed acetic acid.

The polymer product isolated after precipitation and drying corresponded to an overall yield of 83%, which indicates a 70% conversion of ethylene oxide. The size exclusion chromatographic retention volume of this product (see Figure 2) corresponds to an $\overline{M}_n^{GPC}$=3400 with ($\overline{M}_w/\overline{M}_n$)$_{GPC}$=1.04. Several salient features of these results deserve specific comment. Both the narrow molecular weight distribution of the product, PS-PEO, diblock polymer and the absence of any observable peak corresponding to the original polystyrene block (observable by GPC in synthetic mixtures of hydrolyzed A and PS-PEO), indicate that (a) the hydroxyethylation reaction (Equation 2) occurs essentially quantitatively in benzene solution; and (b) no evidence for chain termination or chain transfer is apparent in the polymerization of ethylene oxide with lithium as counterion in a mixture of benzene/dimethylsulfoxide. The 60 MHZ ^{1}H-NMR spectrum of the PS-PEO diblock (Figure 3) clearly indicates the presence of the poly(ethylene oxide) segment. The calculated ratio of the integrated intensities for the aromatic to $-CH_2O-$ protons corresponds to 1.1, while the intensity ratio calculated from the GPC molecular weights corresponds to 1.0.

Thus, a dipolar aprotic solvent such as dimethylsulfoxide provides the necessary solvation and polarity to render lithium alkoxides as effective initiators for ethylene oxide polymerization. Work is underway to further explore the scope and kinetics of this important polymerization system.

Several polymerizations of ethylene oxide with sodium tributyl magnesate have been performed. Reactions at 60°C for 3 days produced polymer in 22% yield. It was necessary to heat the reactions mixtures for 12 days at 60°C to achieve a conversion of 56%. Molecular weights were less than stoichiometric and the molecular weight distributions as determined by GPC were somewhat broad but symmetrical.

Solution Properties of Styrene-Ethylene Oxide Block Polymers. During the course of our studies of synthetic routes to poly(styrene-b-ethylene oxide), we have undertaken an investigation of the solution

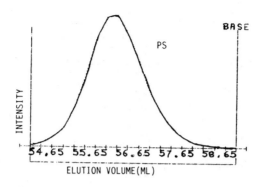

Figure 1. Size exclusion chromatogram of polystyrene.

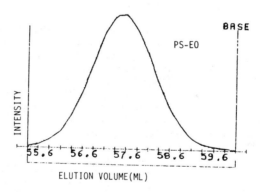

Figure 2. Size exclusion chromatogram of poly(styrene-b-ethylene oxide).

properties of these polymers. We have compared a diblock polymer,
poly(styrene-b-ethylene oxide) (PS-PEO), with the corresponding
triblock polymer, poly(styrene-b-ethylene oxide-b-styrene)(PS-PEO-PS).
The triblock polymer has the same styrene end segment lengths as the
diblock polymer, and the poly(ethylene oxide) center block in the
triblock is twice the poly(ethylene oxide) segment length in the
diblock as shown in Table II. Both PS-PEO and PS-PEO-PS exhibited
the same retention volume in tetrahydrofuran using a three-column set

Table II. Molecular Weight Characterization of PS-PEO and
PS-PEO-PS

Polymer	Stoichiometric Molecular Weight (g/mol)[a]	$\overline{M}_n$ (g/mol)[b]
PS-PEO (Diblock)	83,000 (35 - 48)	82,800
PS-PEO-PS (Triblock)	166,000 (35-96-35)	163,900

[a] The numbers in parentheses correspond to the stoichiometric
molecular weights for the individual block segments $(\times 10^{-3})$
based on the ratio of gm of monomer charged to the moles of
initiator.

[b] Determined by membrane osmometry.

of μ-styragel columns. This surprising result was compounded by the
fact that the measured intrinsic viscosities in tetrahydrofuran were
75 ml/g and 71 ml/g for the diblock and the triblock, respectively.
Thus, the triblock polymer apparently exhibits the same hydrodynamic
volume and viscosity as the diblock polymer which has one-half of the
molecular weight of the triblock. These unusual observations probably
reflect the fact that tetrahydrofuran is listed as non-solvent for
poly(ethylene oxide)(21) and precipitates from a 1% solution at 18°C
(22). This phenomenon was explored further by examining the behavior
of these block polymers in chloroform, a good solvent for polystyrene
and poly(ethylene oxide)(21). The size exclusion chromatograms of
these polymers in chloroform are shown in Figure 4. The retention
volumes were 13.5 ml and 14.0 ml for the triblock and diblock,
respectively. For the polystyrene standards, a retention volume difference of 1.2 ml (versus 0.5 ml observed for the block polymers)
would be expected for a doubling of the molecular weight from 83,000
to 166,000. Further evidence for the unusually small increase in
hydrodynamic volume for the triblock polymer relative to the diblock
in chloroform has been obtained from their intrinsic viscosities and
second virial coefficients as shown in Table III.
       It it noteworthy that the diblock and triblock polymers in
toluene solution could not be separated by ultracentrifugation. A
50/50 mixture of the two polymers in toluene exhibited a single peak
throughout the sedimentation process with the ultracentrifuge operating at a speed of 28,000 rpm. In conclusion, all of the evidence
from solution properties of a PS-PEO-PS block polymer indicates that

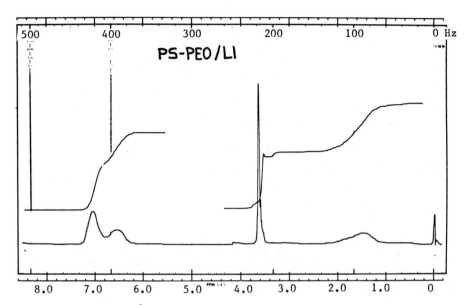

Figure 3.   60 MHz ^{1}H-NMR spectrum of poly(styrene-b-ethylene oxide).

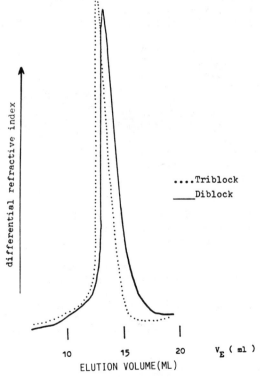

Figure 4.   Size exclusion chromatograms of poly(styrene-b-ethylene oxide-b-styrene) and poly(styrene-b- ethylene oxide).

Table III.   Solution Characterization of PS-PEO
and PS-PEO-PS in Chloroform.

Polymer	$[\eta]$ (ml/g)	Huggins Constant, k_1	Kraemer Constant, k_2	$A_2 \times 10^4$ (ml/mol g^2)
PS-PEO	124	0.40	-0.13	6.63
PS-PEO-PS	132	0.34	-0.14	5.01

this polymer exhibits unique hydrodynamic properties when compared to
the corresponding diblock polymer with one-half of the molecular
weight of the triblock.   Further work is in progress to characterize
the solution properties of these polymers.

Acknowledgments

The authors would like to acknowledge the able assistance of
Mr. Dennis McFay who carried out the initial synthetic and solution
property experiments at Michigan Molecular Institute.

Literature Cited

1.   Sawada, H. J. Macromol. Sci.-Rev. Macromol. Chem., 1970, (C5(1), 151.
2.   Morton, M.   "Anionic Polymerization: Principles and Practice";
     Academic Press: New York, 1983; p. 52.
3.   Boileau, S. in "Anionic Polymerization: Kinetics, Mechanisms, and
     Synthesis"; McGrath, J.E., Ed.; ACS Symposium Series No. 166,
     American Chemical Society: Washington, D.C., 1981; p. 283.
4.   Bailey, F.E.; Koleske, J.V.   "Poly(ethylene Oxide)"; Academic
     Press: New York, 1976.
5.   St. Pierre, L.E.; Price, C.C. J. Am. Chem. Soc., 1956, 78, 3432.
6.   Lebedev, N.N.; Baranov, Yu.I. Polym. Sci. USSR, 1966, 8, 211.
7.   Doroshenko, N.P.; Spirin, Yu.L. Polym. Sci. USSR, 1970, 12, 2812.
8.   Cabasso, F.; Zilkha, A. J. Macromol. Sci.-Chem., 1974, A8(8),
     1313.
9.   Guilbert, Y.; Brossas, J. Polym. Bull., 1979, 1, 293.
10. Chang, C.J.; Kiesel, R.F.; Hogen-Esch, T.E. J. Am. Chem. Soc.,
     1973, 95, 8446.
11. Solov'yanov, A.A.; Kazanskii, K.S. Polym. Sci. USSR, 1970, 12,
     2812.
12. Wakefield, B.J.   "The Chemistry of Organolithium Compounds";
     Pergamon Press: Elmsford, N.Y., 1974; p. 199.
13. Young, R.N.; Quirk, R.P.; Fetters, L.J. Adv. Polym. Sci., 1984,
     56, 1.
14. Dudek, T.J.   Ph.D. Thesis, University of Akron, 1961, p. 74.
15. Kobayashi, S.; Kaku, M.; Mizutani, T.; Saegusa, T. Polym. Bull.,
     1983, 9, 169.
16. Ziegler, K.; Dislich, H. Chem. Ber., 1957, 90, 1107.
17. Morton, M.M.; Fetters, L.J. Rubber Chem. Tech., 48, 359 (1975).
18. Halaska, V.; Lochmann, L.; Lim, D.; Coll. Czech. Chem. Commun.,
     1968, 33, 3245.
19. Figueruelo, J.E.; Worsfold, D.J. Eur. Polym. J., 1968. 4, 439.

20. Anderson, B.C.; Andrews, G.D.; Arthur, P., Jr.; Jacobson, H.W.; Melby, L.R.; Playtis, A.J.; Sharkey, W.H. Macromolecules, 1981, 14, 1599.
21. Fuchs, O.; Suhr, H.-H. in "Polymer Handbook," Second ed.; Brandrup, J.; Immergut, E.H., Eds.; Wiley-Interscience: New York, 1975; p. IV-241-265.
22. Stone, F.W.; Stratta, J.J. in "Encyclopedia of Polymer Science and Technology"; N. Bikales, Ed.; John Wiley and Sons, Inc.: New York, 1967; Vol. 6; p. 114.

RECEIVED September 14, 1984

Free Radical Ring-Opening Polymerization

WILLIAM J. BAILEY

Department of Chemistry, University of Maryland, College Park, MD 20742

Free radical ring-opening polymerization has previously been quite rare with the only examples being
cyclopropane derivatives and o-xylylene dimer. This
fact is surprising in view of the fact that ionic ring-
opening polymerization is very common. Since a carbon-
oxygen double bond is about 50 kcal more stable than a
carbon-carbon double bond, it was found that introducing an oxygen atom into an unsaturated cyclic
monomer would permit free radical ring-opening polymerization. Thus it was shown that cyclic ketene acetals, cyclic ketene aminals, cyclic vinyl ethers,
unsaturated spiro ortho carbonates, and unsaturated
spiro ortho esters, would all undergo such polymerization. Furthermore, all of these monomers would copolymerize with a wide variety of vinyl monomers with the
introduction of functional groups, such as esters,
thioesters, amides, ketones, and carbonates, into the
backbone of the addition polymers. This copolymerization makes possible the synthesis of biodegradable
polymers, functionally terminated oligomers, polymers
with enhanced thermal stability, and monomer mixtures
which expand upon polymerization.

In a research program to find monomers which expand upon polymerization it was desirable to have available monomers which would undergo
double ring-opening polymerization by a free radical mechanism.
However, a search of the literature revealed that there were very
few examples of any free radical ring-opening polymerization. For
example Takahashi (1) reported that during the free radical polymerization of vinylcyclopropane the cyclopropane ring opened to give
a polymer containing about 80% 1,5-units and about 20% of undeter-
mined structural units but no cyclopropane rings. Apparently the
radical adds to the vinyl group to give the intermediate cyclopropylmethyl radical which opens at a rate faster than the addition to
the double bond of another monomer. The driving force for the polymerization is the relief of the strain of the three-membered ring.
Somewhat similar results were obtained with the chloro derivatives.
Very recently, Cho and Ahn (2) studied the related malonic ester
derivative, which underwent free radical ring-opening polymerization
to produce a high molecular weight polymer containing only the
1,5-units.

0097–6156/85/0286–0047$06.00/0

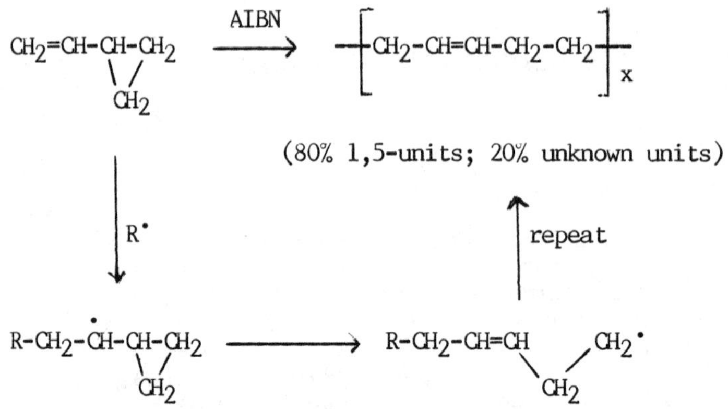

$$CH_2=CH-CH-CH_2 \quad \xrightarrow{\text{AIBN}} \quad \left[CH_2-CH=CH-CH_2-CH_2 \right]_x$$
$$\diagdown / $$
$$CH_2$$

(80% 1,5-units; 20% unknown units)

$R^\bullet$ ↓   ↑ repeat

$$R-CH_2-\overset{\bullet}{CH}-CH-CH_2 \quad \longrightarrow \quad R-CH_2-CH=CH \quad\quad CH_2^\bullet$$
$$\diagdown / \quad\quad\quad\quad\quad\quad\quad\quad\quad \diagdown /$$
$$CH_2 \quad\quad\quad\quad\quad\quad\quad\quad\quad\quad\quad CH_2$$

Hall and coworkers (3) demonstrated that derivatives of bicyclo[1.1.0]butane would polymerize by free radicals by cleavage of the highly strained central bond.

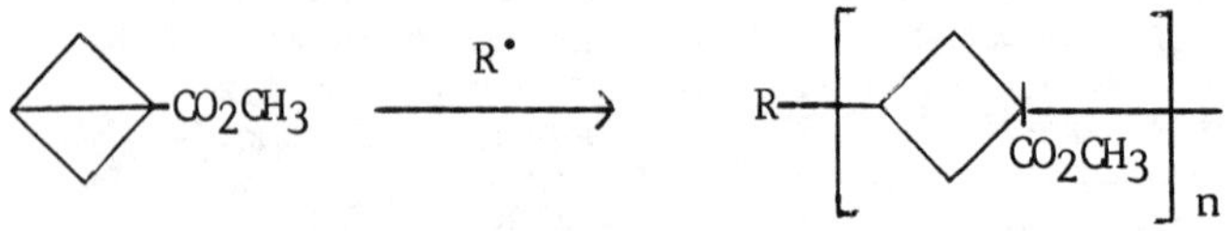

Errede (4) showed that the dimer of o-xylylene would undergo free radical ring-opening polymerization to give the corresponding poly-o-xylylene.

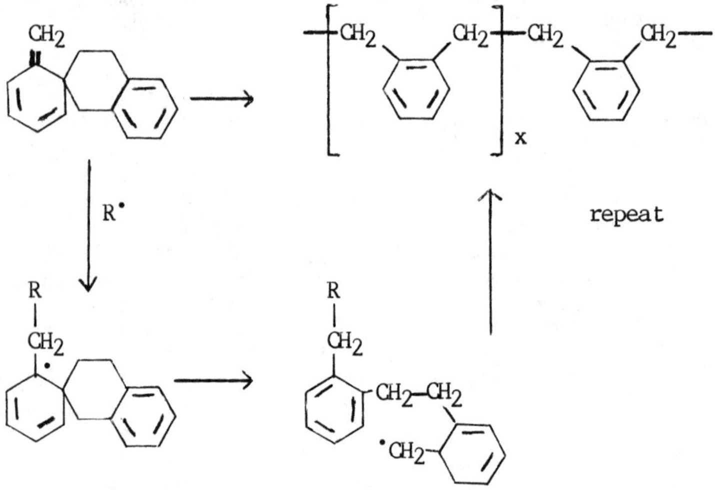

In this case the driving force for the ring-opening step is the formation of the aromatic ring. Finally the ring-opening polymerization of S_8 has been postulated to involve free radicals (5).

The dearth of examples of free radical ring-opening polymerization is rather surprising in view of the fact that the ionic ring-opening polymerization of heterocyclic compounds, such as ethylene oxide, tetrahydrofuran, ethylenimine, β-propiolactone, and caprolactam, as well as the Ziegler-Natta ring opening of cyclic olefins, such as cyclopentene and norbornene, are quite common. One explanation is that unstrained five- and six-membered carbocyclic rings usually are involved in ring-closing reactions rather than ring opening. For example, Butler and Angelo (6) in 1957 found that, when diallyldimethylammonium bromide was polymerized by a free radical mechanism, a soluble polymer containing five-membered rings was obtained by an inter-intramolecular polymerization.

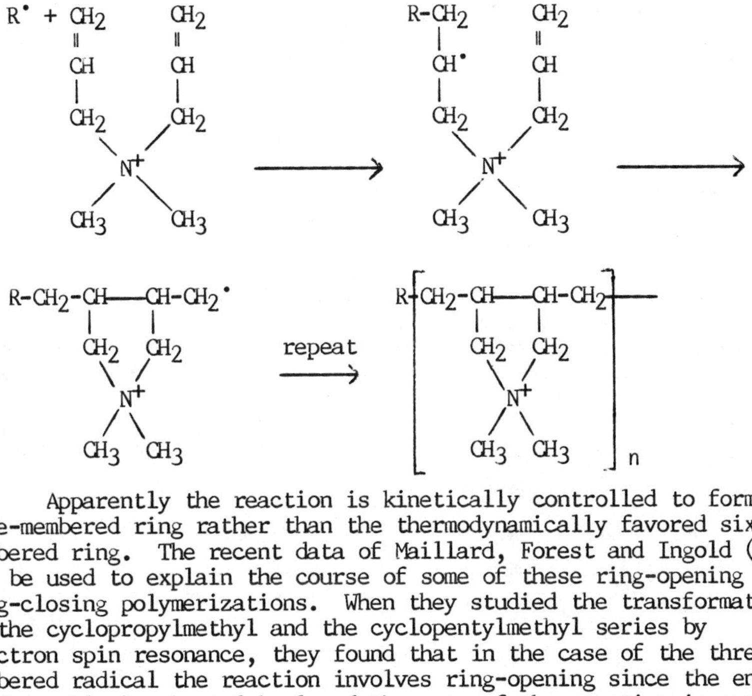

Apparently the reaction is kinetically controlled to form the five-membered ring rather than the thermodynamically favored six-membered ring. The recent data of Maillard, Forest and Ingold (7) can be used to explain the course of some of these ring-opening and ring-closing polymerizations. When they studied the transformations in the cyclopropylmethyl and the cyclopentylmethyl series by electron spin resonance, they found that in the case of the three-membered radical the reaction involves ring-opening since the energy is favorable by about 6 kcal and the rate of the reaction is very high. In the case of the five-membered ring system they found that the reaction proceeds in the direction of ring-closure since the energetics of that reaction is favorable by about 8 kcal and the rate of the ring closure is also moderately high.

Free Radical Ring-Opening of Cyclic Ketene Acetals

Since the carbon-oxygen double bond is at least 50-60 kcal/mole more stable than the carbon-carbon double bond (8), we estimated that the introduction of an oxygen atom in place of the carbon atom in the cyclopentylmethyl radical would favor the reverse reaction or the

ring opening. In other words the ring-opening reaction would be
favored by at least 40 kcal/mole by producing the more stable carbonyl double bond. A search of the literature revealed several ring
systems containing an oxygen atom that would undergo a ring-opening
reaction in the presence of free radical catalysts. One such case
was the cyclic formal, 1,3-dioxolane, which Maillard, Cazaux, and
Lalande (9) found rearranged to ethyl formate when heated at 160°
C.

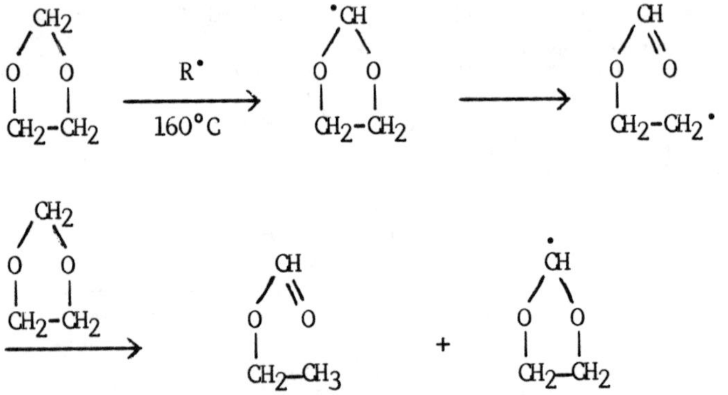

The reaction could be rationalized as indicated where the
driving force for the ring-opening step in the chain reaction was
the formation of the stable carbon-oxygen double bond in the final
ester. With the knowledge that such a ring system would undergo
cleavage, it seemed to be a fairly straight forward process to
synthesize a monomer that would undergo ring-opening polymerization
by introducing a double bond at the carbon atom flanked by the two
oxygens.

The monomer desired for this ring-opening polymerization had
indeed been prepared by McElvain and Curry (10) in 1948. Although

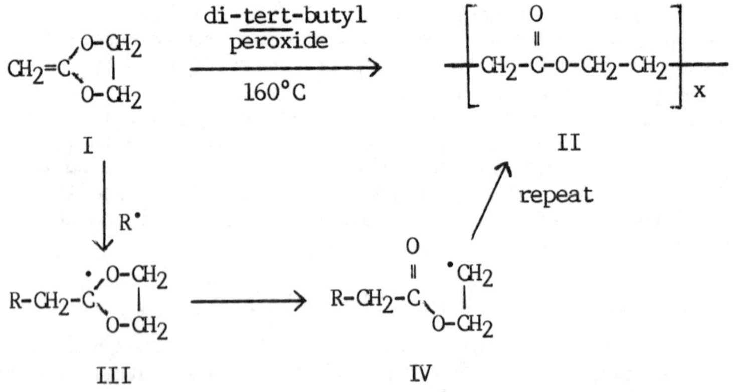

Johnson, Barnes, and McElvain (11) had treated diethyl ketene acetal
with peroxide and had reported that there was no reaction, no such

study was reported for the 2-methylene-1,3-dioxolane (I). A rein-
vestigation of the cyclic ketene acetal I was therefore undertaken.
This polyester II is difficult to synthesize with high molecular
weight from the γ-hydroxybutyric acid because of the stability of
the competing lactone.

When the polymerization is carried out at lower temperatures,
the ring opening is not complete. Thus at 60° C only 50% of the
rings are opened to give a random copolymer of the following structure (12):

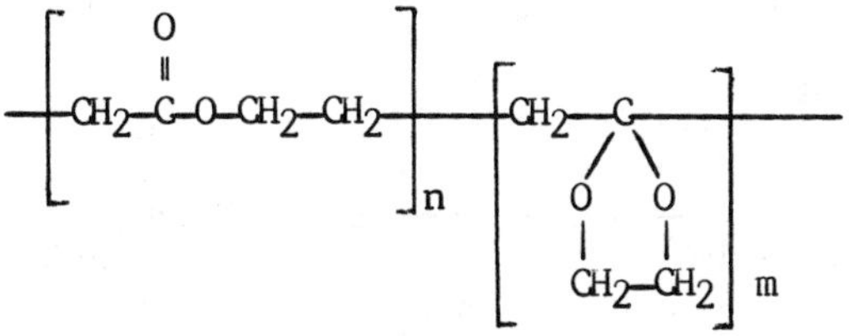

Even at 120° C only 87% of the rings are opened. The uno-
pened radical III apparently can add directly to the monomer I in
competition with the ring-opening process to form the open chain

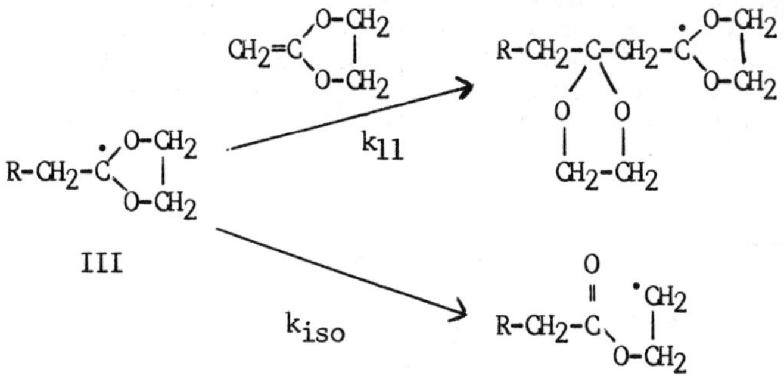

IV

radical IV. High dilution was found to favor the ring-opening process since the addition of III to the monomer I is a second order
reaction while the conversion of III to the open chain radical is
first order. The extent of ring opening is kinetically controlled
with a direct competition between the rate of direct addition, k_{11},
and the rate of ring opening, k_{iso}.

In a program to find other cyclic ketene acetals that would
undergo quantitative ring-opening even at room temperature we prepared the seven-membered ketene acetal, 2-methylene-1,3-dioxepane
(V), which underwent essentially complete ring-opening at room temperature (13-15). This process makes possible the quantitative
introduction of an ester group in the backbone of an addition
polymer.

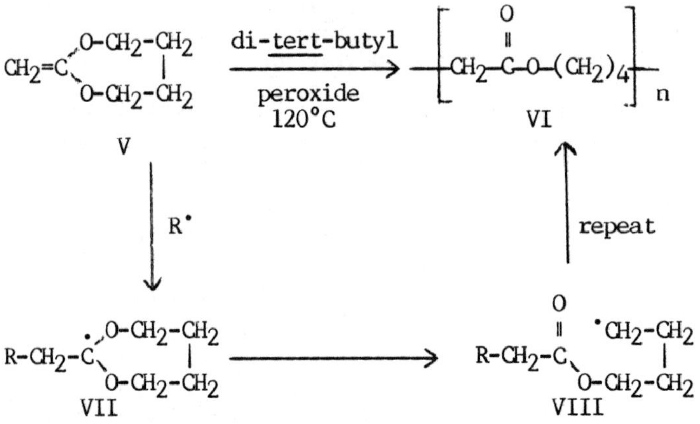

$$CH_2=C \overset{O-CH_2-CH_2}{\underset{O-CH_2-CH_2}{\diagdown}} \quad \xrightarrow[\substack{\text{peroxide} \\ 120^\circ C}]{\text{di-}\underline{\text{tert}}\text{-butyl}} \quad \left[CH_2-\overset{\overset{\textstyle O}{\|}}{C}-O-(CH_2)_4 \right]_n$$

V   VI

R• (down)   repeat (up)

$$R-CH_2-\overset{\bullet}{C}\overset{O-CH_2-CH_2}{\underset{O-CH_2-CH_2}{\diagdown}} \quad \longrightarrow \quad R-CH_2-\overset{\overset{\textstyle O}{\|}}{C}\overset{\bullet CH_2-CH_2}{\underset{O-CH_2-CH_2}{\diagdown}}$$

VII   VIII

Apparently the seven-membered ring increases the steric hindrance in the intermediate free radical VII to eliminate practically all of the direct addition and also introduces a small amount of strain so that the ring-opening to the radical VIII is accelerated.

Additional cyclic ketene acetals (16-18) that have been studied have included the 4-phenyl-2-methylene-1,3-dioxepane (IX) which undergoes quantitative ring-opening to give the polyester X. Apparently the ring-opening step from XI to XII is greatly enhanced

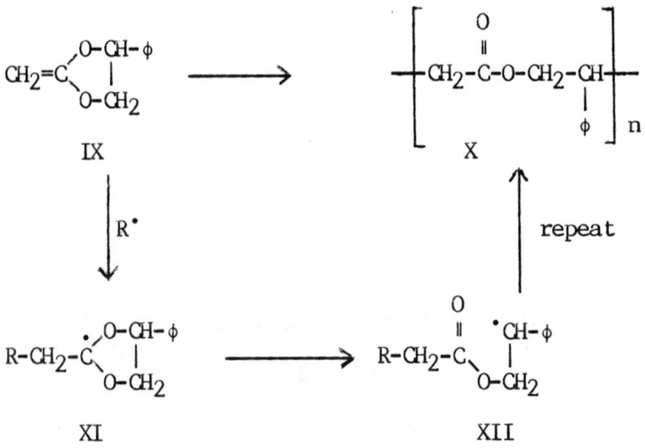

$$CH_2=C \overset{O-CH-\phi}{\underset{O-CH_2}{\diagdown}} \quad \longrightarrow \quad \left[CH_2-\overset{\overset{\textstyle O}{\|}}{C}-O-CH_2-\underset{\underset{\textstyle \phi}{|}}{CH} \right]_n$$

IX   X

R• (down)   repeat (up)

$$R-CH_2-\overset{\bullet}{C}\overset{O-CH-\phi}{\underset{O-CH_2}{\diagdown}} \quad \longrightarrow \quad R-CH_2-\overset{\overset{\textstyle O}{\|}}{C}\overset{\bullet CH-\phi}{\underset{O-CH_2}{\diagdown}}$$

XI   XII

by the formation of the relatively stable benzyl radical in XII even though XI is a five-membered ring analogous to the radical III.

Nitrogen and Sulfur Analogs of Cyclic Ketene Acetals

Since an amide group is more stable than an ester group, the nitrogen analog XIII of the cyclic ketene acetal was synthesized and polymerized to give the polyamide XIV.

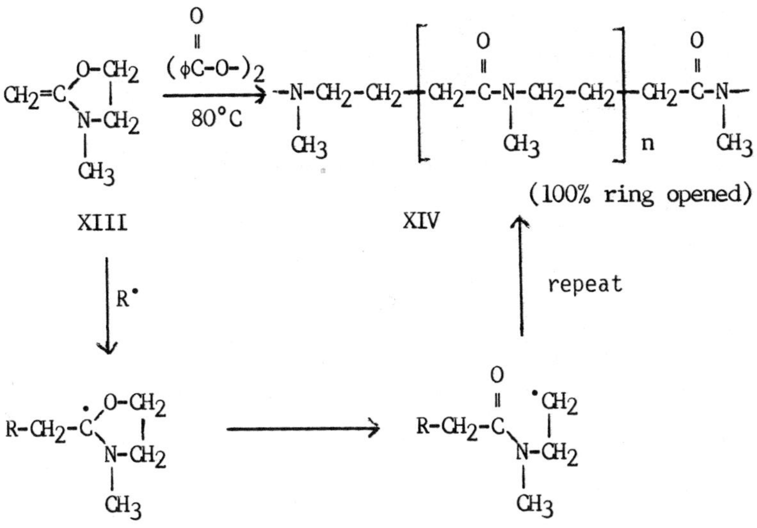

$$CH_2=C \underset{N-CH_2}{\overset{O-CH_2}{\big<}} \quad \xrightarrow[80°C]{(\phi C-O-)_2} \quad -N-CH_2-CH_2-\left[CH_2-\overset{O}{\overset{\|}{C}}-N-CH_2-CH_2\right]_n-CH_2-\overset{O}{\overset{\|}{C}}-N-$$

XIII   XIV

(100% ring opened)

$$\downarrow R^•$$ $$\uparrow \text{repeat}$$

$$R-CH_2-\overset{•}{C}\underset{N-CH_2}{\overset{O-CH_2}{\big<}} \quad \longrightarrow \quad R-CH_2-\overset{O}{\overset{\|}{C}}\underset{N-CH_2}{\overset{•CH_2}{\big<}}$$

XV   XVI

In contrast with the 2-methylene-1,3-dioxolane (I) the nitrogen analog XIII undergoes essentially quantitative ring opening even at room temperature.

Although the sulfur analog of the cyclic ketene acetal I was prepared and polymerized, apparently the resulting thioester is higher energy than the ordinary ester and therefore retards the extent of ring opening. Even at 120°C only 45% of the rings were opened (19-20).

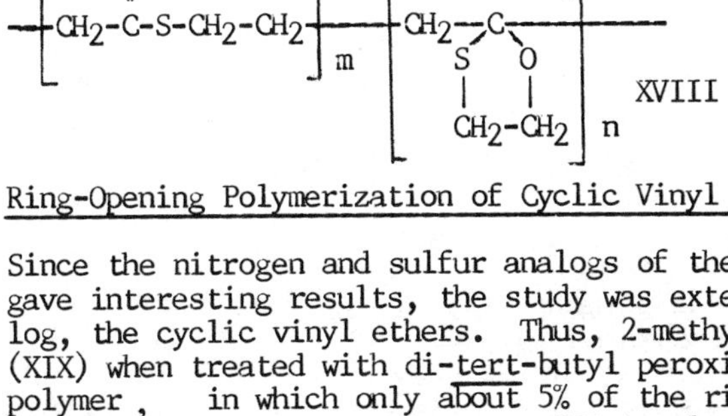

$$CH_2=C\underset{S-CH_2}{\overset{O-CH_2}{\big<}} \quad \xrightarrow[(CH_3)_3-C-O-O-C(CH_3)_3]{120°C}$$

XVII

$$-\left[CH_2-\overset{O}{\overset{\|}{C}}-S-CH_2-CH_2\right]_m \left[\underset{CH_2-CH_2}{\overset{CH_2-C}{\big<}}\underset{O}{\overset{S}{|}}\right]_n \quad \text{XVIII}$$

Ring-Opening Polymerization of Cyclic Vinyl Ethers

Since the nitrogen and sulfur analogs of the cyclic ketene acetal I gave interesting results, the study was extended to the carbon analog, the cyclic vinyl ethers. Thus, 2-methylenetetrahydrofuran (XIX) when treated with di-tert-butyl peroxide at 120° C gave a polymer, in which only about 5% of the rings had opened (19). Apparently the ketone group is sufficiently less stable than the ester group that the extent of ring opening decreases from 87% for I to 5% for XIX. Even when the cyclic vinyl ether XIX was diluted with an equal volume of benzene, the ring opening increased only to 7%.

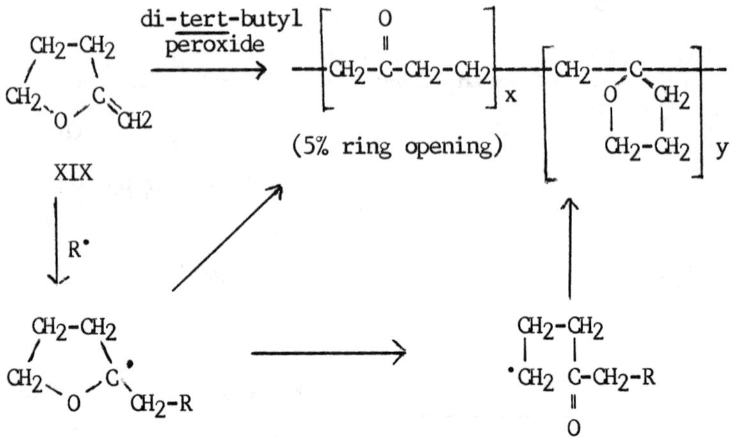

Since the extent of ring opening of the cyclic vinyl ethers is less than that of the cyclic ketene acetals, this series appeared to be an ideal system to study the effect of steric hindrance and the presence of radical-stabilizing substituents on the extent of ring opening. Thus the following monomers were prepared and poly-merized at 120° C in the presence of di-tert-butyl peroxide. The number in parentheses indicate the extent of ring opening.

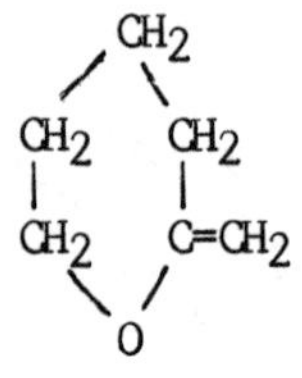

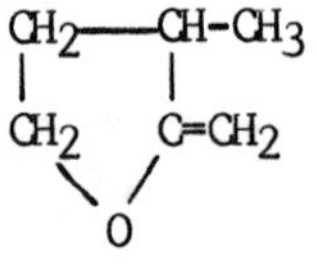

The highest extent of ring opening in this series was observed for the 4-phenyl-2-methylenetetrahydrofuran (XX). In this case the formation of relatively stable benzyl radical helps promote ring cleavage so that about one half of the rings are opened (19).

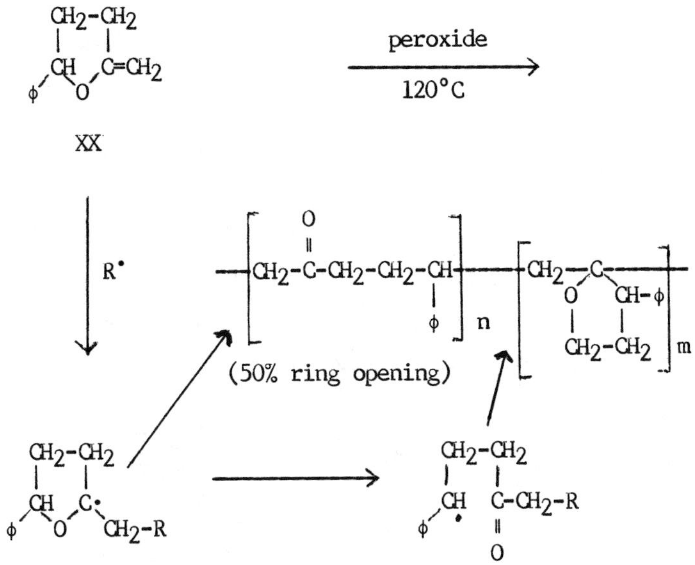

Even ring strain will not produce quantitative ring opening in this series. Although the oxetane ring possesses considerable ring strain, the corresponding radical does appear to open at a rapid rate. For example, 2-oxetanyl gives no signals in the ESR spectrum for the ring-opened product, which is in contrast to the five- and six-membered acetals which give the ring-opened radical signals at room temperature (21-22). For example, the polymerization of 2-methyleneoxetane (XXI) (23) at 120° C gave a copolymer in which only about 40% of the rings were opened (20). Apparently the small size of the ring reduces the steric hindrance so that the

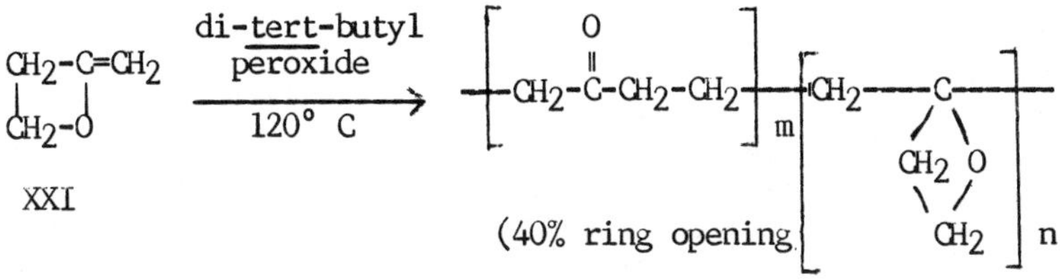

direct addition can effectively compete with the sluggish ring opening step.

Ketene dimer (XXII), which is an analog of the 2-methylene-oxetane (XXI), will undergo a higher extent of ring opening on polymerization to give copolymers with different structures depending on whether the reaction is run in a sealed tube or open to the air.

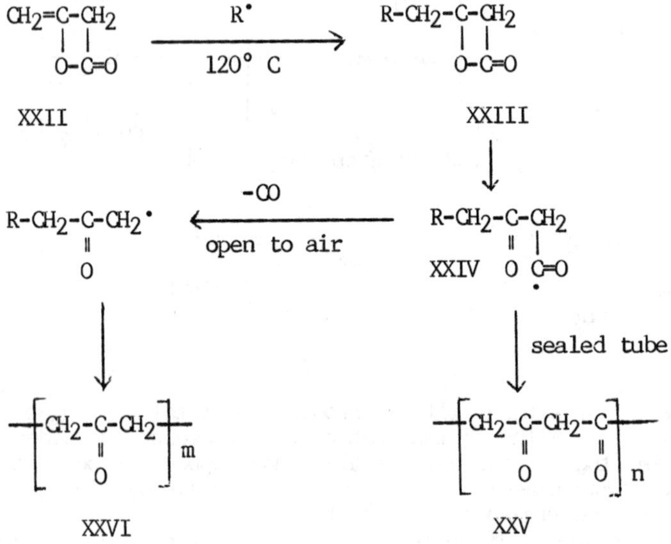

$$CH_2=C-CH_2 \quad \xrightarrow[\quad 120°\ C \quad]{R^\bullet} \quad R-CH_2-C-CH_2$$

O-C=O	O-C=O
XXII	XXIII

$$R-CH_2-C-CH_2^\bullet \quad \xleftarrow[\text{open to air}]{-CO} \quad R-CH_2-C-CH_2$$

XXIV

XXIV

sealed tube

$$\left[\begin{array}{c} CH_2-C-CH_2 \\ \| \\ O \end{array}\right]_m \qquad \left[\begin{array}{c} CH_2-C-CH_2-C \\ \| \quad\quad \| \\ O \quad\quad O \end{array}\right]_n$$

XXVI   XXV

The extent of ring opening is higher because the acyl radical XXIV is more stable than the primary radical that one would expect to obtain from the 2-methylene-1,3-oxetane (XXI). In the sealed tube the acyl radical XXIV does not lose very much carbon monoxide to give a copolymer XXV which contains a large amount of the 1,3-diketone structure. On the other hand the acyl radical XXIV, when the reaction is open to the air, can lose a substantial proportion of the acyl groups to give the copolymer XXVI which is largely the 1,4-diketone.

A surprising result was obtained from the polymerization of the 2-methylene-3,4-benzotetrahydrofuran (XXVII); XXVII gives a polymer XXVIII with little or no ring opening.

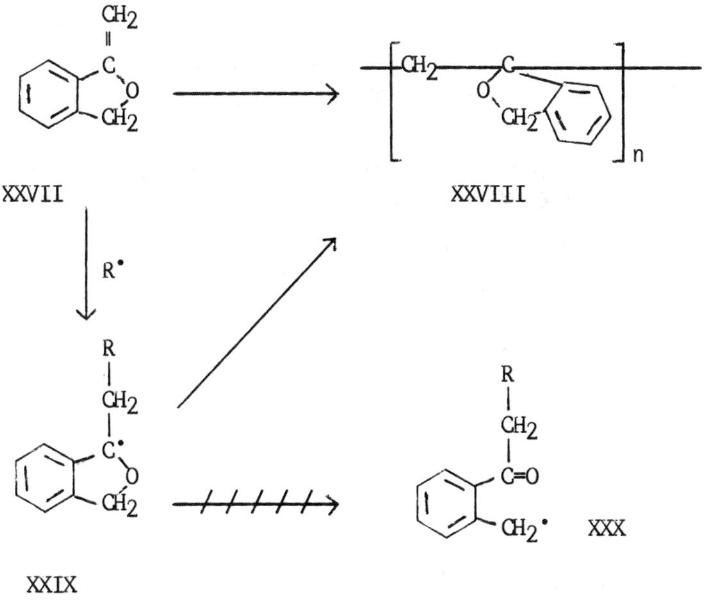

Apparently the intermediate radical XXXIX, which is a tertiary benzyl radical with additional stabilization from the oxygen, is sufficiently more stable than the open chain primary benzyl radical XXX that only direct addition takes place.

Double Ring-Opening in Free Radical Polymerization

In a program to develop monomers that expand upon polymzerization, we had prepared a series of spiro ortho esters, spiro ortho carbonates, trioxabicyclooctanes, and ketal lactones that could be polymerized ionicaly with ring opening expansion in volume (24). Since a large proportion of industrial polymers are prepared by free-radical polymerization, it was desirable to have available a series of monomers that would polymerize by a free radical process with no change in volume or slight expansion. Such monomers could be added to common monomers to produce copolymers with reduced shrinkage or expansion in volume. It was reasoned that the introduction of unsaturation into a spiro ortho carbonate would permit double ring opening with expansion in volume. For that reason we undertook the synthesis of 3,9-dimethylene-1,5,7,11-tetraoxaspiro-[5.5]undecane (XXXI) by the following set of reactions (25).

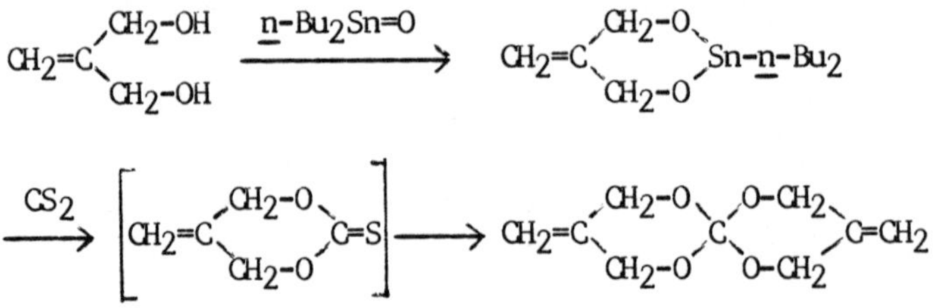

$$CH_2=C\overset{CH_2-OH}{\underset{CH_2-OH}{\big<}} \xrightarrow{\underline{n}-Bu_2Sn=O} CH_2=C\overset{CH_2-O}{\underset{CH_2-O}{\big<}}Sn-\underline{n}-Bu_2$$

$$\xrightarrow{CS_2} \left[CH_2=C\overset{CH_2-O}{\underset{CH_2-O}{\big<}}C=S\right] \longrightarrow CH_2=C\overset{CH_2-O}{\underset{CH_2-O}{\big<}}C\overset{O-CH_2}{\underset{O-CH_2}{\big>}}C=CH_2$$

When this monomer was treated with peroxide, it underwent double ring opening by the following mechanism:

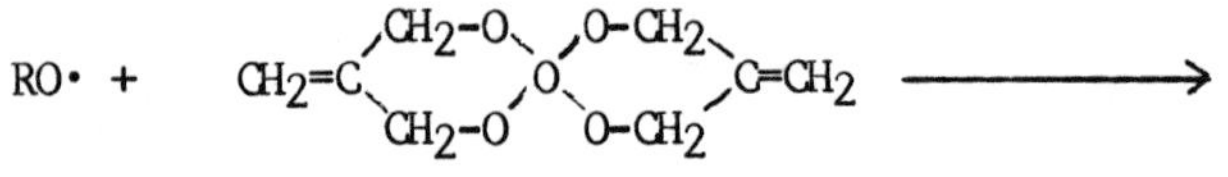

$$RO\cdot + \quad CH_2=C\overset{CH_2-O}{\underset{CH_2-O}{\big<}}\overset{O}{C}\overset{O-CH_2}{\underset{O-CH_2}{\big>}}C=CH_2 \longrightarrow$$

XXXI

$$R-O-CH_2-\overset{\bullet}{C}\overset{CH_2-O}{\underset{CH_2-O}{\big<}}C\overset{O-CH_2}{\underset{O-CH_2}{\big>}}C=CH_2 \longrightarrow$$

$$RO-CH_2-\overset{\overset{CH_2}{\|}}{C}\overset{\cdot O}{\underset{CH_2-O}{\big<}}C\overset{O-CH_2}{\underset{O-CH_2}{\big>}}C=CH_2 \longrightarrow$$

$$RO-CH_2-\overset{\overset{CH_2}{\|}}{C}\overset{O}{\underset{CH_2-O}{\big<}}C\overset{O-CH_2}{\underset{\cdot O-CH_2}{\big>}}C=CH_2 \xrightarrow{repeat}$$

$$RO-\left[\overset{\overset{CH_2}{\|}}{CH_2-C}-CH_2-O-\overset{\overset{O}{\|}}{C}-O-CH_2-\overset{\overset{CH_2}{\|}}{C}-CH_2-O\right]_x$$

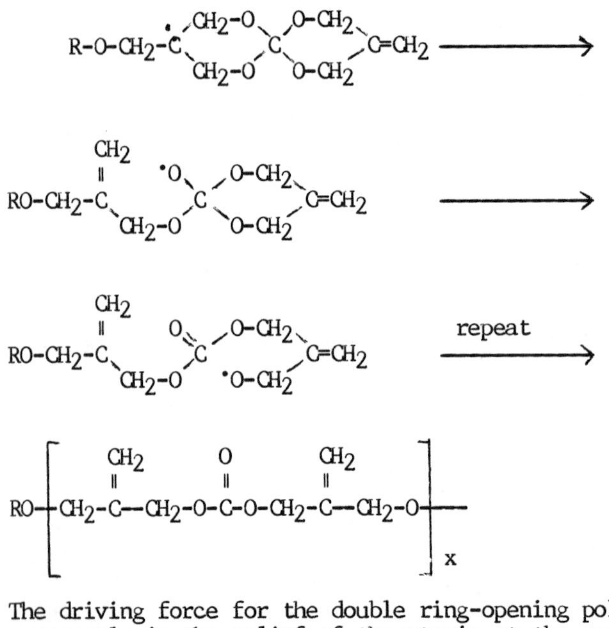

The driving force for the double ring-opening polymerization apparently is the relief of the strain at the central spiro atom as well as the formation of the stable carbonyl group (26).

The chemistry of a wide variety of unsaturated spiro ortho carbonates and unsaturated spiro ortho esters have been discussed in detail previously (14,19,24,27-30). They not only are useful for monomers that expand on polymerization (24,32-37) but are useful as dental filling materials (38), for the synthesis of oligomers capped with hydroxyl groups, production of biodegradable polymers, and enhancement of thermal stability of copolymers (39-45).

Copolymerization with Cyclic Ketene Acetals

One of the intriguing characteristics of the ethylene ketene acetal
is its ability to copolymerize with a wide variety of common monomers, including styrene and methyl methacrylate. One should note
that this process introduces an ester group into the backbone of an
addition polymer. Although the copolymerization of oxygen would
introduce a peroxide linkage into the backbone, this is the first
time that a relatively stable but yet hydrolizable functional group
has been introduced into the backbone of an addition polymer ($\underline{12}$).
     For example, when the seven-membered ketene acetal V was
copolymerized with styrene, a styrene copolymer containing ester
groups in the backbone was obtained (r_1=0.021 and r_2=22.6).

$$CH_2=C\begin{smallmatrix}O-CH_2-CH_2\\ \\O-CH_2-CH_2\end{smallmatrix} \;+\; CH_2=CH\!\!\begin{smallmatrix}\\ \\ \phi\end{smallmatrix} \quad \dfrac{(CH_3)_3-C-O-O-C-(CH_3)_3}{120°C} \longrightarrow$$

V

$$—CH_2-\overset{\overset{O}{\parallel}}{C}-O-(CH_2)_4-\!\left[CH_2-\underset{\phi}{CH}\right]_n\!-CH_2-\overset{\overset{O}{\parallel}}{C}-O-(CH_2)_4-\!\left[CH_2-\underset{\phi}{CH}\right]_m\!-$$

XXXII

     Hydrolysis of these copolymers makes possible the synthesis
of oligomers of styrene capped with a hydroxyl on one end and a carboxylic acid group on the other (27,43-45).

$$\text{XXXII} \quad \xrightarrow[\text{the } H^+]{OH^-} \quad HO-(CH_2)_4-\!\left[CH_2-\underset{\phi}{CH}\right]_n\!-CH_2-\overset{\overset{O}{\parallel}}{C}-OH$$

XXXIII

     When a mixture containing about 80% V and 20% styrene was
used, a copolymer containing 90 mole-% styrene and 10 mole-% ester-
containing units was obtained. The hydrolysis of this copolymer
gave an oligomer XXXIII of styrene containing an average of about
nine styrene units end-capped with the hydroxyl and carboxylic acid
groups. Thus a very general method has been developed for the
synthesis of a wide variety of oligomers with any desired molecular
weight range. Of course, since the copolymers are random, the molecular weight distribution of the oligomers is quite broad. However,
these oligomers should prove quite useful for the synthesis of
polyurethanes and block polyesters.

Synthesis of Biodegradable Addition Polymers

In spite of the fact that most synthetic polymers are nonbiodegradable since they have not been on the earth long enough for microorganisms or enzyme systems to have evolved to utilize them as food, polyesters that are relatively low molecular weight and rather low melting are biodegradable (46). This observation is related to the fact that poly(β-hydroxybutyric acid) occurs widely in nature and many micro-organisms use this polyester to store energy in the same way that animals use fat. On the other hand, no synthetic addition polymer was known that was readily biodegradable (28). In an effort to produce a biodegradable addition polymer the 2-methylene-1,3-dioxepane (V) and ethylene were copolymerized at 120°C for 30 minutes at a pressure of 1800 psi to give a low conversion of copolymers XXXIV with ester-containing units varying from 2.1 to 10.4 mole-%. The copolymers were in fact biodegradable with the copolymers containing the high amount of ester groups being rapidly degraded and the copolymers containing only 2.1% comonomer only slowly degraded (29). Apparently there are enzymes in the micro-organisms that are capable of hydrolyzing the ester linkages in the ethylene copolymer to produce the oligomers with terminal carboxylic acid groups; these oligomers are then degraded as analogs of fatty acids by the normal metabolic processes.

$$CH_2=CH_2 \ + \ CH_2=C \overset{O-CH_2-CH_2}{\underset{O-CH_2-CH_2}{<}} \quad \xrightarrow[\ 120°C\ ]{\text{peroxide}}$$

V

$$-CH_2-\overset{O}{\overset{\|}{C}}-O-(CH_2)_4-\left[CH_2-CH_2\right]_n-CH_2-\overset{O}{\overset{\|}{C}}-O-(CH_2)_4-\left[CH_2-CH_2\right]_m-$$

XXXIV

$$\xrightarrow[\text{then } H^+]{OH^-} \quad HO-(CH_2)_4-\left[CH_2-CH_2\right]_m-CH_2-\overset{O}{\overset{\|}{C}}-OH$$

XXXV

In a separate step the ethylene-2-methylene-1,3-dioxepane copolymer was hydrolyzed to give oligomers XXXV that were capped with a hydroxyl group at one end and a carboxylic acid group at the other. When the ester-containing unit was 2.1 mole-%, the value of n was approximately 47 and when it was 10.4 mole-%, the value of n was approximately 9. The copolymers with 6 or less mole-% of the ester-containing units had melting points in excess of 90°C (30-31).

   Copolymerization of several of the unsaturated cyclic ethers,

particularly the 2-methylene-5-phenyltetrahydrofuran (XXI) with
various monomers, is a convenient way to introduce a ketone group
into the backbone of the addition polymer. Such copolymers containing the keto groups can be expected to be photodegradable. The
competitive process involves the copolymerization of the monomer
with carbon monoxide under high pressure (47).

Ketene Acetals as New Chain Transfer Agents

Since the cyclic ketene acetals undergo free radical addition-
elimination, we wondered whether open chain ketene acetals would
also undergo a very similar free radical addition-elimination.
Johnson, Barnes, and McElvain (11) treated diethyl ketene acetal
with a peroxide and reported that nothing happened. It appeared
more likely that a free radical rearrangement occurred and no
increase in viscosity was observed. For this reason we treated a
1:1 mixture of diethyl ketene acetal and styrene with a peroxide at
120°C and obtained a low molecular weight oligomer of styrene capped
with an ethyl group at one end and an ethyl ester group at the other
but containing some copolymerized diethyl ketene acetal units (48).

Since the 4-phenyl-2-methylene-1,3-dioxolane (XXI) underwent much
more extensive ring opening than did the unsubstituted
2-methylene-1,3-dioxolane (I), it was reasoned that a benzyl ketene
acetal would be a more effective chain transfer agent than diethyl
ketene acetal. Thus when methyl benzyl ketene acetal was used with
styrene, a complete addition-elimination occurred to produce an oligomer of styrene.

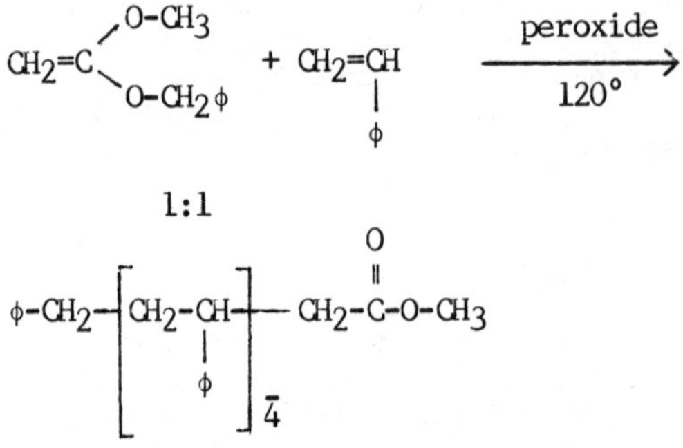

$$CH_2=C \begin{smallmatrix} O-CH_3 \\ \\ O-CH_2\phi \end{smallmatrix} \quad + \quad CH_2=CH \Big|_{\phi} \quad \xrightarrow[120°]{\text{peroxide}}$$

1:1

$$\phi-CH_2-\left[CH_2-CH \Big|_{\phi} \right]_{\overline{4}} \!\!\!- CH_2-\overset{O}{\overset{\|}{C}}-O-CH_3$$

The mechanism of the reaction is very probably as follows:

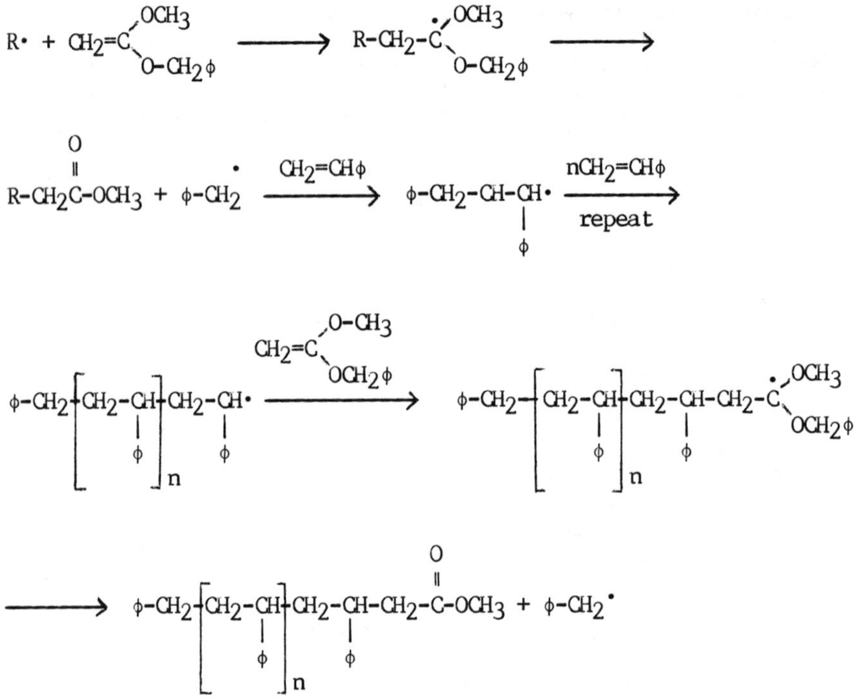

$$R\cdot \; + \; CH_2=C \begin{smallmatrix} OCH_3 \\ \\ O-CH_2\phi \end{smallmatrix} \quad \longrightarrow \quad R-CH_2-\overset{\cdot}{C} \begin{smallmatrix} OCH_3 \\ \\ O-CH_2\phi \end{smallmatrix} \quad \longrightarrow$$

$$R-CH_2\overset{O}{\overset{\|}{C}}-OCH_3 \; + \; \phi-\overset{\cdot}{C}H_2 \quad \xrightarrow{CH_2=CH\phi} \quad \phi-CH_2-CH-\overset{\cdot}{C}H \Big|_{\phi} \quad \xrightarrow[\text{repeat}]{nCH_2=CH\phi}$$

$$\phi-CH_2\left[CH_2-CH \Big|_{\phi} \right]_{n}\!\!\!CH_2-\overset{\cdot}{C}H \quad \xrightarrow{\quad CH_2=C \begin{smallmatrix} O-CH_3 \\ OCH_2\phi \end{smallmatrix} \quad} \quad \phi-CH_2-\left[CH_2-CH \Big|_{\phi} \right]_{n}\!\!\!CH_2-CH-CH_2-\overset{\cdot}{C} \begin{smallmatrix} OCH_3 \\ \\ OCH_2\phi \end{smallmatrix}$$

$$\longrightarrow \quad \phi-CH_2\left[CH_2-CH \Big|_{\phi} \right]_{n}\!\!\!CH_2-CH-CH_2-\overset{O}{\overset{\|}{C}}-OCH_3 \; + \; \phi-\overset{\cdot}{C}H_2$$

Thus the ketene acetals are a new class of chain transfer agents that do not have a sulfur or a halogen present. The introduction of functional groups, such as hydroxyl groups, into the benzyl and the alkyl group makes possible the direct synthesis of functionally capped oligomers without the necessity of hydrolysis.

Acknowledgments

This work was supported in part by the Polymer Program of the

National Science Foundation, the Frasch Foundation, and the Goodyear Tire and Rubber Company.

Literature Cited

1. Takahashi, T. J. Polym. Sci. 1968, A6, 403.
2. Cho, I.; Ahn, K. D. J. Polym. Sci., Polym. Letters Ed. 1977, 15, 751.
3. Errede, L. A. J. Polym. Sci. 1961, 49, 253.
4. Hall, Jr., H. J.; Ykman, P. Macromol. Rev. 1976, 11, 1.
5. Tobolski, A. V.; Eisenberg, A. J. Am. Chem. Soc. 1959, 81, 780.
6. Butler, G. B.; Angelo, R. J. J. Am. Chem. Soc. 1957, 79, 3128.
7. Maillard, B.; Forrest, D.; Ingold, K. U. J. Am. Chem. Soc. 1976, 98, 7024.
8. Streitwieser, Jr., A.; Heathcock, C. H. "Introduction to Organic Chemistry," 2nd ed.; Macmillan Publishing Company: New York, 1981; p. 1195.
9. Maillard, B.; Cazaux, M.; Lalande, R. Bull. Soc. Chim. Fr. 1973, 1368.
10. McElvain, S. M.; Curry, M. J. J. Am. Chem. Soc. 1948, 70, 3781.
11. Johnson, P. R.; Barnes, H. M.; McElvain, S. M. J. Am. Chem. Soc. 1940, 62, 964.
12. Bailey, W. J.; Yamazaki, N. Unpublished Results.
13. Bailey, W. J. Am. Chem. Soc., Div. Polym. Chem., Preprints 1984, 27(1), 210.
14. Bailey, W. J.; Chen, P. Y.; Chen, S. -C.; Chiao, W. -B.; Endo, T.; Gapud, B.; Lin, Y. -N.; Ni, Z.; Pan, C. -Y.; Shaffer, S. E.; Sidney, L.; Wu, S. -R.; Yamamoto, N.; Yamazaki, N.; Yonezawa, K. In "Symposium Honoring Professor Carl S. Marvel"; Mark, H.; Pearce, E., Eds. Marcel Dekker: New York, 1984.
15. Bailey, W. J.; Ni, Z.; Wu, S. -R. J. Polymer Sci, Polym. Chem. Ed. 1982, 20, 2420.
16. Bailey, W. J.; Wu, S. -R.; Ni, Z. J. Macromol. Sci. - Chem. 1982, A18(6), 973.
17. Bailey, W. J.; Ni, Z.; Wu, S. -R. Macromolecules 1982, 15, 711.
18. Bailey, W. J.; Wu, S. -R.; Ni, Z. Makromol. Chem. 1982, 183, 1913.
19. Bailey, W. J.; Arfaei, A.; Chen, P. Y.; Chen, S. -C.; Endo, T.; Pan, C. -Y.; Ni, Z.; Shaffer, S. E.; Sidney, L.; Wu, S. -R.; Yamazaki, N. Proc. IUPAC 28th Macromolecular Symposium, Amherst, MA, July 12-16, 1982, p. 214.
20. Sidney, L.; Shaffer, S. E.; Bailey, W. J. Am. Chem. Soc., Div. Polym. Chem., Preprints 1981, 22(2), 373.
21. Dobbs, A. J.; Gilbert, B. C.; Norman, K. U. C. J. Chem. Soc. 1971, A, 124.
22. Dobbs, A. J.; Gilbert, B. C.; Norman, K. U. C. J. Chem. Soc., Perkin Trans. 1972, 2, 786.
23. Hudrlik, P.; Hudrlik, A.; Wan, C. J. Org. Chem. 1975, 40, 1116.

24. Bailey, W. J; Sun, R. L.; Katsuki, H.; Endo, T.; Iwama, H.;
     Tsushima, R.; Saigo, K.; Bitritto, M. M. In "Ring-Opening
     Polymerization," Saegusa, T.; Goethals, E., Eds.; ACS
     SYMPOSIUM SERIES No. 54, American Chemical Society:
     Washington, D. C., 1977; p. 38.
25. Bailey, W. J.; Katsuki, H.; Endo, T. Am. Chem. Soc., Div.
     Polym. Chem., Preprints 1974, 15, 445.
26. Endo, T.; Bailey, W. J. J. Polym. Sci., Polym. Letters Ed.
     1975, 13, 193.
27. Bailey, W. J.; Endo, T.; Gapud, B.; Lin, Y. -N.; Ni, Z.; Pan,
     C. -Y.; Shaffer, S. E.; Wu, S. -R.; Yamazaki, N.; Yonezawa, K.
     "U. S./Japan Seminar on Synthesis and Reactions of Oligomers
     and End Reactive Polymers," Osaka, Japan, October 17-20, 1983,
     p. 74.
28. Bailey, W. J. "Proc. Third International Conference on
     Advances in the Stabilization and Controlled Degradation of
     Polymers," Lucerne, Switzerland, June 1, 1981, p. 12.
29. Bailey, W. J; Gapud, B. Am. Chem. Soc., Div. Polym. Chem.,
     Preprints 1984, 27(1), 58.
30. Bailey, W. J. "Proc. Sixth International Conference on
     Advances in the Stabilization and Controlled Degradation of
     Polymers," Lucerne, Switzerland, May 22-24, 1984, p.38
31. Bailey, W. J.; Gapud, B. Annals of the New York Academy of
     Sciences In Press.
32. Bailey, W. J. Kobunshi 1981, 30(5), 331.
33. Endo, T.; Bailey, W. J. J. Polym. Sci., Polym. Chem. Ed. 1975,
     13, 2525.
34. Bailey, W. J.; Endo, T. J. Polym. Sci., Polym. Sym. 1978, 64,
     17.
35. Endo, T.; Bailey, W. J. Makromol. Chem. 1975, 176, 2897.
36. Bailey, W. J.; Endo, T. J. Polym. Sci., Polym. Chem. Ed.
     1976, 14, 1735.
37. Endo, T.; Cai-Song, M.; Okawara, M.; Bailey, W. J. Polym. J.
     1982, 14(6), 485.
38. Thompson, V. P.; Williams, E. F.; Bailey, W. J. J. Dental
     Res. 1979, 58, 522.
39. Endo, T.; Bailey, W. J. J. Polym. Sci., Polym. Letters Ed.
     1980, 18, 25.
40. Fukuda, H.; Hirota, M.; Endo, T.; Okawara, M.; Bailey, W. J.
     J. Polym. Sci., Polym. Chem. Ed. 1982, 20, 2935.
41. Endo, T.; Okawara, M.; Yamazaki, N.; Bailey, W. J. J. Polym.
     Sci., Polym. Chem. Ed. 1981, 19, 1283.
42. Bailey, W. J. Polymer J. In Press.
43. Endo, T.; Okawara, M.; Bailey, W. J.; Azuma, K.; Nate, K.;
     Yokono, H. J. Polymer Chem., Polym. Letters Ed. 1983, 21,
     373.
44. Bailey, W. J.; Chen. P. Y.; Chiao, W. -B.; Endo, T.; Sidney,
     L.; Yamamoto, N.; Yamazaki, N.; Yonezawa, K. In "Contemporary
     Topics in Polymer Science," Vol. 3; Shen, M., Ed.; Plenum
     Publishing Corporation: New York, 1979; p. 29.
45. Bailey, W. J.; Gapud, B.; Lin, Y. -N.; Ni, Z.; Wu, S. -R. Am.
     Chem. Soc., Div. Polym. Chem., Preprints 1984, 27(1), 142.
46. Potts, J. E.; Clendinning, R. A.; Ackart, W. B.; Niegisch, W.

D. In "Polymers and Ecological Problems"; Guillet, J., Ed.;
Plenum Press: New York, 1973; p. 61.
47. Columbo, P.; Steinberg, M.; Fontana, J. J. Polym. Sci. 1973,
B1, 447.
48. Bailey, W. J; Endo, T.; Gapud, B.; Lin, Y. -N.; Ni, Z.; Pan, C.
-Y.; Shaffer, S. E.; Wu, S. -R.; Yamazaki, N.; Yonezawa, K. J.
Macromol. Sci. - Chem. 1984, A21, p. 979.

RECEIVED April 1, 1985

Mechanism of N-Carboxy Anhydride Polymerization

H. JAMES HARWOOD

Institute of Polymer Science, The University of Akron, Akron, OH 44325

Mechanisms proposed for strong base initiated polymerizations of N-carboxy anhydrides (NCA's) and evidence
bearing on their validity are critically reviewed. A
kinetic analysis is presented to show that the maximum
D.P. obtainable in polymerizations proceeding by the
"activated monomer" mechanism is given by $\sqrt{k_p/k_i}$ and it
is argued that $\sqrt{k_p/k_i}$ values that can reasonably be
expected for the "activated monomer" mechanism are too
low to account for the molecular weights obtained in NCA
polymerizations. It is shown that the activated monomer
mechanism for NCA polymerization bears little similarity
to the mechanism presently accepted for the alkali-ini-
tiated polymerization of lactams. Arguments favoring
the Idelson-Blout mechanism for strong base initiated
NCA polymerizations are presented.

The polymerization of N-carboxy anhydrides (NCA's) is a complicated
process that is difficult to study. The sensitivity of NCA's to
moisture and other impurities, the limited solubility of the products
of NCA polymerizations in most solvents that are suitable for anionic
polymerizations, the tendency of NCA's to associate with polypeptides,
leading to enhanced monomer concentrations in the vicinity of growing
polypeptide chains, the general complexity of ionic reactions in non-
aqueous solvents and the diversity of possible mechanisms for amide
bond formation or destruction, including catalysis by CO_2, acids,
bases, etc., collectively make it difficult to establish mechanisms
for NCA polymerizations.
    Much of our present understanding of the mechanisms of NCA polym-
erization is summarized in several recent reviews (1-4). In addition,
several papers (5-9) and references cited therein provide more current
information. Although there is agreement about the mechanism of NCA
polymerizations that are initiated by primary amines (weak base mechanism), there is considerable controversy concerning the mechanisms of
NCA polymerizations that are initiated by tertiary amines, hindered
secondary amines or basic salts. Three mechanisms have been invoked
to explain these reactions. These are the mechanism of Wieland (10-11),

0097-6156/85/0286-0067$06.00/0

which involves the formation and subsequent polycondensation of zwitterion intermediates, the mechanism of Blout and Idelson (12), which
assumes propagation by carbamate ions, and the "activated monomer" or
"a.m." mechanism, which was originally proposed by Ballard and
Bamford (13), was modified somewhat by Bamford and Block (14-15), and
was elaborated further by Szwarc (16). Experimental evidence, some
of which is contradictory, and theoretical considerations stand
against the acceptance of all three mechanisms. In his excellent review, Sekiguchi (1,9) has argued that the Blout-Idelson and the
Bamford mechanisms may operate simultaneously in some polymerization
systems and that in other systems, one or the other of these mechanisms may predominate. The presence of alcohol, for example, is proposed to favor the Blout-Idelson path. For reasons to be discussed
later in this paper, however, it seems desirable to resist this compromise interpretation.

The principal purposes of this paper are to comment on the evidence for and against these various mechanisms, to present kinetic
considerations that make the "a.m." mechanism inappropriate for aprotic base initiated NCA polymerizations and to dispel an analogy that
is frequently used to enhance the plausibility of the "a.m." mechanism. In addition, the results of experiments done to check experimental results cited in favor of the "a.m." mechanism will be mentioned and discussed.

Background and Critique

NCA polymerizations that are initiated by hindered secondary amines,
tertiary amines and basic salts yield much higher molecular weight
polymers than are obtained in primary amine initiated polymerizations.
The polymerizations initiated by aprotic bases such as sodium methoxide or sodium hydride occur considerably faster than those initiated
by amines. The molecular weights of the polymers are often considerably larger than expected based on the monomer/initiator (M/I) ratios
employed; they do not increase linearly with conversion and the ultimate molecular weight obtainable in a polymerization system may be
reached at conversions below 50 percent. The highest molecular
weights are obtained in tertiary amine initiated polymerizations and
in such cases there is no relationship of molecular weight to M/I.
The aprotic base initiated NCA polymerizations exhibit complex kinetic
behavior. These findings and others (1-4), make this class of NCA
polymerizations very complicated and difficult to understand. Three
general mechanisms have been proposed for these polymerizations.

The Wieland mechanism. Wieland (10,11) suggested that tertiary amine
initiated NCA polymerizations proceed via the polycondensation of
zwitterion intermediates. Such intermediates can more properly be
represented as betaines as proposed by Kricheldorf (17) and as represented below. If such intermediates also participate in polyaddition
reactions with NCA's, this mechanism may come close to explaining tertiary amine initiated NCA polymerizations (vide infra). This mechanism has not been given adequate attention, partly because zwitterion
intermediates such as II and III were considered more prone to cyclize
and form diketopiperazines or hydantoin derivatives than to propagate
further, partly because some investigators have claimed that tertiary

$$I + R_3N \rightleftharpoons {}^{\ominus}O-\overset{O}{\overset{||}{C}}-NHCHR\overset{O}{\overset{||}{C}}-\overset{\oplus}{N}R_3$$

II

$$(I + II) \text{ or } 2\,II \longrightarrow {}^{\ominus}O-\overset{O}{\overset{||}{C}}-NHCHR\overset{O}{\overset{||}{C}}-O-\overset{O}{\overset{||}{C}}-NHCHR\overset{O}{\overset{||}{C}}-\overset{\oplus}{N}R_3 + R_3N$$

$$\downarrow -CO_2$$

$$ {}^{\ominus}O-\overset{O}{\overset{||}{C}}-NHCHR\overset{O}{\overset{||}{C}}NHCHR\overset{O}{\overset{||}{C}}-\overset{\oplus}{N}R_3 \xrightarrow[-CO_2]{I} {}^{\ominus}O-\overset{O}{\overset{||}{C}}-(NHCHR\overset{O}{\overset{||}{C}})_3-\overset{\oplus}{N}R_3$$

III

$$\downarrow \text{II or III} \qquad\qquad\qquad \text{etc.}$$

$$ {}^{\ominus}O-\overset{O}{\overset{||}{C}}-[NHCHR\overset{O}{\overset{||}{C}}]_n-\overset{\oplus}{N}R_3 + CO_2 + NR_3$$

$$\downarrow$$

etc.

amines cannot initiate the polymerizations of N-substituted NCA's and partly because of efforts to develop a common mechanism for tertiary amine and basic salt-initiated NCA polymerizations. The general acceptance of zwitterion mechanisms for other ionic polymerizations (18) may render the Wieland mechanism more palpable today than it has been in the past. It should be noted, however, that this mechanism involves mixed carbamic acid-carboxylic acid intermediates, which are claimed to decompose too slowly to be intermediates in aprotic base initiated NCA polymerizations. Additional comments on this aspect will be discussed in the next section.

The Idelson-Blout mechanism. Idelson and Blout (12) proposed that hydroxide ion and methoxide ion initiated polymerizations of NCA's are initiated by nucleophilic attack of these ions at the 5-carbonyl group of an NCA. This leads to a carbamate ion V that can similarly react with another NCA in a propagation step that involves the formation and subsequent decarboxylation of a mixed carbamic acid-carboxylic acid anhydride.

Based on the failure of the research groups of Goodman (19-20) and Peggion (21) to find adequate amounts of end groups derived from radioactive NaOCH₃, sodium N-benzylcarbamate or diisopropylamine in polypeptides prepared by NCA polymerizations that were initiated with these materials, many workers abandoned the Idelson-Blout mechanism. However, a number of other studies have shown that initiator fragments can be found in polypeptides derived from NCA polymerizations that were initiated by alkoxide or carbamate salts (22-28). It seems that the Idelson-Blout mechanism should not be discounted based on the early radiotracer analyses.

The strongest evidence against the Idelson-Blout mechanism comes from the work of Kopple and Thursack (29) who estimated the rate of decarboxylation of a mixed carbamic acid-carboxylic acid anhydride, similar to that postulated as an intermediate in the Idelson-Blout mechanism. They reported that the decarboxylation rate was two orders of magnitude too slow to be compatible with polymerization rates reported by Blout and Karlson for NaOH initiated polymerizations of γ-benzyl-L-glutamate NCA. We have confirmed Kopple's work in our laboratory (30) but are now concerned about the interpretation of the data. First of all, the polymerization rates estimated from the Blout-Karlson data were based on assumptions concerning the concentrations of active polymerization centers present in the polymerization experiments. There is an opportunity for error here since many aspects of the polymerization are uncertain. Secondly, Kopple's experiments were conducted under acidic conditions. (A small amount of a tertiary amine was present in one experiment, but this was not enough to render the system basic). In contrast, NCA polymerizations are conducted under basic conditions. Should the decarboxylation of mixed carbamic acid-carboxylic acid anhydrides be catalyzed by bases, then the Kopple-Thursack experiment may have underestimated the decarboxylation rate of the mixed anhydride intermediates. Preliminary NMR studies we have conducted on the reaction of ethylisocyanate with phenylacetic acid in the presence and absence of sodium phenylacetate, using dimethoxyethane as a solvent, indicate a very low steady state concentration of the mixed anhydride in the presence of sodium phenylacetate (30). This indicates that decarboxylation of the mixed anhydride is catalyzed by sodium phenylacetate. Kricheldorf has shown that polymeric carbamic acid-carboxylic acid anhydrides are rapidly decarboxylated in the presence of pyridine in triethylamine at 0° (31). Kinetic studies should be done on this decomposition.

At the present time, it seems advisable to retain the Idelson-Blout mechanism, or a modified form of it. In this connection, it should be noted that many workers (1,12,31,33) have suggested that decarboxylation may accompany carbamate ion reactions in which the carbamate nitrogen acts as a nucleophile, viz.

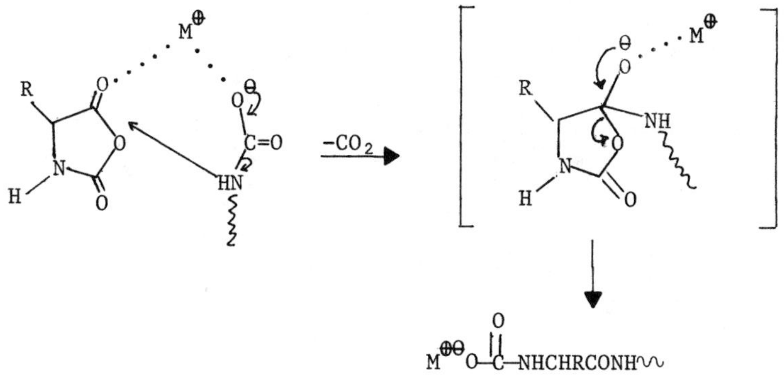

In such a case, the kinetic data of Kopple and Thursack would not be relevant to the polymerization mechanism.

The "activated monomer" mechanism. The "a.m." mechanism emphasizes the basic rather than the nucleophilic properties of aprotic initiators. The principal nucleophile considered in the mechanism is the anion obtained by abstracting a proton from a NCA, this anion being termed an "activated monomer," VI. If an aprotic base is represented by B (B=$^{\ominus}$OCH$_3$, R$_3$N, etc.), the initiation step can be represented as follows:

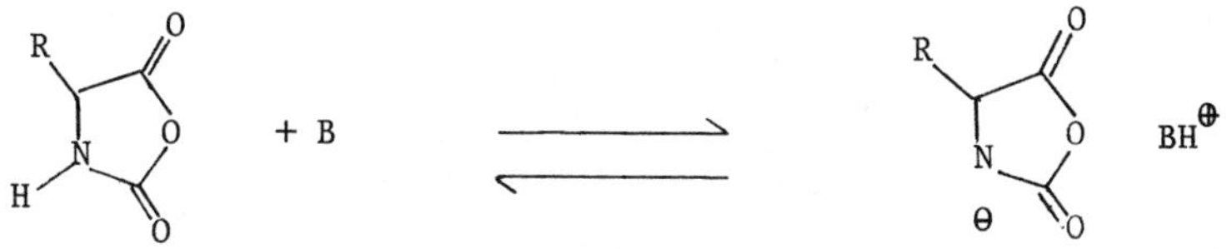

A second aspect of the initiation process involves reaction of VI with NCA to yield VII, which contains carbamate ion and N-acyl NCA moieties.

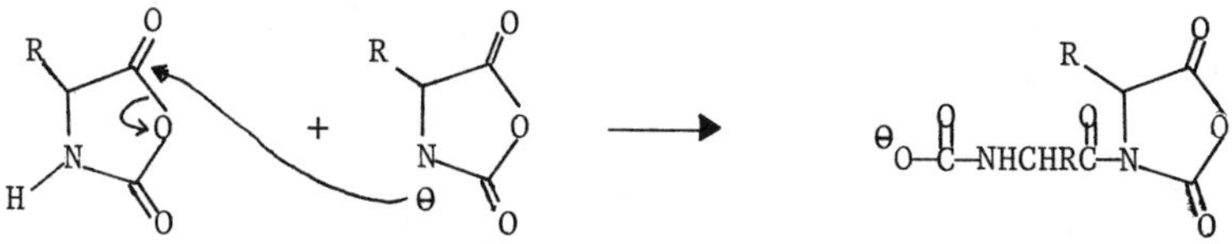

It is important to make two important observations before proceeding. First, these two reactions are completely reasonable and can occur whenever a base with a pKa greater than that of an NCA is employed. Most, if not all, of the evidence that has been cited in favor of the "a.m." mechanism (1-4,34) does nothing more than demonstrate that

species such as VI and VII can be reactive intermediates in NCA
polymerizations, at least in the early stages. The question of significance is what happens to VII once formed. It could conceivably
self-condense or initiate an Idelson-Blout polymerization process,
but it is not likely to undergo the reactions proposed for the propagation step in the "a.m." mechanism. The second point is that if VI
is present during the entire course of a polymerization reaction, it
can continually react with NCA to generate VII. It will be shown
later in this paper that this changes the character of a NCA polymerization from a polyaddition process to a polycondensation process.
Based on the considerable difference in the pKa's of NCA's and carbamic acids ($\Delta pKa \sim 6$) it seems reasonable to expect any VI generated
in the initial stages of an NCA polymerization to be extensively converted into species such as VII, and thereby be reduced to an ineffective concentration. This is not what is proposed in the "a.m."
mechanism, however.

The propagation step postulated in the "a.m." mechanism involves
reaction of the carbamate ions in VII or larger n-mers (e.g., VIII)
with NCA to generate <u>additional</u> "activated monomers." This is then
proposed to react as a nucleophile with acyl NCA moieties of n-mers
to yield carbamate ions of (n + 1)-mers, and so forth.

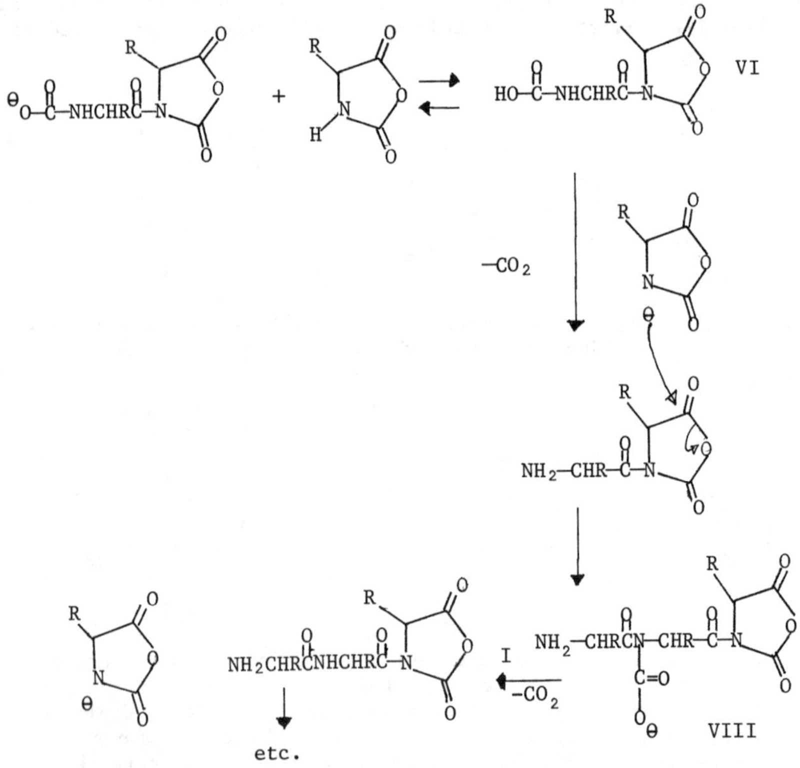

Polymer chain growth can also occur by condensation of terminal amino groups with the acyl NCA moieties of n-mers, but this is considered to occur more slowly than polyaddition.

Many workers (1,4,12,35-37) have raised serious criticisms of the reactions proposed for the propagation step in this "a.m." mechanism. The large difference in the pKa's of carbamic acids and NCA's ($\Delta pKa \sim 6$) makes the deprotonation of a NCA by a carbamate ion an extremely unreasonable step. The propagation mechanism also assumes that "activated monomer" once generated, will react almost selectively with N-acyl NCA chain ends of n-mers rather than with NCA, so that the polymerization has the character of a polyaddition process. This is an utterly unreasonable expectation. These deficiencies in the "a.m." mechanism have been pointed out in numerous papers (1,4,12,22,35-37) and it is astonishing that this mechanism continues to be accepted by so many research groups. In later portions of this paper, these arguments will be placed on a more quantitative basis, in the hope that they will thereby become more convincing.

Compromise mechanism. Sekiguchi (1,9) has proposed recently a compromise mechanism that assumes operation of the "a.m." mechanism under highly aprotic conditions and the operation of the Idelson-Blout mechanism under conditions where proton donors such as alcohols are present in appreciable concentration. His concepts appear to have support from investigators studying stereochemical aspects of NCA polymerizations, since polymerizations done under conditions that would favor the "a.m." mechanism are more stereospecific than those done under conditions that would favor the Idelson-Blout mechanism (2,38,41). However, the objections that can be raised against the "a.m" mechanism are so serious that it should not be retained as part of even a compromise mechanism.

Acid-base Considerations

The propagation step in the "a.m." mechanism involves a very unfavorable proton transfer step in which a carbamate ion abstracts a proton from an NCA to form an "activated monomer." Based on the pKa's of carboxylic acids and succinimide in water, Shalitin estimated that the pKa's of carbamic acids and NCA's differ by at least 5 (4). Similarly, based on the pKa (18.9) determined by Sekiguchi's group for α-aminoisobutyric acid NCA in dimethoxyethane (42) and on the pKa's (~ 13) determined for benzoic acid and acetic acid in non-aqueous solvents such as DMF, DMSO, dimethylacetamide and N-methylpyrrolidone (43,44), it seems that there is a six-fold difference in the ionization constants of NCA's and carbamic acids. This makes the reaction of NCA's with carbamate ions very unfavorable thermodynamically. This objection to the "a.m." mechanism has been raised many times (1,4,12, 22,35-37). It has been countered with the argument (16) that the NCA anions, are so highly nucleophilic that their reactivity toward NCA's compensates for their low concentration. With such an argument one can justify almost any intermediate for any reaction, at least until the rate constants for the postulated reactions exceed the limits imposed by diffusion processes. Unfortunately, we are not quite at that point. Calculations based on data reported by Idelson and Blout for a γ-benzyl-L-glutamate NCA polymerization (M/I=400) initiated by NaOCH$_3$ in dioxane, assuming $\Delta pKa=6$ and a carbamate ion concentration

equal to the initial initiator concentration, indicate a propagation
rate constant of the order of 10^4 $1M^{-1}sec^{-1}$. This is too low to be
limited by diffusion. Although the counter argument cannot thus be
negated, it nevertheless seems unreasonable to base a mechanism on
such an unfavorable equilibrium. The estimates of ΔpKa discussed
above are crude and there seems to be a need for better electrochemi-
cal data concerning equilibria involved in NCA polymerization systems.
It is unfortunate, for example, that Sekiguchi's group did not measure
the pKa of a carboxylic acid in dimethoxyethane when they studied the
NCA of α-aminoisobutyric acid.

Kinetic Considerations

Should the "a.m." mechanism for NCA polymerizations be correct, these
polymerizations would belong to a class of polymerizations that might
be called "monomer-promoted" polymerizations. Such polymerizations,
which include the thermal polymerization of styrene (45-47), and the
alkali metal initiated polymerization of lactams (48-52), have the
common characteristic that monomers mutually react to form the species
required for propagation. It can be shown that the maximum molecular
weight obtainable in such polymerizations is given by $\sqrt{k_p/k_i}$, where
k_i and k_p are the specific rate constants for initiation and propagation. The ratio k_p/k_i must therefore be very large if high molecular weight polymers are to be obtained in such polymerizations.
This requirement is easily met when styrene is polymerized thermally
(45-47) and when lactams are polymerized in the presence of alkali
metals (48-52) but it is doubtful that a large k_p/k_i ratio can be
justified for the "a.m." mechanism.

The result discussed above will be derived for the specific case
of the "a.m." mechanism. This derivation is similar in some respects
to one developed by Shalitin (4) for an NCA polymerization that
involves the continuous, slow generation of growing polypeptide chains
by an unspecified mechanism.

According to the "a.m." mechanism, initiation involves the following reaction.

$$[M] \qquad\qquad [M^-] \qquad\qquad\qquad\qquad [P] \qquad\qquad\qquad (1)$$

Defining P as a growing chain, the rate of initiation (dP/dt) is
then proportional to the product of the concentrations of monomer (M)
and activated monomer (M⁻), <u>viz.</u>

$$\frac{dP}{dt} = k_i \; (M^-) \; (M) \qquad\qquad\qquad (2)$$

Monomer is consumed by this initiation step and by propagation, which
occurs as follows:

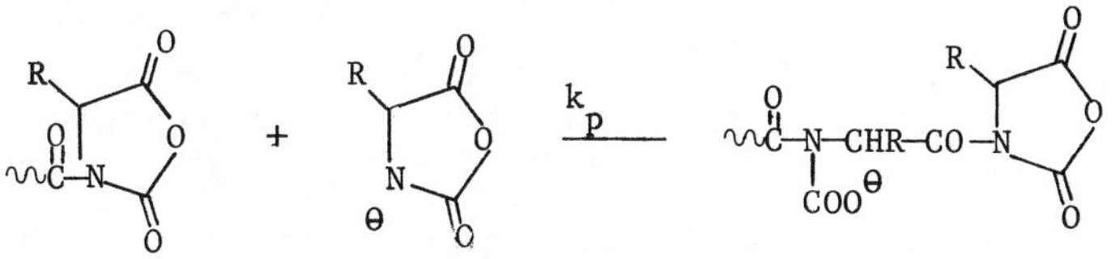

$$\text{(3)}$$

II (P)   (M⁻)   (P)

The rate of monomer consumption is given by the sum of the rates of its consumption by reactions (1) and (3), <u>viz.</u>

$$\frac{-d(M)}{dt} = 2k_i(M)(M^-) + k_p(P)(M^-) \tag{4}$$

Dividing Equation (4) by Equation (2) yields an expression for the instantaneous ratio of the rate of monomer consumption to the rate of polymer chain generation, an "instantaneous D.P."

$$\frac{-d(M)}{d(P)} = 2 + \frac{k_p(P)}{k_i(M)} \tag{5}$$

Integration of this expression between the initial limits of $M=M_0$ and $P=0$ to the final limits of $M=0$ and $P=P_f$ yields an expression for the maximum D.P. obtainable in the reaction, which is as follows where $R=k_p/k_i$.

$$\text{D.P. (Maximum)} = \sqrt{R} \, \exp\left\{ -\frac{1}{\sqrt{R-1}} \left(\tan^{-1}\frac{1}{\sqrt{R-1}} - \frac{\pi}{2} \right) \right\} \tag{6}$$

Although this equation is exact, a more meaningful approximate solution can be obtained by assuming $k_p(M^-)(P) \gg 2k_i(M^-)(M)$. Making this assumption, Equations (4) and (5) become Equations (7) and (8).

$$\frac{-d(M)}{dt} = k_p(M^-)(M) \tag{7}$$

$$\frac{-d(M)}{d(P)} = \frac{k_p(P)}{k_i(M)} \tag{8}$$

Equation (8) can be easily integrated by separating the variables, as follows:

$$-\int_0^{M_0}(M)d(M) = \frac{k_p}{k_i}\int_{P_f}^{0}(P)d(P)$$

$$\frac{-(M_0)^2}{2} = \frac{-k_p}{k_i} \cdot \frac{(P_f)^2}{2}$$

$$\boxed{\text{D.P.} = \frac{(M_0)}{(P_f)} = \sqrt{k_p/k_i} = \sqrt{R}} \tag{9}$$

According to Equation 9, <u>the maximum D.P. that can be obtained</u>
<u>by a mechanism such as the "a.m." mechanism is</u> given by $\sqrt{k_p/k_i}$.
Table I lists D.P.'s calculated according to equations (6) and (9) as
well as values obtained by CSMP (53) and Monte-Carlo simulations of
"a.m." polymerization. All show that the maximum D.P. is very close
to $\sqrt{k_p/k_i}$.

Since D.P.'s greater than 300 are not uncommon in synthetic
polypeptides prepared by strong base-initiated NCA polymerizations,
k_p/k_i must be at least 10^5 if the "a.m." mechanism <u>prevails</u>. It is
unreasonable to expect the 5-carbonyl group in an N-acyl NCA (II in
Equation (3)) to be 10^5 times more reactive than the 5-carbonyl group
in an NCA (I in Equation (1)). Based on data presented by Kricheldorf
(54), the reactivities of glycine NCA or alanine NCA and N-acetyl-
glycine NCA can be estimated to differ by factors of 30-90. Even
these factors seem higher than what would seem reasonable for the
enhancement of the electrophilicity of 5-carbonyl group that would
result from acylation of a nitrogen atom two bonds away, with no con-
jugation possible. N-acyl NCA's are simply not reactive enough,
relative to NCA's for the "a.m." mechanism to be correct.

Table I.   Maximum D.P. For "A.M." Polymerizations

$R = k_p/k_i$	Maximum D.P.		
	$\sqrt{R}$	Exact Integral	Monte Carlo Simulation	CSMP
5	2.32	3.89	3.89	3.89
50	7.07	8.67	8.67	8.67
100	10.00	11.59	11.57	11.59
500	22.36	23.94	23.83	23.97
1000	31.62	33.20	33.02	33.23
5000	70.71	72.28	72.30	72.33
10,000	100.00	101.57	101.21	101.54
50,000	223.60	225.17	223.41	225.14
100,000	316.22	317.79	312.11	317.77
500,000	707.11	708.67	705.55	708.63

<u>Concerning the Analogy Between the "a.m." Mechanism for NCA Polymeri-</u>
<u>zation and The Mechanism for The Anionic Polymerization of Lactams.</u>

A large number of authors (1,3,4,16,55-58) casually equate the "a.m."
mechanism for NCA polymerization to the generally accepted mechanism
for the anionic polymerization of lactams (45-52). The two mechanisms share only the common features that cyclic amide ions are
involved and that N-acylamides, which promote propagation reactions,
are created in the initiation step. There is a fundamental difference between the two mechanisms, however; k_p/k_i <u>can be</u> as large as
10^5 in the case of the anionic polymerization of lactams because the
leaving group in the propagation step is a much weaker base than the
leaving group in the initiation step (which is very unfavorable).

Initiation

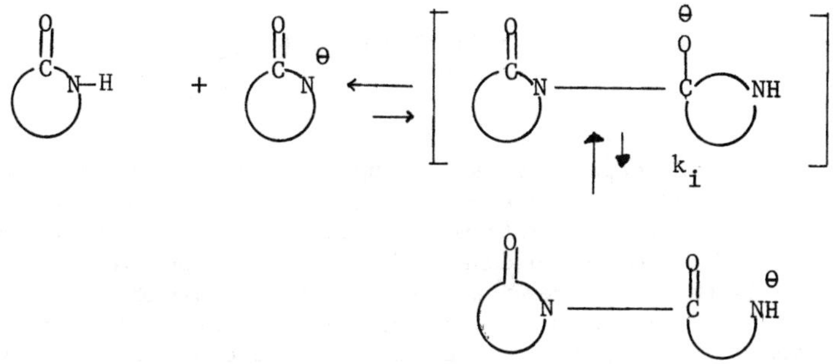

Propagation

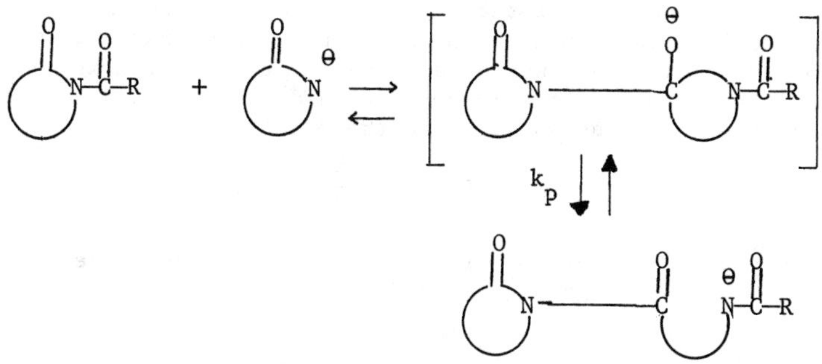

Such a difference in the basicity of the leaving groups for initiation and propagation reactions does not prevail in the case of the "a.m." mechanism, since carbamate ions are generated in both initiation (Equation 1) and propagation (Equation 3) reactions. In this case k_i and k_p should have similar values. Failure of many workers to appreciate this fundamental difference between the two mechanisms may be one of the reasons for the widespread acceptance of the "a.m." mechanism. Only when the bond between the 3-nitrogen and the 2-carbonyl carbon in an acyl NCA is broken during a rate determining propagation step can NCA polymerization bear any resemblance to anionic lactam polymerization. Even in such a case k_i and k_p may not differ by a large amount, since resonance stabilized ions would result from both initiation (carbamate ion) and propagation (amide ion) reactions. Perhaps if decarboxylation occurred in the transition state for propagation, k_p might be substantially greater than k_i. This might be a way to salvage the "a.m." mechanism, but it is clearly unacceptable in its present form.

Role of Self-Condensation Reactions

According to the "a.m." mehcanism, polymer chains are formed that contain acyl NCA moieties at one end and amino groups or carbamate ions at the other end, viz.

$$\text{NH}_2\text{CHRCO[NHCHRCO]}\underset{n}{\text{—N}}$$

Such chains are capable of self-condensation to form polymers of higher molecular weight. Peggione and coworkers (59), have obtained evidence that can be interpreted in terms of such a process. We have verified their experimental work and shown that coupling does indeed involve chain ends (60). However the coupling effects are observed only when DMF or N-formylpiperidine are employed as solvents. It does not occur when benzene or dioxane are used, so one should be cautious about generalizing this result. In addition, coupling is not observed in polymerizations of D,L-leucine NCA.

To obtain some perspective concerning the possible influence of such coupling reactions on NCA polymerizations that proceed by the "a.m." mechanism, allowance was made for the possibility of self-condensation in the kinetic scheme outlined above. This was done by adding the following equation to the model.

$$P_n + P_m \xrightarrow{k_c} P_{(n+m)} \tag{10}$$

Incorporation of this feature into the kinetic model required that Equation (2) be changed to Equation (11).

$$\frac{d(P)}{dt} = k_i(M^-)(M) - k_c(P)(P) \tag{11}$$

Dividing Equation (4) by Equation (11) yielded an expression that could not be integrated analytically. It was possible, however, to use CSMP programming (53) to simulate the process defined by these equations. The parameters required for the simulations, k_i, k_p and $x = k_c/k_p(M^-)$ were varied to determine their relative influence on D.P. vs conversion curves. Figure 1 shows the results of a typical simulation, where $R = k_p/k_i = 10$ and where x was varied from 1 to 500. The arrow points to the conversion at which monomer was depleted. The x values provide a measure of the relative importance of condensation and propagation processes in the model. It can be seen that the D.P. vs conversion curve resembles that of a condensation reaction when x is small (condensation much slower than propagation); high molecular polymer is obtained only at very high conversions, after monomer has disappeared and end-linking reactions must be used to increase M.W. When condensation is more favorable (x = 100–500), reasonable molecular weights are obtained at moderate conversions. This same general behavior holds for R values as high as 5000, as can be seen in Figures 2–4. Based on the belief that R must be below 1000, condensation reactions must accompany NCA polymerizations that proceed according to the "a.m." mechanism, if such reactions yield high molecular weight polymers. In fact, condensation must be a major feature of the polymer-forming process. This is not in accord with what is known about NCA polymerizations. The simulated D.P. vs conversion results generated in this study are not similar to any

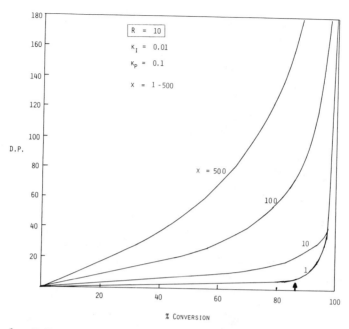

Figure 1:   D.P. vs. conversion plot for CSMP simulation of a monomer
            promoted polymerization with accompanying polycondensation
            where R=10 and X varies from 1.0 to 500.

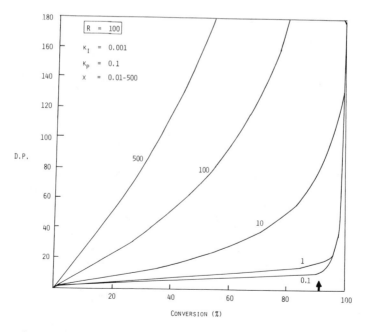

Figure 2:   D.P. vs. conversion plot for CSMP simulation of a monomer
            promoted polymerization with accompanying polycondensation
            where R=100 and X varies from 0.1 to 500.

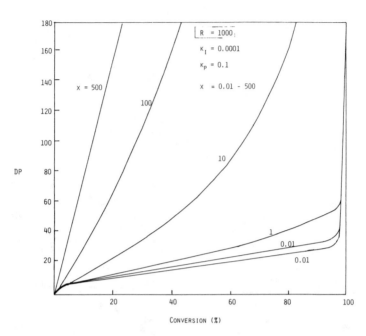

Figure 3: D.P. vs. conversion plot for CSMP simulation of a monomer
promoted polymerization with accompanying polycondensation
where R=1000 and X varies from 0.01 to 500.

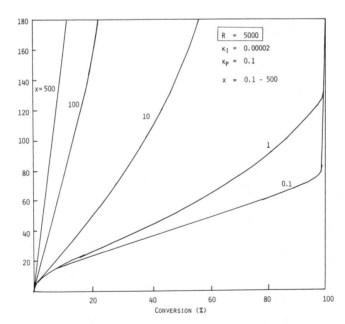

Figure 4: D.P. vs conversion plot for CSMP simulation of a monomer
promoted polymerization with accompanying polycondensation
where R=5000 and X varies from 0.1 to 500.

published results with one exception. Curves generated with R=500–2000 resemble D.P. vs conversion curves reported by Kricheldorf (61) for polymerizations of sarcosine NCA initiated by tertiary amines. Such polymerizations may proceed by propagating zwitterions according to a carbamate mechanism. Coupling of the zwitterions is expected to be a significant feature of such reactions, which may account for the agreement of the simulated curves with Kricheldorf's data. The "a.m." mechanism, if appropriate for NCA polymerization, can only account for the formation of difunctional species that subsequently condense to yield high molecular weight polymers. For this reason and many others, we do not believe that this mechanism can be valid for NCA polymerization.

The coupling effects reported by Peggion, et al. (59) and confirmed by Mackey (60) should not be taken as evidence for the "a.m." propagation mechanism in any case. As discussed in an earlier section, they may provide evidence for the "a.m." initiation mechanism but even this is not certain. A number of other routes to polypeptide chains having chain ends with acylating capability can be envisaged. Tertiary amines and hindered amines were used as initiators in Peggion's work. These might have yielded chain ends identical to those proposed by Kricheldorf (17) which would have acylating capability.

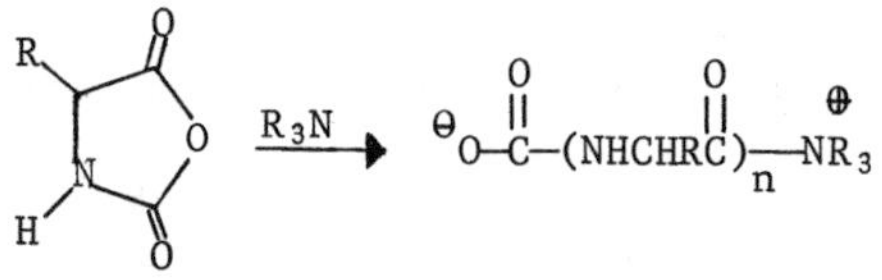

In addition, the fact that chain coupling has been mostly observed for polymerizations conducted in DMF or N-formylpiperidine suggests that Vilsmeyer-like reactions (62,63) may be responsible for the acylating capability of the chain ends, viz.

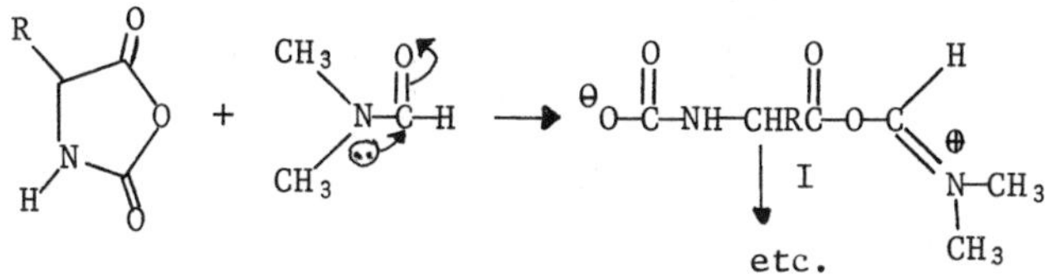

Furthermore, if carboxylic acid salts are generated in the reaction, these can react with the 5-carbonyl of an NCA to generate a carboxylic acid–carboxylic acid chain end. As has already been shown by Miwa and Stahman (64), such structures have acylating capability. The formation of hydantoin acetic acid derivatives from NCA's in DMF solution is a rather favorable process (65) and this is therefore also a likely source of acylating capability in the systems studied by Peggion.

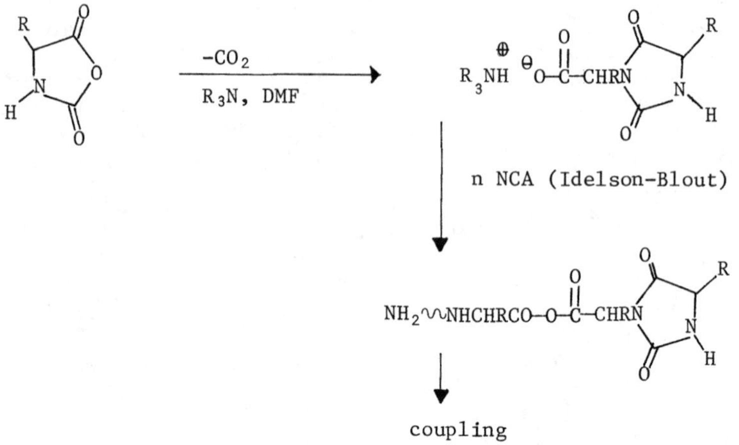

n NCA (Idelson-Blout)

coupling

In those instances where coupling is observed in basic salt-initiated polymerizations (<u>66</u>) it may be explained by reaction of a portion of the initiator with the 2-carbonyl of an NCA. This would generate a carboxylic acid salt that can react with an NCA to yield a carbamate ion that contains a carboxylic acid anhydride function. Such a species could function in the same manner proposed for N-acyl NCA's.

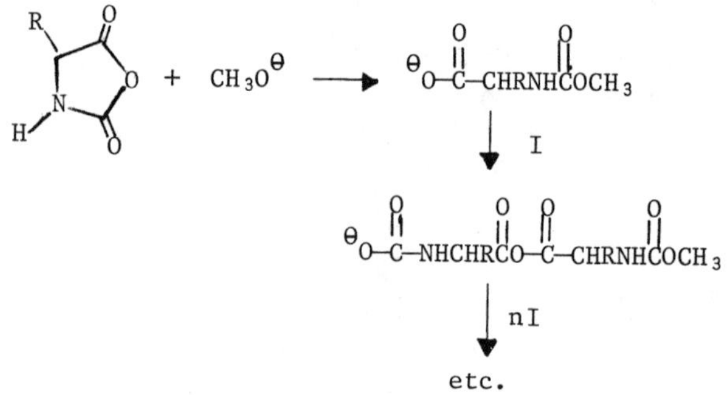

etc.

It is clear that more work must be done before the interesting results of Peggion, <u>et al</u>. can be explained unequivocally. And it must be remembered that even if Peggion's interpretation is correct, his results demonstrate the validity of only the reactions postulated for the initiation step in the "a.m." mechanism; they cannot provide evidence for the controversial propagation step.

The Carbamate Ion Mechanism as a Viable Alternative

The propagating carbamate ion mechanism can explain most features of NCA polymerizations without the "hocus pocus" that attends many aspects of the "a.m." mechanism. The principal objections against it are the low nucleophilicity of the carbamate ion and the argument of Kopple (29) that mixed anhydrides of carbamic and carboxylic acids do not decompose at rates high enough to be compatible with rates observed in NCA polymerizations. The carbamate ion should certainly be nucleophilic enough to equilibrate with an NCA or any other carboxylic acid anhydride, since the displacing and displaced groups are both carboxylate ions having similar basicity. In fact, carboxylic acids and their anhydrides are known to equilibrate readily. Independent measurements of the rates of decarboxylation of mixed carbamic acid-carboxylic acid anhydrides need to be made under a variety of conditions before the Idelson-Blout mechanism is abandoned. Finally, the proposal that carbamate ions can react with NCA's without the intermediate formation of mixed carbamic acid-carboxylic acid anhydrides seems very attractive and should be tested further.

Although a universally accepted mechanism for strong base initiated NCA polymerization remains to be established, it is hoped that the comments made herein will help clarify some aspects of the subject.

Acknowledgments

This work was supported in part by the National Science Foundation. The author is indebted to Mrs. Louise Hutchison for neat and accurate typing.

Literature Cited

1. Sekiguchi, H., Pure and Appl. Chem. 1981, 53, 1689.
2. Imanishi, Y., Pure and Appl. Chem. 1981, 53, 715.
3. Goodman, M.; Peggion, E., Pure and Appl. Chem. 1981, 53, 699.
4. Shalitin, Y., in "Kinetics and Mechanisms of Polymerization, Vol. 2, Ring Opening Polymerization"; Frisch, K.C.; Reegen, S.L., Eds.; Marcel Dekker: New York, 1969, p. 421.
5. Kricheldorf, H.R.; Mulhaupt, R., J. Macromol. Sci.-Chem. 1980, A14, 349.
6. El-Sabbah, M.M.B.; Elias, H.G., Makromol. Chem. 1981, 182, 1617.
7. Hashimoto, Y.; Imanishi, Y., Biopolymers 1981, 20, 507.
8. Kricheldorf, H.R.; Hull, W.E., Biopolymers 1982, 21, 1635.
9. Amouyal, H.; Coutin, B.; Sekiguchi, H., J. Macromol. Sci.-Chem. 1983, A20, 675.
10. Wieland, T., Angew. Chem. 1951, 63, 7.
11. Wieland, T., Angew. Chem. 1954, 66, 507.
12. Idelson, M.; Blout, E.R., J. Am. Chem. Soc. 1958, 80, 2387.
13. Ballard, D.G.H.; Bamford, C.H., J. Chem. Soc. 1956, 381.
14. Bamford, C.H.; Block, H., J. Chem. Soc. 1961, 4989.
15. Bamford, C.H.; Block, H., in "Polyamino Acids, Polypeptides and Proteins"; Stahmann, M.A., Ed., Wisconsin University Press: Madison, 1962; Chapter 7.
16. Szwarc, M., Adv. Polym. Sci. 1965, 4, 1.
17. Kricheldorf, H.R., Makromol. Chem. 1976, 177, 1243.
18. Johnston, D.S., Adv. Polym. Sci. 1982, 42, 53.
19. Goodman, M.; Arnon, U., J. Am. Chem. Soc. 1964, 86, 3384.
20. Goodman, M.; Hutchison, J., J. Am. Chem. Soc. 1966, 88, 3627.

21. Peggion, E.: Terbojevich, M.; Cosani, A.; Colombini, C.,
    J. Am. Chem. Soc. 1966, 88, 3630.
22. Seeney, C.E.; Harwood, H.J., J. Macromol. Sci.-Chem. 1975, A9,
    779.
23. Avny, Y.; Zilkha, A., Isr. J. Chem. 1965/1966, 3, 307.
24. Avny, Y.; Migdal, S.; Zilkha, A., Europ. Polym. J. 1966, 2, 355.
25. Avny, Y.; Zilkha, A., Europ. Polym. J. 1966, 2, 367.
26. Giannakidis, D.; Harwood, H.J., J. Polym. Sci., Polym. Lett. Ed.
    1978, 16, 491.
27. Sawan, S.P.; Suzuki, T.; Mackey, D.E.; Harwood, H.J., Am. Chem.
    Soc., Div. Polym. Chem., Preprints 1979, 20, 657.
28. Goodman, M.; Peggion, E., Vysokomol. Soed. 1967, A9, 247.
29. Kopple, K.D.; Thursack, R.A., J. Chem. Soc. 1962, 2065.
30. Nyeu, T.K.; Niknam, M.; Harwood, H.J., unpublished work.
31. Kricheldorf, H.R., Makromol. Chem. 1971, 149, 127.
32. Miwa, T.K.; Stahmann, M.A., in "Polyamino Acids, Polypeptides
    and Proteins"; Stahmann, M.A., Ed.; The University of Wisconsin
    Press: Madison, 1962; pg. 94.
33. Walker, E.E., Proc. Internat. Colloq. on Macromolecules,
    Amsterdam 1949; pp 381-382.
34. Goodman, M.; Peggion, E.; Szwarc, M.; Bamford, C.H.,
    Macromolecules 1977, 6, 1299.
35. Katchakki, E.; Sela, M., Adv. Protein Chem. 1958, 13, 243.
36. Seeney, C.; Harwood, H.J., J. Macromol. Sci.-Chem 1975, A9, 779.
37. Kopple, K.D., J. Am. Chem. Soc. 1957, 79, 6443.
38. Hashimoto, Y.; Imanishi, Y., Biopolymers 1981, 20, 507, 489.
39. Imanishi, Y., J. Polym. Sci., Macromol. Rev. 1979, 14, 1.
40. Kricheldorf, H.R., Makromol. Chem. 1982, 183, 2113.
41. Kricheldorf, H.R.; Mang, T., Makromol. Chem. 1981, 182, 3077.
42. Sekiguchi, H.; Froyer, G., J. Polymer Sci., Symposium No. 52
    1975, 157.
43. Kolthoff, I.M.; Elving, P.J., "Treatise on Analytical Chemistry,
    Part 1, Volume 2," J. Wiley & Sons: New York 1979, pg. 278.
44. Tremillon, B., "Chemistry in Non-aqueous Solvents," D. Reidel
    Pub. Co.: Dordrecht Holland/Boston, 1974.
45. Mayo, F.R., J. Am. Chem. Soc. 1953, 75, 6133.
46. Pryor, W.A.; Coco, J.H., Macromolecules 1970, 3, 500.
47. Pryor, W.A.; Lasswell, L.D., in "Advances in Free Radical
    Chemistry"; Williams, G.J., Ed., 1975, 5, p. 27.
48. Sebenda, J., "Progress in Polymer Science, Volume 6"; Jenkins,
    A.D., Ed.; Pergamon Press: London, 1978; p. 123.
49. Sebenda, J., in "Comprehensive Chemical Kinetics, Vol. 15, Non-
    radical Polymerization"; Bamford, C.H.; Tipper, C.F.H., Eds.;
    Elsevier: New York, 1976; Chapter 6.
50. Reimschuessel, H.K., in "Kinetics and Mechanisms of Polymeri-
    zation, Volume 2, Ring Opening Polymerization"; Frisch, K.C.;
    Reegen, S.L., Eds.; Marcel Dekker: New York, 1969; Chapter 7.
51. Wichterle, O.; Sebenda, J.; Kradicek, U., Adv. Polym. Sci.,
    1961, 2, 578.
52. Millich, F.; Seshardi, K.V., in "Cyclic Monomers"; Frisch, K.C.,
    Ed.; Wiley Interscience: New York, 1971; Chapter 3.
53. Harwood, H.J.; Dworak, A.; Nyeu, T.K.L.; Tong, S.-N., in
    "Computer Applications in Applied Polymer Science"; Provder, T.,
    Ed.; ACS SYMPOSIUM SERIES NO. 197, American Chemical Society:
    Washington, D.C. 1982; p. 65.

54. Kricheldorf, H.R., Makromol. Chem. 1977, 178, 905.
55. Odian, G., "Principles of Polymerization, 2nd Edition," J. Wiley
    & Sons: New York, 1981; pp. 543-545.
56. Elias, H.G., "Macromolecules 2, Synthesis and Materials,"
    Plenum Press; New York, 1977; Section 18.24, pp. 637-638.
57. Bamford, C.H.; Block, H., in "Cyclic Monomers; Frisch, K.C., Ed.;
    Wiley Interscience: New York, 1972; Chapter 7, pg. 729.
58. Bamford, C.H.; Block, H., in "Comprehensive Chemical Kinetics,
    Volume 15"; Bamford, C.H.; Tipper, C.F.H., Eds.; Elsevier: New
    York, 1976; Chapter 8.
59. Terbojevich, M.; Pizziola, G.; Peggion, E.; Cosani, A.; Scoffone,
    E., J. Am. Chem. Soc. 1967, 89, 2733.
60. Mackey, D.E., Ph.D. Dissertation, The University of Akron,
    Akron, Ohio, 1980.
61. Kricheldorf, H.R.; Bosinger, K., Makromol. Chem. 1976, 177,
    1243.
62. Vilsmeier, A.; Haack, A., Ber deutsch. Chem. Ges. 1927, 60, 119.
63. Spiewak, J.W., private communication, for which we are grateful.
64. Miwa, T.K.; Stahmann, M.A., in "Polyamino Acids, Polypeptides,
    and Proteins"; Stahmann, M.A., Ed.; Univ. of Wisconsin Press:
    Madison, 1962; p. 81.
65. Bamford, C.J.; Elliott, A.; Hanby, W.E., "Synthetic Polypeptides,"
    Academic Press: New York, 1956; pp. 92-93.
66. Choi, N.S.; Goodman, M., Biopolymers 1972, 11, 67.

RECEIVED October 26, 1984

Ring-Opening Polymerization in the Synthesis of Block Copolymers

D. H. RICHARDS

PERME, Ministry of Defence, Waltham Abbey, Essex, EN9 1BP, United Kingdom

A survey is given of the methods developed to synthesize novel block copolymers in which at least one of the polymer segments is polytetrahydrofuran (poly THF). The techniques covered include the use of transformation reactions, direct reaction between living polymers and quaternization of performed polymers. The relative efficiencies of these synthetic routes are discussed.

The synthesis of block copolymers of well defined structures and chain lengths, and uncontaminated with the component homopolymers, relies heavily on the use of 'living' polymer techniques. These are defined as techniques in which transfer and termination reactions are absent, and the propagating end therefore remains available for subsequent reaction even after polymerization has ceased.

Such living conditions are found principally in anionically initiated systems and involve common monomers such as styrene, α-methylstyrene, butadiene and isoprene (1,22). They are far less common in cationically initiated systems, there being virtually no established example involving vinyl monomers, but some cyclic monomers such as tetrahydrofuran (THF) and the oxetanes may be polymerized under carefully specified conditions to yield living polymers (2). Although living free radical systems have also been described in which radicals have been preserved on surfaces, in emulsion, or by precipitation before termination occurs, these are special conditions not easily adapted for clean block copolymer synthesis.

Block copolymers are therefore most readily prepared by the sequential addition of monomers to systems polymerizing under living ionic conditions. However, this approach is limited in its applicability by the necessity of meeting two requirements, viz (1) that the monomers involved must all be capable of clean polymerization by the selected propagating mechanism, and (2) that the order of monomer addition must be such that the polymer anion generated by the preceding monomer must be capable of initiating rapidly the polymerization of the succeeding monomer. The former

0097–6156/85/0286–0087$06.00/0

condition limits the combination of monomers which can be used to
make block copolymers since, for example, block copolymers of
styrene (polymerized anionically) and THF (polymerized cationically)
cannot be prepared in this way. The latter conditions restrict the
order in which sequences of the polymer segments can be synthesized;
for example, living polystyrene can rapidly initiate the
polymerization of ethylene oxide (3), but the alcoholate ion of
living polyethylene oxide cannot initiate styrene.

It appeared, therefore, that considerable development work was
necessary in order to devise new synthetic techniques capable, at
least in principle, of extending the combinations of monomers which
could be converted into block copolymers of the desired compositions
and structures. This paper summarizes the results of attempts by
the author's group to increase the number of copolymers in which one
of the component monomers is THF.

Results and Discussion

The ring opening polymerization of THF can be effected only by
cationic initiators, but under stringently defined conditions this
polymerization may be carried out with the minimum of transfer, and
essentially living polymers are formed. We have made use of the
conditions described by Groucher and Wetton (4), and have developed
techniques for preparing homopoly THF which involve the reaction in
THF of a selected organic halide with the silver salt of a strong
acid (Equation 1) (5).

$$\langle O \rangle - CH_2Br \xrightarrow[AgPF_6]{THF} \langle O \rangle - CH_2 - O \oplus \langle O \rangle \quad PF_6^{\ominus} \quad + 2AgBr \qquad (1)$$

Living monofunctional poly THF is prepared using p-methylbenzyl
bromide (6) as the co-initiator, and living difunctional poly THF is
prepared using m-xylylydibromide as coinitiator (5). These
procedures or adaptations thereof have been employed in the
techniques described below.

Transformation Reactions

If the mechanism of propagation can be transformed at will from one
best suited to the first monomer to that best suited to the second
and so on, then the restriction that all monomers must be
polymerizable by the same mechanism is removed. Further
consideration of this concept indicates that each transformation
requires three distinct steps; the first involving the living
polymerization of the first monomer using the appropriate mechanism,
the second involving termination of the polymer with an appropriate
reagent capable of effecting the subsequent transformation,
isolating the polymer and then dissolving it in an appropriate
solvent containing the second monomer, and the third step carrying
out the reaction on the metastable polymer end group to transform it
into an active specie capable of polymerizing the second monomer.

The following transformation reactions involving the ring
opening polymerization of THF have been examined.

Anion to Cation

Two related procedures have been developed to effect this transformation. Both involve the initial synthesis of mono- or difunctional living anionic polymers of styrene, butadiene, or block copolymers of both. They are then reacted via Grignard intermediates (7) with either excess bromine or with excess m-xylylyl dibromide (8-10) to yield polymers with reactive halide terminal groups (benzylic or allylic depending upon the polymer and terminating agent). The reactions for polystyrene are shown in equations 2 and 3.

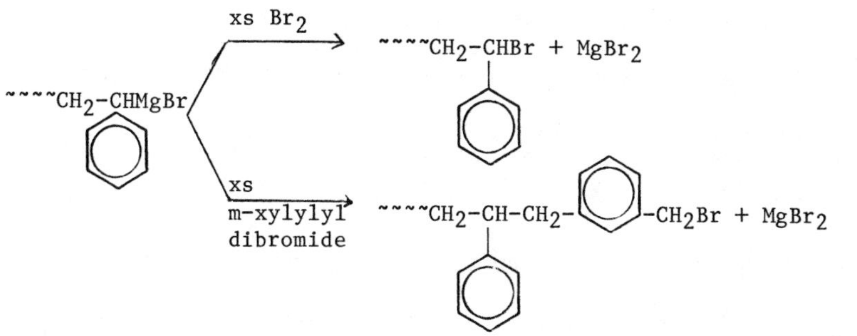

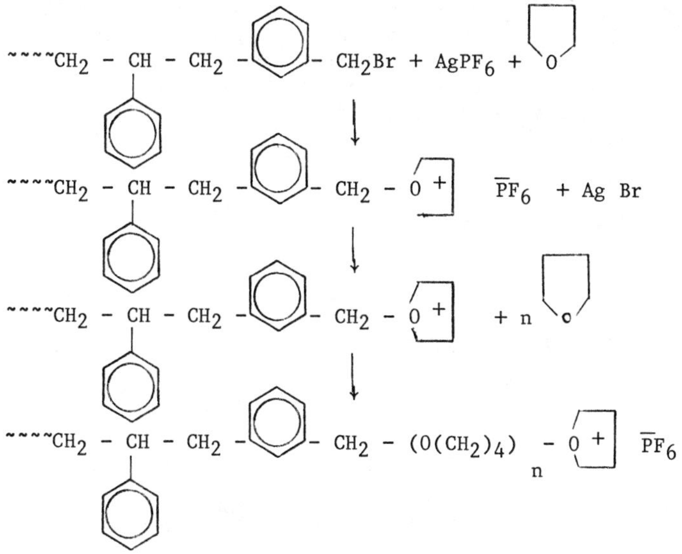

These materials may be isolated and stored if required, and then redissolved in THF before reaction with a silver salt, such as $AgPF_6$, to generate an oxonium ion capable of polymerizing THF under living conditions (Equations 4 and 5). Of the two brominating agents used, m-xylylyl dibromide is favored since it may be used in the presence of unsaturated polymer without side reactions, the reactivity of the terminal bromide is independent of the polymer chain and, most importantly, the absence of hydrogen on the penultimate carbon prevents the competing chain stopping proton elimination reaction from occurring (Equation 6, with product from Equation 2).

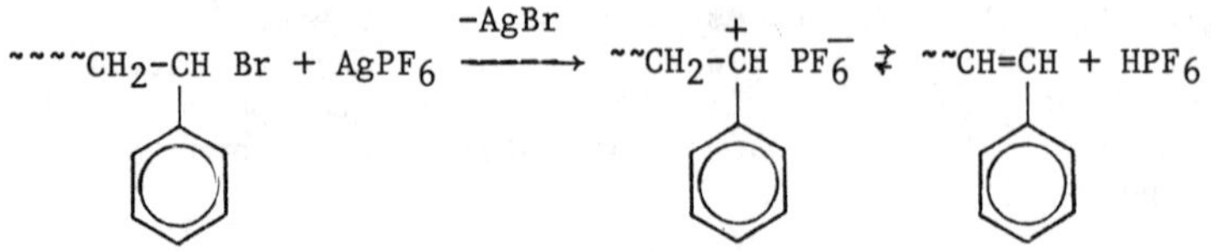

$$\sim\sim\sim\sim CH_2\text{-}CH\ Br\ +\ AgPF_6\ \xrightarrow{\ -AgBr\ }\ \sim\sim CH_2\overset{+}{\text{-}}CH\ \overline{PF_6}\ \rightleftharpoons\ \sim\sim CH\text{=}CH\ +\ HPF_6 \qquad (6)$$

This transformation yields block copolymers of low polydispersity (M_w/M_n = 1.1 to 1.2) with an efficiency of about 90% (10). AB, ABA, ABC and ABCBA block copolymers have been prepared in this way involving styrene and butadiene as monomers with mono- or difunctional anionic initiators. More complex structures may also be made by titrating the living cationic block copolymers with difunctional reagents such as the disodium salt of resorcinol (Equation 7).

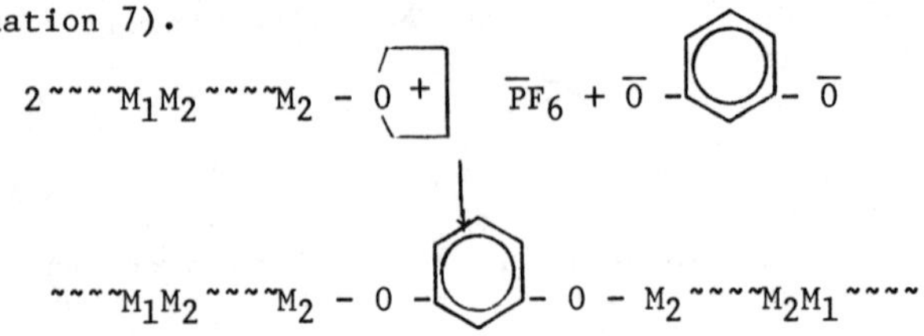

$$2\sim\sim\sim\sim M_1 M_2 \sim\sim\sim\sim M_2 - \left[O + \right]\ \overline{PF_6}\ +\ \overline{O} - \bigcirc - \overline{O}$$

$$\qquad (7)$$

$$\sim\sim\sim\sim M_1 M_2 \sim\sim\sim\sim M_2 - O - \bigcirc - O - M_2 \sim\sim\sim\sim M_2 M_1 \sim\sim\sim\sim$$

Cation to Anion

Attempts to effect this transformation efficiently have not been very successful. Mono-or difunctional living poly THF was prepared using the appropriate cationic initiator and subsequently reacted with the sodium salt of cinnamyl alcohol or of 1-phenyl-1-buten-4-ol to yield styryl groups on terminal ligands (Equation 8) in which n = 1 or 2 respectively (11). In both cases reaction was shown to

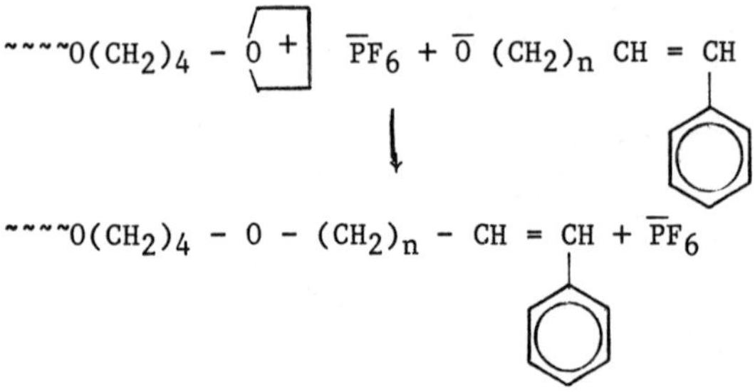

$$\sim\sim\sim\sim O(CH_2)_4 - \left[O + \right]\ \overline{PF_6}\ +\ \overline{O}\ (CH_2)_n\ CH = CH$$

$$\qquad (8)$$

$$\sim\sim\sim\sim O(CH_2)_4 - O - (CH_2)_n - CH = CH + \overline{PF_6}$$

be quantitative, and the materials were subsequently isolated, redissolved in THF or benzene and reacted with n-butyl lithium to generate styryl anions (Equation 9) which should then have co-

$$\sim\sim\sim O(CH_2)_4 - O - (CH_2)_n - CH = CH + n\ BuLi$$

$$\downarrow \quad\quad nBu$$

(9)

$$\sim\sim\sim O(CH_2)_4 - O - (CH_2)_n - CH - \overline{C}H\ Li^+$$

polymerized freshly added styrene monomer. Although the red color characteristic of styryl anions was generated rapidly on introduction of n–butyllithium and the styrene subsequently added was polymerized, molecular weight determinations on the isolated polymer indicated that only about 20% of the poly THF chains had participated in the copolymerization process. The reason for this low efficiency of block copolymer formation is not known at present.

A second approach to effecting this transformation was attempted by reaction of living poly THF with a primary amine to generate a terminal secondary amine ligand (12) (Equation 10), and

$$\sim\sim\sim O(CH_2)_4 - O + \quad \overline{P}F_6 + RNH_2 \longrightarrow \sim\sim\sim O(CH_2)_4 - NRH + HPF_6 \quad (10)$$

by its subsequent reaction with potassium to generate the amide anion. The amination reaction can be made quantitative (13), as can the metallation. These macromolecular initiators polymerized added styrene quantitatively, but the rate of initiation versus that of polymerization was slow. Consequently only about 30% of the poly THF chains produced block copolymers, and the molecular weight distributions were broad (typically M_w/M_n = 2 to 3).

In summary, the cation to anion transformation has not as yet been made an efficient process, and further development is required in this area.

Cation to Free Radical

This transformation has been effectively achieved by combining techniques devised within our Group with those developed at the Liverpool, and the hitherto unpublished results described here were obtained cooperatively at the latter institution (14).

Living poly THF was terminated by the lithium salt of bromoacetic acid to yield a polymer possessing terminal bromide ligands (Equation 11). This reaction can be made essentially

$$\sim\sim O(CH_2)_4 - O + \quad \overline{P}F_6 + Li\ OOCCH_2Br \longrightarrow \sim\sim O(CH_2)_4 - OOCCH_2Br + LiPF_6 \quad (11)$$

quantitative, and the material isolated and dissolved in methyl methacrylate (MMA) into which dimanganese decacarbonyl has been added as photoinitiator (λ = 436 nm). Irradiation resulted in

free radical formation (Equation 12) and the consequent formation of
AB poly (THF-b-MMA) copolymer with some ABA material also present.

$$\sim\sim\sim\sim O(CH_2)_4-OOCCH_2Br+Me(0) \longrightarrow \sim\sim\sim\sim O(CH_2)_4OOCCH_2\bullet + Me(I)Br \qquad (12)$$

The relative proportions of the two copolymers are directly
dependent on the nature of the PMMA radical terminating process
under the experimental conditions, which at 25°C is 70/30 dispro-
portionation/combination. The photoinitiating system has been
studied in great detail with small molecule organic halides, and has
been shown to be very efficient (15). Similar high efficiencies
have been observed with bromide terminated poly THF, with no
evidence of formation of homo PMMA being obtained.

Clearly, this technique can be extended to include other
monomers polymerizable by free radical mechanisms and work has
begun using chloroprene. It must be stressed, however, that the
specificity and purity of the product is largely controlled by the
detailed kinetics of the radical polymerization process, in
particular the prominence of chain transfer reactions and the nature
of the termination step which will, of course, vary from monomer to
monomer.

Reactions of Living Anionic Polymers With Living Poly THF

A potentially simpler approach to the synthesis of block copolymers
containing poly THF is the direct reaction of living cationic
homopoly THF with anionic living polymers (Equation 13). Assuming

$$\sim\sim\sim\sim O(CH_2)_4- \stackrel{+}{\boxed{O}} \quad PF_6^- + Li^{+-}M\sim\sim\sim\sim \longrightarrow \sim\sim\sim\sim O(CH_2)_4-M\sim\sim\sim\sim + LiPF_6 \quad (13)$$

a clean metathetical reaction, this route could lead to AB, ABA, BAB
and $(AB)_n$ block copolymers, depending on the functionalities of the
two reagents.

The feasibility of this approach was first demonstrated by
Berger et al. (16) using living polystyrene but, although they
showed that block copolymer was formed, they did not determine the
efficiency of the reaction. We have studied this reaction in more
detail (17), and have shown that with living polystyrene the process
is essentially quantitative. All combinations of block copolymer
structures listed above were prepared in very high yield, and
calculations based on the molecular weight of $(AB)_n$ obtained when
difunctional reagents were employed gave a reaction efficiency of at
least 95%.

Similar reactions involving living poly (α-methylstyrene) were
not so efficient, the coupling efficiency being of the order of 20%
only (18). The competing reaction in this case was identified as
being one of proton abstraction from living poly THF to yield an
unsaturated terminal group (Equation 14).

The effectiveness of this coupling process in the synthesis of
block copolymers with living poly THF is, therefore, highly

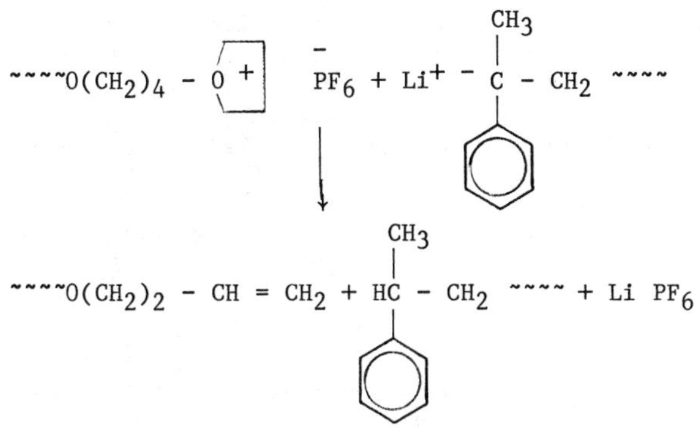

$$\sim\sim\sim O(CH_2)_4 - O^+ \quad PF_6^- + Li^+ \ ^- \overset{\overset{\displaystyle CH_3}{|}}{C} - CH_2 \sim\sim\sim$$

$$\downarrow$$

$$\sim\sim\sim O(CH_2)_2 - CH = CH_2 + H\overset{\overset{\displaystyle CH_3}{|}}{C} - CH_2 \sim\sim\sim + Li\,PF_6 \qquad (14)$$

dependent on the nature of the anionic living polymer involved, and so each system requires individual evaluation. An interesting polymer combination hitherto not examined is that of living poly THF and the living polydienes. If metathesis proves efficient, its development could generate many block copolymers of great morphological interest.

Block Copolymers With Ionic Linking Groups

More recently we have begun to explore the synthesis of block copolymers in which the linking of the polymeric segments is by a stable ionic group. Two such synthetic routes involve living poly THF, and the reaction schemes are outlined below.

In the earlier approach, living poly THF was reacted with a secondary amine to generate a polymer possessing tertiary amine end groups (19) (cf Equation 10). The product was then isolated and subsequently reacted with polystyrene possessing a terminal bromide ligand derived either by direct reaction with bromine (Equation 2) or by reaction with m-xylylyl dibromide (Equation 3). Gpc studies showed no observable quaternisation with the former material, but with the latter material there was evidence of significant reaction (Equation 15), albeit only after a period of several days at ambient temperatures. Indeed, although the gpc traces were complicated by

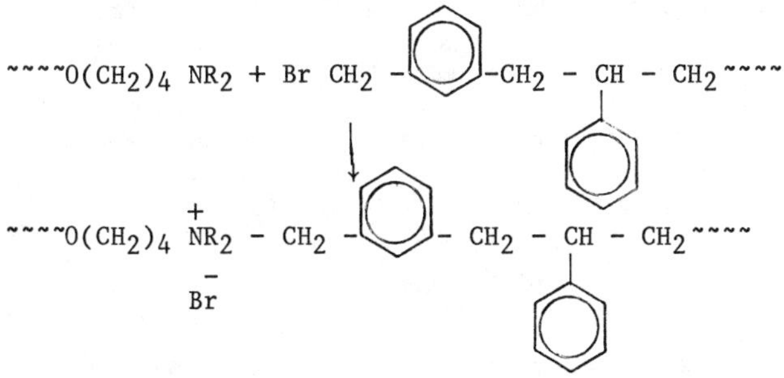

$$\sim\sim\sim O(CH_2)_4 \ NR_2 + Br\ CH_2 - \bigcirc - CH_2 - CH - CH_2 \sim\sim\sim$$

$$\downarrow \qquad\qquad\qquad (15)$$

$$\sim\sim\sim O(CH_2)_4 \ \overset{+}{NR_2} - CH_2 - \bigcirc - CH_2 - CH - CH_2 \sim\sim\sim$$
$$\underset{Br^-}{}$$

the presence of unreactive coupled polystyrene, there was a strong indication of complete consumption of the bromine terminated polystyrene component under the reaction conditions prevailing.

An alternative approach to synthesizing block copolymers with ionic linking groups was later developed. Polystyrene with tertiary amine and groups was prepared by the reaction of living polystyrene with α, w—chloroamines such as 3-(dimethylamino)propyl chloride (20) (Equation (16). This reaction was shown to take place with at least

$$
\sim\sim CH_2\text{-}\overline{C}H \ Li^+ + Cl(CH_2)_3N(CH_3)_2 \rightarrow \sim\sim CH_2\text{-}CH\text{-}(CH_2)_3N(CH_3)_2 + LiCl \quad (16)
$$

95% efficiency, and the product was precipitated, purified and redissolved in THF. A molar equivalent of living poly THF was then added and the resulting material was isolated. Gpc examination showed that block copolymer had been formed in virtually quantitative yield (Equation 17) (21); moreover the reaction was found to be very fast at room temperature. Since the gegen ion can be easily exchanged if desired, the product of reaction 17 can be

$$
\sim\sim CH_2\text{-}CH\text{-}(CH_2)_3\text{-}N(CH_3)_2 + PF_6^- \ \Big[\overset{+}{O}\text{-}(CH_2)_4O\sim\sim \rightarrow
$$

$$
\sim\sim CH_2\text{-}CH\text{-}(CH_2)_3\text{-}\overset{+}{N}(CH_3)_2(CH_2)_4O\sim\sim \ P\overline{F}_6 \quad (17)
$$

made structurally very similar to that of Equation 15, and the reaction involving living poly THF directly, because of its rapidity and efficiency, must therefore be regarded as the preferred synthetic route to such block copolymers.

The relative rapidity of the quaternisation reaction involving living poly THF compared with polystyrene bromide is directly related to the greater electrophilicity of the former reagent. This facile reaction has also been used to prepare homopoly THF with quaternary ammonium salt linkages at predetermined positions along the polymer chain (19). Thus tertiary amine terminated poly THF has been reacted with further living poly THF to generate a quaternised homopolymer (of equation 17). Gpc analysis of the process again indicated that the reaction was both facile and specific. The molecular weight of the product and the position of the quaternary ammonium grouping along the chain are therefore easily and independently controlled by appropriate choice of the molecular weights of the component poly THF reagents.

Literature Cited
  1. Szwarc, M. 'Carbanions, Living Polymers and Electron Transfer
      Processes' Interscience, 1968.
  2. Penczek, S.; Kubiza, P.; Matyjaszewski, K. 'Cationic Ring-
      Opening Polymerization', Advances in Polymer Science 37,
      Springer-Verlag 1980.

3.   Richards, D. H.; Szwarc, M. Trans Faraday Soc., (1959), 58, 1644.
4.   Croucher, T. G.; Wetton, R. E. Polymer (1967), 17, 205.
5.   Burgess, F. J.; Cunliffe, A. V.; Richards D. H.; Thompson, D. Polymer, (1978), 19, 334.
6.   Richards, D. H.; Thompson, D. Polymer, (1979), 20, 1439.
7.   Burgess, F. J.; Cunliffe, A. V.; MacCallum, J. R.; Richards, D. H. Polymer, (1977), 18, 719.
8.   Burgess, F. J.; Cunliffe, A. V.; MacCallum, J. R.; Richards, D. H. Polymer, (1977), 18, 726.
9.   Burgess, F. J.; Cunliffe, A. V.; Dawkins, J. V.; Richards, D. H., Polymer, (1977) 18, 733.
10. Richards, D. H.; Viguier, unpublished results.
11. Abadie, M. J. H.; Schue, F.; Souel, T.; Hartley, D. B.; Richard, D. H. Polymer, (1982), 23, 445.
12. Cohen, P.; Abadie, M. J. M.; Schue, F.; Richards, D. H. Polymer (1982), 23, 1105.
13. Cohen, P.; Abadie, M. J. M.; Schue, F.; Richards, D. H. Polymer, (1982), 23, 1350.
14. Bamford, C. H.; Eastmond, G. C.; Woo, J.; Richards, D. H., unpublished results.
15. Bamford, C. H. "Reactivity, Mechanisms and Structure in Polymer Chemistry' Eds Jenkins, A. D. and Ledwith, A. Wiley, NY (1974), 52.
16. Berger, G.; Levy, M.; Vofsi, D. J. Polymer Sci. (B) (1966), 4, 183.
17. Richards, D. H.; Kingston, S. B.; Souel, T. Polymer (1978), 19, 68.
18. Richards, D. H.; Kingston, S. B.; Souel, T. Polymer (1978), 19, 806.
19. Hartley, D. B.; Hayes, M. S.; Richards, D. H. Polymer (1981), 22, 1081.
20. Richards, D. H.; Service, D. M.; Stewart, M. J. Brit Polym J, in press.
21. Hurley, J. N.; Richards, D. H.; Stewart, M. J., unpublished results.
22. Noshay, A.; McGrath, J. E. "Block Copolymers: Overview and Critical Survey" Academic Press (1977).

RECEIVED March 27, 1985

Metal–Alcoholate Initiators

Sources of Questions and Answers in Ring-Opening Polymerization of Heterocyclic Monomers

P. TEYSSIÉ, J. P. BIOUL, P. CONDÉ, J. DRUET, J. HEUSCHEN, R. JÉRÔME, T. OUHADI, and R. WARIN

Laboratory of Macromolecular Chemistry and Organic Catalysis, University of Liège, Sart Tilman, 4000 Liège-Belgium

Soluble μ-oxo-bimetallic trinuclear alkoxides rank among the best initiators known for oxiranes polymerization into high M.W. polyethers. After an initial rearrangement, coordination propagation proceeds on three types of centers: non-selective ones, producing oligomers of specific lenghts, and other ones generating high M.W. chains either atactic, or stereoregular (up to 80 % isotactic). The relative importance of these three centers, working through similar but competitive pathways, can be determined to a large extent by modifying the structure of the aggregates, implying "topochemical control" in solution. These features are tentatively related to a detailed ^{1}H, ^{13}C and ^{27}Al NMR study, demonstrating the rigidity of these aggregates which offer different coordination sites for the monomer. The initiators also polymerize most lactones, the perfectly living character of the process allowing interesting block copolymerizations. Their activity towards cumulenes, i.e. isocyanates and CO_2, also leads to valuable reactions.

In agreement with a former proposal by E. Vandenbergh (1) and with the mode of preparation of different active systems, polynuclear structures are clearly a key feature for the design of catalysts able to polymerize substituted oxiranes into very high molecular weight polyethers. In an exploratory approach based on these premises and aimed at the synthesis of well characterized and versatile specie of that type, we have shown indeed (2) that soluble μ-oxo-bimetallic trinuclear alkoxides having the general formula $\left[(RO)_x M^1\text{-O-}M^2\text{-O-}M^1 (OR)_x \right]_{\overline{n}}$ rank among the best known

initiators for oxiranes polymerization, yielding products
having molecular weights in the 10^6 range and displaying
in some cases an amazing degree of stereoregularity despi-
tes the perfect solubility of the initiator and the polymer.
These compounds, particularly those where M^1=Al and M^2=
Zn, have proven to be very interesting models, allowing
a better understanding of several determinant phenomena
in ring opening polymerization, as well as in other areas
of chemistry(3). It is the purpose of this paper to sum-
marize the current status of the field, and to indicate
its potential developments .(see also ref. 4b).

Synthesis and properties of active µ-oxo-bimetallic alkoxides

- Preparation : The simplest method to obtain a stable
polynuclear system, containing the metal atoms experimentally recognized to give active systems for the high
M.W. polymerization of oxiranes, is a thermal condensation process between metal alkoxides and carboxylates,
i.e. :

$$2 \ M^1(OR)_x \ + \ 1 \ M^2(OAc)_2 \rightarrow (RO)_{x-1}^- M^1\!\!-\!O\!-\!M^2\!-\!O\!-\!M^1\!\!-\!(OR)_{x-1}^- + 2 \ ROAc \uparrow$$

Conveniently carried out in a hydrocarbon solvent like
decaline, this reaction is characterized by a rapid first
step, followed by a more difficult elimination (at 200°C)
of the second ester molecule; it is important to drive
the equilibrium to completion by distillation, and to
avoid so coordination of by products to the catalyst. It
must also be realized that too high temperatures may
further promote formation of Al-O-Al bonds upon elimination of ether (or alcohol + olefin), a detrimental process in terms of solubility and activity.
Most of the elements of the periodic table can be used
in that type of condensation reaction; additional ver-
satility is offered by the possibility to exchange OR
groups quantitatively (again by displacing equilibrium
through distillation).
Most of the glassy compounds resulting from these reactions have a composition fitting the above formula, based
on elemental and functional analysis as well as on mass
spectrometry.

- Characteristic structural features : A more detailed
analysis of these products using cryoscopic measurements
in conjunction with ^{1}H, ^{13}C and ^{27}Al NMR spectroscopy
has revealed a number of interesting structural charac-
teristics(15):they are present in solution as aggregates

$$\left[(RO)_2 \ Al\text{-}O\text{-}M_2\text{-}O\text{-}Al(OR)_2 \right]_n^-$$

probably of a globular and thermodynamically favored type

of structure (usually low and integer association number
n, from cryoscopic determinations). The NMR spectra indicate of course the presence of bridged and non-bridged
(free) OR groups, and of Al atoms with at least two different coordination environments, while electronic spectra
(visible range), indicate different types of coordination
geometries for transition metals (blue or red cobalt derivatives) depending on the nature of the R group (normal
or branched). The most important structural indication
though is obtained[6] from ^{13}C NMR spectra of $Al_2ZnO_2(On.Bu)$,
which reveal an unexpected rigidity in these aggregates :
the spectra display indeed broad diffuse resonances not
only for the α but also the β carbon of the alkyl groups,
rather unsensitive to shift reagents, dilution and temperature; more narrow bonds (with slight shift) appear only
upon addition of alcohols (known to dissociate the aggregates), oxiranes or very polar solvents. These features
obviously imply a very rigid core for the aggregate in
solution, and a number of similar but slightly non-equivalent coordination situations : the name "tecto-complexes"
is accordingly proposed for this class of compounds, which
also exhibit a particular electronic behaviour as put in
evidence by e.p.r. and visible -I.R. spectra.
Although this complex situation probably justifies the
amorphous glassy character of these products, a 1:1
acetato-alkoxide has been crystallized $(Al_2Mo_2(Oi.Pr)_8(OAc))$,
and its structure established by XR crystallography. (16)

Polymerization of oxiranes

- The catalyst : although most compounds having the
composition indicated above initiate oxiranes polymerization, their activity as well as the properties (i.e.
M.W. and stereoregularity) of the resulting polymers depend
critically on their structural characteristics : these
include not only the composition, but also the size
(degree of association) and the shape (coordination number
and geometry around the metals) of these coordinatively
aggregated compounds. These parameters, obviously difficult to determine directly and independently, depend
in a very sensitive manner on slight variations in the
synthesis conditions, as shown by changes in kinetic
data and product characteristics.
The $(n.BuO)_4Al_2O_2Zn$ compound, with a mean degree of association $n = 6$ in non polar solvents is particularly
efficient : the half-polymerization time of methyloxirane (MO) in heptane at 50°C amounts to 5 minutes (with
$[MO] = 1M$, and $[Zn] = 10^{-2}$ .M); the reaction follows a simple
kinetics $(R=k'.[C][M])$ and $\bar{M}_n$ over 10^6 are currently
obtained.
These features, together with the stability of the
catalyst, make of it one of the best candidates for a

possible industrial production of high M.W. polyether
rubbers.

- The active groups : that polymerization proceeds by
insertion into the Al-OR bonds is clearly demonstrated
by the fact that linear chains produced contain (after
hydrolysis) one OR and one OH end-group. Furthermore,
the direct relationship between the catalytic properties
and a μ-oxo-bridged multinuclear bimetallic structure
has been ascertained by using oxo-alkoxides obtained from
hydrolysis of Meerwein complexes[4] : $Al_2(OR)_6 \cdot M^2(OR)_2$
+ $H_2O \leftrightarrows Al_2M^2O_2(OR)_4$ + 2 ROH. Although the starting
Meerwein complex only produces (slowly) low oligomers,
the hydrolyzed structure display exactly the same structural and kinetic charasteristics as the product obtained
from the thermal condensation reaction described above.

- Type of mechanism : a typical flip-flap coordination
mechanism (obviously involving also a $C^{\delta-} \ldots Al^{\delta+}$ pola-
rization) is strongly supported by the following features:
the reaction rate is extremely sensitive to the bulkiness
and ligand character of the oxiranes substituents (i.e.
$t^{0.5}$ are 1600 and over 2000 minutes for epichlorohydrin
(ECH) and t.butyloxirane (t BO) respectively), decrease
with increasing solvent polarity, and obeys to more
complex kinetic laws for strongly coordinating monomers[17].
But the more interesting indication comes from solvent
competition experiments in the copolymerization of ECH
and MO[4b]: in non-polar solvents, the slower but strongly
coordinating comonomer ECH is incorporated preferentially
(in a globally slow reaction) while in polar solvents
wiping off this coordination preference, MO largely takes
over in a high rate process (apparent r_{ECH}/r_{MO} from 16.3
to 0.01).

- The initiation period, extending up to a few percent
conversion, is a typical autoacceleration stage during
which the mean M.W. of the product increases and broadens
significantly. It has been shown that in fact, the catalytic aggregate which produces initially oligomers of a
preferred specific lenght (depending on factors such as
monomer : catalyst ratio, solvent, temperature), soon
develop the ability to build up high MW chains, the oligomer content decreasing from more than 80 % down to as
low as 3 % under suitable conditions (high (M):(C) ratio).
More or less rapidly (depending on the structure) after
that initial rearrangement, one enters an apparent steady-
state situation obeying simple kinetics.

- The chain propagation proceeds on 3 different types
of active sites[4,5]:

a) non-selective ones, producing oligomers by random opering of the oxirane ring(as indicated by NMR analysis and use of optically active monomer); these oligomers, similar to those obtained in the initiation period, exhibit particular lenghts depending on the conditions, and sometimes multimodal distributions which are believed to reflect somehow the size and distribution of aggregates.

b) non-selective ones, but producing high M.W. atactic polymers, and stereoselective ones yielding isotactic high M.W. polyether. The degree of isotacticity, which can vary from 5 to 80 % depending on the conditions and structural parameters, relates clearly to the size and shape of the catalytic aggregate as indicated by experiments with $(RO)_4 Al_2O_2Co^{II}$ compounds having different degrees of association and coordination geometries. It seems accordingly probable that these different but simultaneous pathways, take place through the same insertion propagation scheme yielding open-chain products of different lenghts and stereoregularities; these differences will depend on the steric environment of the particular center involved , the relative amount and activity of each one being a sensitive function of the aggregate structure. Selective poisoning of the catalyst with Li Cl allowed an evaluation of the maximum relative content in active sites promoting high polymer formation, i.e. 4 môle % of the total amount of $Al_2ZnO_2(OR)_4$.

It must also be stressed that with some catalytic systems (again depending sensitively on their synthesis conditions), a "quasi-living" behaviour is observed : $\overline{M}_n$ increases with conversion and (M) : (C) ratio, while resumption of polymerization with an additional crop of monomer also leads to a corresponding chain lenghtening.

- Correlation with the aggregate structure. Although tBO does not polymerize at a significant rate at room temperature, it changes significantly the MO polymerization characteristics, even when added in a 1:1 (tBO): (C) ratio. That observation prompted a ^{13}C NMR study[6] of that oxirane interaction with the catalyst, in other words its use as a molecular probe for the behaviour of an oxirane monomer on the catalytic aggregate. Chemical shifts by 18 and 27ppm. for the CH and CH_2 resonances respectively, confirm a very strong coordination of the t BO on $Al_2ZnO_2 (On.Bu)_4$. The first stoechiometric equivalent is quantitatively bound, while the second is in partial equilibrium; the corresponding i-propyl derivative, which has been shown to be less rigid and less associated by NMR and cryoscopy, induces apparently a weaker coordination. Interestingly, asymmetric doublets are observed for

both CH and CH_2 resonances of t.BO, strongly suggesting
at least two different types of coordination sites. In
the presence of small amounts of n-butanol, known to
partly dissociate the aggregate, the bands accordingly
sharpen but still indicate strong coordination of the
oxirane, thus a successfully competing ligand; at the
same time, the shape and shift of both CH and CH_2 resonances are significantly modified, indicating an envi-
ronmental change in the coordination sites.

Obviously, these observations might be correlated at
least in part to the results reported above in terms
of products distribution. They certainly illustrate
the structural complexity of these systems, and support
the general idea of a "topochemical control" of such polymerizations in solution, a borderline case between
homogeneous and heterogeneous catalysis.

Polymerization of other monomers

- Lactones : the fast and perfectly living polymerization
of lactones by complex aluminum alkoxides has now been
well documented(7), where it is possible to control the
catalyst efficiency by the nature of its ligands and the
degree of association. This behaviour has opened the way
to new block copolymerization methods : two of them are
worthwhile to be mentioned here, although many other ones
are to be expected :

a) the synthesis of caprolactone block copolymers with
   styrenes or dienes (8), extremely useful as emulsifiers
   for the stabilization of different polymer blends(9);

b) the preparation of lactide block copolymers with other
   lactones, potentially interesting as biodegradable
   implants (10).

- Cumulenes : interestingly, μ-oxo-bimetallic alkoxides
are also active towards heterocumulenes. Two important
examples involve :

a) isocyanates, which are able to polymerize(11) to very
high M.W. (10^6) polyamides-1 $\left[N-\overset{\displaystyle O}{\underset{\displaystyle R}{C}} \right]_n$. Again, the resulting

product, formed by insertion into the Al-O bonds, contains an oligomer (cyclic trimer isocyanurate, being the
sole product above a T_C of ca. 70°C), and a bimodal high
polymer (main product at low temperature and conversion).

b) carbon dioxide, which inserts slowly into all of the
Al-OR bonds to produce the corresponding mixed carbonate.
Under pressure (3B) and in the presence of methyloxirane,
one obtains a copolymer containing ca. 25 % carbonate
units (12).

- <u>Olefins</u> : Some structural modifications of similar
oxoalkoxides containing transition metals (Ti plus
another metal) have lead to complexes which, in the
presence of additional aluminium alkyls, polymerize
ethylene and propylene[13]. In that latter case,
interesting variations in stereoregularity can be
precisely controlled.

    <u>In conclusion</u>, complex metal alkoxides and par-
ticularly those of aluminum, are very active and significant model compounds which can be tailored to better
understand and control a variety of interesting reactions.
In particular, they represent a rather unique example
of surface-like behaviour in solution, and are certainly
prone to broad extensions in different areas of chemistry:
sufficeit to mention here their binding to OH groups-
bearing supports (polymeric or inorganic) in a controlled
state of dispersion, and the fixation of molecular oxygen
(14 b) under the active form of peroxo or even superoxo
complexes when $M^2 = Fe^{2+}$ or MO^{2+}.
Other important questions, i.e. the understanding and
control of termination reactions in ring-opening polymerization, certainly deserve further detailed study.

Literature Cited

(1) Vandenberg E.J., Polymer Sci., (1960), <u>47</u>, 489 and
     (1969) <u>A1,7</u>, 525.
(2) a) Osgan M. and Teyssié Ph., Polymer Letters, (1967)
     <u>B5</u>, 789 and (1970), <u>B8</u>, 319, with J.J. Pasero ;
     b) Fr. Pat 1.478.334 (1965) and U.S. Pat. 3.432.445
     (1967).
(3) Teyssié Ph., Bioul J.P., Hocks L. and Ouhadi T.
     Chemtech (1977), 7, 192.
(4) a) Bioul J.P., Ph. D. Thesis, University of Liège
     (1973);
     b) Teyssié Ph., Ouhadi T. and Bioul J.P., Int. Rev.
     Sci., Phys. Chem. Ser., (1975) <u>II8</u>, 191, Butter-
     worths, London.
(5) Ouhadi T., Ph. D. Thesis, University of Liège
     (1973).
(6) Warin R., Condé P. and Teyssié Ph., to be published.
(7) Hamitou A., Jérôme R. and Teyssié Ph., J. Polymer
     Sci., Chemistry Ed., (1977) <u>15</u>, 1035; ibid (1977),
     <u>15</u>, 865 with Ouhadi T.
(8) a) Heuschen J., Jérôme R. and Teyssié Ph.(1981),
     Macromolecules, <u>14</u>, 242;
     b) Fr. Pat. 2.383.208 (1978), U.S. Pat. 4. 281.087
     (1981)

(9) Teyssié Ph. et al. in "Ring-opening Polymerization"
     Saegusa T. and Goethals E., Ed.,A.C.S.Symposium
     Series, N° 59,p.174 (1977); see also 8b.
(10) Feng X.D., Song C.X. and Chen W.Y., J. Polymer Sci.,
     Polymer Letters Ed., (1983), 21, 593.
(11) Druet J.P., Ph. D. Thesis, University of Liège
     (1976).
(12) Condé P., Ouhadi T., Warin R. and Teyssié Ph., in
     preparation.
(13) Herrmann C. and Streck R., BadNauheim Meeting,
     Fachgruppe Makromolekulare Chemie, Gesellschaft
     Deutscher Chemiker, April 1980.
(14)a)Durbut P., Teyssié Ph., and Marbehant L.,
      in preparation;
    b)Hocks L. Teyssié Ph. et al., J. Molecular Cat.,
      (1977), 3, 135, and (1980) 7, 75.
(15) Ouhadi T. et al., Inorganica Chimica Acta, (1976),
     19, 203.
(16) Lamotte J., Dideberg O., Dupont L. and Durbut P.,
     Cryst. Struct., Comm., (1981),10, 59.
(17) Xie H. , personal communication.

RECEIVED October 23, 1984

Solvent and Substituent Effects in the Anionic Polymerization of α,α-Disubstituted β-Propiolactones

ROBERT W. LENZ, EILEEN M. MINTER[1], DOUGLAS B. JOHNS[2], and SØREN HVILSTED[3]

Department of Chemical Engineering, University of Massachusetts, Amherst, MA 01003

Solvent and substituent effects in the anionic polymerization of α-ethyl-α-n-butyl-β-propiolactone, EBPL, were investigated in dimethyl sulfoxide and in N-methylpyrrolidone. In dimethyl sulfoxide, DMSO, gelation of the growing polymer resulted in decreased polymerization rates and formation of bimodal molecular weight distributions, but polymerization in N-methylpyrrolidone, NMP, continued at a constant rate despite precipitation of the growing polymer. The polymerization rate was less in DMSO than in NMP although the former has a higher dielectric constant. It is suggested that solvation of the carboxylate ion endgroup by DMSO may be responsible for the lower rate. The propagation rate constant for EBPL in DMSO was lower than that for the α-methyl-α-butyl monomer indicating that the steric acceleration observed in the α-methyl-α-n-alkyl monomer series did not continue when two large alkyl groups are present. It is believed that living polymers are formed in all of the polymerization reactions of EBPL although somewhat broad molecular weight distributions are obtained because of the heterogeneity of the reactions.

The anionic polymerization reactions of α,α-disubstituted-β-propiolactones, and the properties of the resulting polyesters, with the structures shown on the next page, have been under investigation in this laboratory for over ten years (1,2).

[1] Current address: Rochester, NY 14612
[2] Current address: Ethicon, Somerville, NJ 08876
[3] Current address: Scandinavian Paint and Printing Ink Research Institute, Denmark

0097-6156/85/0286-0105$06.00/0
© 1985 American Chemical Society

$$R_1 \overbrace{}^{R_2\,O} \longrightarrow \left[OCH_2 \underset{R_2}{\overset{R_1\ O}{C}} C \right] \qquad R_1, R_2 = CH_3, C_2H_5, \underline{n}\text{-}C_3H_7, \underline{n}\text{-}C_4H_9$$

Unusual and unexpected solvent (3) and substituent (4) effects
on the polymerization rates have been observed in these studies,
and quite surprisingly, the polymers obtained from racemic monomers
(i.e., those in which R_1 and R_2 are different) have all been found
to be crystalline (1,2,5), although they are apparently atactic (6).
For the unexpected substituent effects on propagation rates, it was
observed that the rate constants for propagation, k_p, increased with
increasing substituent size within a series of α-methyl-α-$\underline{n}$-alkyl-β-
propiolactones (R_1 = CH_3, R_2 = C_2H_5, C_3H_7, C_4H_9), which were polymerized in either tetrahydrofuran, THF, or dimethyl sulfoxide, DMSO,
with tetraethylammonium benzoate, TEAB, initiator (4). In DMSO the
activation energies also increased with increasing size of the $\underline{n}$-
alkyl group as might be expected because this substituent on the
α-position could cause steric hindrance to attack at the β-position
in the S_N2 reaction. That is, nucleophilic substitution not carbonyl
attack is the mechanism of the propagation reaction, as follows:

$$\sim\sim CH_2 \underset{(CH_2)_x}{\overset{CH_3\ O}{\underset{|}{C}}} C \overset{O}{\underset{O^{\ominus}}{{}}} N Et_4^{\oplus} \quad + \quad CH_3(CH_2)_x - \underset{CH_2\!-\!O}{\overset{CH_3}{\underset{|}{C}}} - C \overset{O}{} \xrightarrow{\ k_p\ } I$$

It is of interest to note also that the increase in activation
energy was offset by an increase in activation entropy with increasing alkyl group size along the series of ethyl (x = 1), propyl (x =
2) and butyl (x = 3) for R_2.

For the unexpected solvent effects referred to above, it was
observed that the rate of propagation, as indicated by the measured
value of k_p, was higher for the reaction in THF than in DMSO, even
though the latter is a much higher polarity solvent. We suggested
that the reduced reactivity in DMSO could be attributed to solvation
of the carboxylate anion by this solvent with formation of a stable,
relatively unreactive solvated ion pair by this endgroup nucleophile.

The present report describes a continuation of these rate investigations with another monomer in this series and also the
preparation of polymers with other α-substituents.

Results and Discussion

Anionic Polymerization of α-Ethyl-α-Alkyl-β-Propiolactones.   For comparison with the previously studied series of α-methyl-α-alkyl substituted monomers and polymers, three monomers containing α-ethyl-α-n-alkyl substituents (R_1 = C_2H_5, R_2 = C_2H_5, C_3H_7, C_4H_9) were prepared and polymerized in DMSO at 30°C with TEAB initiator (see Table I for conditions), and the polymers obtained were characterized for solubility, molecular weight and melting points, with the results shown in Tables I and II. Rate studies were also made on one member of this series, α-ethyl-α-n-butyl-β-propiolactone at five different temperatures with the results shown in Table III.

The molecular weight distributions reported in Table I are seen to be quite broad for a "living polymer" polymerization reaction, although for two of the three polymerization reactions in this table (for R_2 = C_2H_5 and C_3H_7), the observed, $\overline{M}_{n,GPC}$, and calculated, $\overline{M}_{n,Calc}$, number average molecular weights (see the footnotes in Table I for definitions) are in good agreement. For the third reaction in which R_2 is the n-C_4H_9 group, the observed $\overline{M}_n$ was less than half of that expected on the basis of the monomer-to-initiator mole ratio.

In the first two polymerizations of this series in which R_2 is either C_2H_5 or n-C_3H_7, the polymers precipitated as they formed during the reactions at fairly low conversions. Quite likely, therefore, the living endgroups in these growing polymer chains, while still reactive, had variable accessibilities to the monomer and, hence, the polymers grew at different rates. As a result, a broadened molecular weight distribution was obtained, but the average molecular weight after complete conversion of the monomer was still close to that predicted because there was no termination or transfer reaction.

In contrast, for the polymerization of the α-ethyl-α-butyl monomer, the polymer solution became an immobile gel at an intermediate conversion, and as discussed later, a large fraction of the living polymer present at that stage became inactive because the monomer was inaccessible to the endgroup. This fraction of the polymer stopped growing entirely while the other portion of the polymer in the gel which remained active, for some reason, continued to grow until all monomer was consumed. As a result, a bimodal distribution of molecular weights was obtained in the final polymer, and because the $\overline{M}_n$ and $\overline{M}_w$ values reported in Table I for this polymer are for the entire distribution, the values are meaningless in relation to the living character of the polymerization reaction.

Precipitation and gelation of these polymers from solution are believed to be caused by their crystallization, and the melting points, T_m, of the polymers determined by differential scanning calorimetry, DSC, both before, $T_{m,o}$, and after, $T_{m,ann}$, thermal annealing, are given in Table II. The melting points reported in Table II are very similar to those previously observed by Thiebaut and coworkers (5). The melting points $T_{m,o}$ listed in Table II were obtained by plotting the temperature of the final endothermic peak

Table I. Polymerization of α-Ethyl-α-alkyl-β-propiolactones Initiated by TEAB in DMSO at 30°C

R_2[a]	Reactants Monomer g	mmol (M)	TEAB mg	mmol (I)	Solvent DMSO ml	Yield[b] %	Polymer Molecular Weight $\overline{M}_n$,Calc.	$\overline{M}_{n}$,GPC[c][d]	$\overline{M}_w$,GPC[d]	$\overline{M}_w/\overline{M}_n$
C_2H_5	1.276	9.96	28.2	0.112	30.0	99	11,400	10,900	13,400	1.23
$n\text{-}C_3H_7$	1.238	8.71	24.6	0.098	30.0	100	12,700	13,800	19,300	1.41
$n\text{-}C_4H_9$	1.236	7.91	21.6	0.086	31.0	98	14,400	5,700[e]	12,800[e]	2.24[e]

a. $R_1 = C_2H_5$.

b. Polymer isolated after 65 hours of polymerization.

c. $\overline{M}_n$ calculated from [M]/[I] ratio.

d. $\overline{M}_n$ and $\overline{M}_w$ measured by gel permeation chromatography relative to polystyrene sample in THF solution.

e. Bimodal molecular weight distribution.

Table II. Elemental Analysis and Melting Points of Poly(α-ethyl-α-alkyl-β-propiolactones)

R_2	Repeating Unit Composition	Elemental Analysis of Polymers C Calc.	C Found	H Calc.	H Found	N Calc.	N Found	Polymer Melting Point, °C $T_{m,o}$[a]	$T_{m,ann}$[b]	$T_{m,lit}$[c]
C_2H_5	$(C_7H_{12}O_2)_n$	65.60	65.35	9.44	9.46	0.0	<0.1	223	231	245
$n\text{-}C_3H_7$	$(C_8H_{14}O_2)_n$	67.57	67.59	9.92	9.91	0.0	<0.1	199	209	205
$n\text{-}C_4H_9$	$(C_9H_{16}O_2)_n$	69.19	69.12	10.32	10.55	0.0	<0.1	165	175	180

a. Melting point of precipitated polymer obtained by DSC analysis extrapolated to zero heating rate.

b. Melting point by DSC analysis after annealing polymer in instrument.

c. Melting points reported in Reference 5.

(multiple peaks were observed in all cases) against the square of the heating rate (at rates from 10 to 160°C/min) and extrapolation to zero heating rate to eliminate the possible effects of superheating on T_m. The melting points, $T_{m,ann}$, are for polymers annealed at temperatures just below their melting points for 30 minutes, then cooled at a rate of 20°C/min to 50°C and reheated at the same rate to measure T_m. By this procedure, only a single melting peak was obtained.

Kinetic Parameters. The propagation rate constants which were measured directly for the α-ethyl-α-n-butyl monomer in DMSO at temperatures of 30.3, 34.0, 40.4, 45.1 and 49.9 are reported in Table III and compared in Table IV with those previously obtained in this laboratory for the α-methyl-α-alkyl series of monomers (4). The rate constants for the α-ethyl-α-butyl monomer are lower than those we previously observed for the α-methyl-α-butyl monomer within the temperature range studied, so the observed increase in k_p with increase in alkyl group size for the α-methyl series of monomers does not continue for the present monomer in which two larger alkyl groups are present. However, the increase in activation energy, ΔE_p, along the series with increasing alkyl group size is followed for the present monomer as seen in Table IV.

Gelation During Polymerization. When the data for monomer conversion vs. time in the polymerization of the α-ethyl-α-butyl monomer at 40.4°C is plotted as shown in Figure 1, a sharp change in rate is seen to occur after about 20 minutes of reaction time. A similar effect is also observed for the reaction carried out at 34°C, but the abrupt change in rate could be prevented by reducing the monomer concentration in the polymerization reaction as seen in Figure 2. These unusual rate effects were found to be caused by the formation of a gel by the reaction mixture during the polymerization reaction, and the onset of gelation occurred at approximately the same time as that for the decrease in rate of propagation. Gelation was preceded by a sharp increase in the viscosity of the polymerization solution over a period of a few minutes before the rigid gel was formed, so it was not possible to assign an exact gelation time to the reaction.

By decreasing monomer and initiator concentration it is possible to eliminate gelation because polymers which form gels by crystallization from solution do so at a critical concentration and molecular weight (7). The effect of monomer and initiator concentration for the polymerization reaction at 34°C is also shown by the data in Table V.

The molecular weights and molecular weight distributions of the polymers present in the reaction solution before and after gelation were measured by gel permeation chromatography, GPC, with the results shown in Figure 3. The molecular weight distributions of polymer samples isolated before gelation were somewhat broader but single-valued, while the distributions of samples quenched after gelation were bimodal as shown in Figure 3. These results suggest, as discussed above, that at gelation a portion of the growing chains becomes inaccessible to further growth, while the remainder continues

Table III. Rate Constants for the Polymerization of
α-Ethyl-α-n-butyl-β-propiolactone Initiated by TEAB in DMSO

Temperature °C	$[M]_o^a$ $10^2 \cdot mol/l$	$[I]_o^a$ $10^3 \cdot mol/l$	k_p^b $M^{-1} min^{-1}$
30.3	4.94	2.36	9.5
34.0	7.14	0.421	11
34.0	5.41	0.592	12
40.4	3.75	1.96	13.8
45.1	4.36	1.87	16.8
49.9	4.26	1.70	20.0

a. Initial monomer, M, and initiator, I, concentrations.

b. Observed propagation rate constant.

Table IV. Propagation Rate Constants and Activation Energies
for Polymerization of α,α-Disubstituted-β-propiolactones
in DMSO Initiated by TEAB

Monomer Substituents α,α	k_p, $M^{-1} min^{-1}$ at Temp., °C							ΔE_p^c kJ/mol	ΔS_p^c cal/ °mole
	22°	30°	34°	37°	40°	45°	50°		
$CH_3,CH_3{}^a$	10.2								
$CH_3,C_2H_5{}^a$	11.2			19.0				26.8	-42.3
$CH_3,C_3H_7{}^a$	11.8			20.4				27.6	-41.5
$CH_3,C_4H_9{}^a$	13.0			22.9				28.9	-40.3
C_2H_5,C_4H_9	9.5	11.5^b			13.8	16.8	20.0	31.3	-33.6

a. Reference 4.

b. Average of two measurements, see Table 1.

c. Calculated activation energy, ΔE_p, and activation entropy,
    ΔS_p, for propagation.

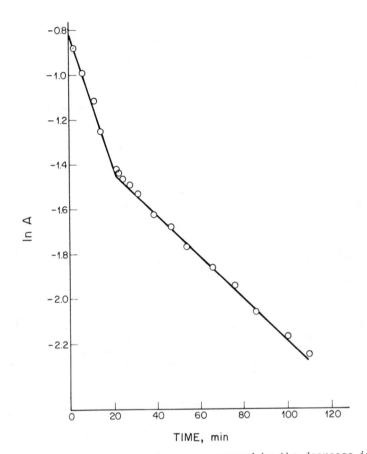

Figure 1.   Monomer conversion, as measured by the decrease in absorbance, A, of the lactone carbonyl group (4), as a function of reaction time for the polymerization of α-ethyl-α-butyl-β-propiolactone, EBPL, in DMSO at 40.4°C.

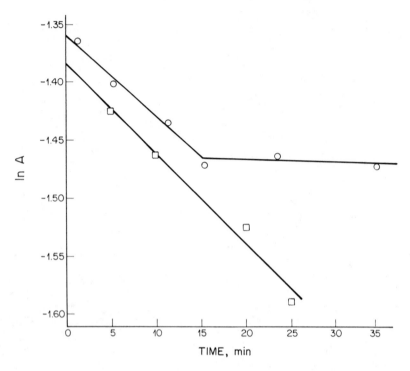

Figure 2. Monomer conversion (see Figure 1 title) <u>vs</u>. time for
           the polymerization of EBPL in DMSO at two different
           initial monomer concentrations: (a) -O- for $[M]_o$ =
           0.0636M, and (b) -□- for $[M]_o$ = 0.0541M.

Table V. Dependence of Gelation Time on Reactant Concentrations[a]

$[M]_o$ M	$[I]_o \times 10^2$ M	k_p M^{-1} min^{-1}	Approx. Gelation Time, min[b]
0.0714	0.0421	11	15
0.0541	0.0592	12	10
0.060	0.056		13

a. At 34°C.

b. Extrapolated from solution viscosity measurements.

to grow. As a result, the molecular weight of the inactive polymer fraction remains unchanged with time after gelation, while the molecular weight of the active polymer fraction continues to increase, and a bimodal distribution is formed whose breadth increases with time.

The fraction of growing polymer chains that continues propagating after gelation can be estimated from the relative areas of the high and low molecular weight polymer fractions in each distribution. From this analysis, it is concluded that about 40% of the polymer continues to propagate under these conditions, as shown in Table VI.

The change in molecular weight with time can also be used to estimate the rate constant of propagation after gelation, which was found to be 13 M^{-1} min^{-1}, although this calculation has a large error associated with it because only two data points were available. The value of k_p determined prior to gelation is 11 M^{-1} min^{-1}, so apparently the propagation rate of those polymer chains that continue to grow remains essentially unchanged by gelation. The decrease in overall polymerization rate upon gelation, therefore, is caused by a sudden decrease in the number of chains capable of growth.

For comparison, the α-ethyl-α-butyl monomer was also polymerized under the same conditions in N-methylpyrrolidone, NMP, as the reaction solvent. Gelation did not occur in this solvent, although the polymer precipitated almost immediately after initiation. Despite the system's heterogeneity, however, propagation continued at apparently constant rate as shown in Figure 4. The propagation rate constant measured for this polymerization reaction was 57 $\pm$ 3 M^{-1} min^{-1} at 34°C. The molecular weight distributions of this polymer as determined by GPC contained only a single peak, but the peak was quite broad, as might be expected from a heterogeneous system.

The much higher rate of polymerization of this monomer in NMP compared to DMSO, even with precipitation in the former polymer, again suggests that an unusual effect occurs for reactions in the latter solvent in these polymerization systems. As discussed earlier, the lower values of k_p for the α-methyl-α-propyl monomer in DMSO (which has dielectric constant, ε, of 46.4) compared to THF (ε of 7.6) suggests that DMSO may solvate the carboxylate anion, thereby reducing its reactivity. In contrast, the much higher k_p values in NMP suggest that this type of interaction or anion solvation does not occur, but instead this solvent (ε of 33.0) simply acts as a highly polar medium for the S_N2 reaction of the carboxylate ion endgroup in the polymerization. Hence, NMP facilitates the dissociation of this tetraethylammonium carboxylate ion pair much more effectively than does DMSO, even though the former has a lower dielectric constant than the latter.

Additional studies on the rate of polymerization of the α-methyl-α-propyl monomer in other solvents, as well as detailed

Table VI.   Molecular Weight as a Function of Polymerization Time[a]

$[I]_o$	Polym. time, min	$\overline{M}_n$ of Inactive Polymer	$\overline{M}_n$ of Active Polymer	Fraction of Active Chains
0.000282 M	25.9	800	2400	0.4099
0.000283 M	35.3	800	3600	0.3588

a. Polymerization reaction initiated by tetraethylammonium
    benzoate at 34°C.

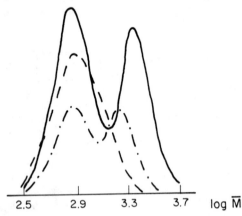

Figure 3.   Molecular weight distributions of EBPL polymers pre-
            pared in DMSO at 34°C after different reaction times:
            (a) dashed line at 12 minutes, (b) dot-dashed line
            at 26 minutes, and (c) solid line at 35 minutes.

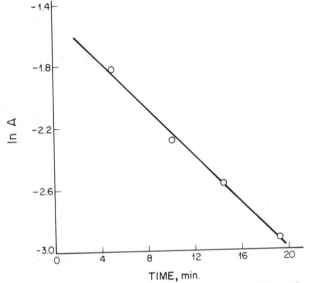

Figure 4.   Monomer conversion vs. time for EBPL polymerized in
            NMP at 34°C.

investigations on the ion pair contributions to the propagation reaction in these systems, are now in progress.

Experimental

Polymerization Reactions. All polymerization reactions were conducted in 50 ml double-wall glass vessels which allowed precise temperature control by rapid circulation of a thermostatted liquid. The vessel was designed to ensure effective magnetic stirring and to allow the polymerization reaction to be performed under dry argon by fitting it with septum caps. Before charging, the vessel was evacuated and flushed with dry argon at least five times with intermittent heating and transferred to a "dry bag" which was maintained with an argon atmosphere. The initiator was then added either as a solid or a solution (prepared in the dry bag) from a gas-tight syringe. The solvent was added from a syringe and temperature equilibration was established before addition of the monomer. The polymerization reaction was terminated by addition of 5-10 ml of methanol. The polymer was isolated by adding this mixture to a tenfold excess of methanol, storing the system overnight in a refrigerator, filtration and drying under vacuum at 50°C for at least 24 hours.

Rate Measurements. A continuous flow detecting system was designed by connecting the polymerization vessel, through a constant flow pump, to an adjustable path-length, flow-through IR cell with sodium chloride windows (Harrick Scientific Corporation) with double tipped needles returning the polymerization mixture. The IR cell mounted in a Perkin-Elmer Model 257 Spectrometer was filled in about one minute after starting the pump, and from that time on allowed repetitive recording of the partial spectrum. The solvent amount totaled approximately 40 ml by weight in these runs.

Initiator. Tetraethylammonium benzoate, TEAB, (Eastman Kodak) was recrystallized from dry benzene, dried under vacuum at 35°C (m.p. 75-76°C), and stored in a desiccator over P_2O_5 under dry argon.

Solvents. All solvents used in the polymerization experiments, including dimethyl sulfoxide, DMSO, tetrahydrofuran, THF, and N-methylpyrrolidone, NMP, were spectrophotometric grades from Aldrich. The solvents were dried by stirring overnight under dry argon with calcium hydride. After 1-2 hours of refluxing, the dry solvents were collected in storage flasks fitted with stopcocks by distillation under dry argon.

Polymer Melting Points. The thermograms of the polymers were obtained under dry nitrogen on a Perkin-Elmer DSC-2 with aluminum pans containing 3-5 mg samples of polymers precipitated from methanol and dried as described. The instrument was temperature calibrated with indium and lead standards. For $T_{m,o}$ determinations, separate samples were recorded at 10, 20, 40, 80, and 160°C/min heating rates.

Molecular Weights. Molecular weight distributions were determined with a Waters 201 HPLC instrument, and molecular weights were measured using polystyrene standards on THF solutions.

Acknowledgments

The authors are grateful to the National Science Foundation for the support of this work under Grant No. GH-38848 and for the use of the facilities of the Materials Research Laboratory of the University of Massachusetts. We also gratefully acknowledge the financial support provided to Søren Hvilsted by the Otto Mønsted Foundation of Denmark.

Literature Cited

1. Cornibert, J.; Marchessault, R. H.; Allegrezza, A. E. Jr.; Lenz, R. W. Macromolecules 1973, 6, 676.
2. Lenz, R. W. Bull. Soc. Chem. Beograd. 1974, 39, 395.
3. Bigdeli, E.; Lenz, R. W. Macromolecules 1978, 11, 493.
4. Eisenbach, C. D.; Lenz, R. W. Makromol. Chem. 1976, 177, 2539-45.
5. Thiebaut, R.; Fischer, N.; Etienne, Y.; Costa, J. Ind. Plast. Mod. 1962, 14, 1.
6. Leborgne, A.; Spassky, N.; Sigwalt, P. Polymer Bulletin 1979, 1, 825.
7. Rogovina, L. Z.; Slonimskii, G. L. Russ. Chem. Rev. 1974, 43(6), 503-23.

RECEIVED October 4, 1984

Structure–Reactivity Relationships in Ring-Opening Polymerization

STANISLAW PENCZEK, PRZEMYSLAW KUBISA, STANISLAW SLOMKOWSKI, and
KRZYSZTOF MATYJASZEWSKI

Center of Molecular and Macromolecular Studies, Polish Academy of Sciences, 90-362 Łódź,
Poland

Correlations of structures and reactivities
for anionic and cationic ring-opening polymerization are reviewed. The following topics
are discussed: chemical structure of active
species and their isomerism, determination
of active centers concentration, covalent vs
ionic growth and correlations between structures of active centers or monomers and
their reactivities.

Correlations of structures and reactivities require for
the ring-opening polymerization as well as for other ionic
polymerizations approaches differing from these in radical
polymerization of the unsaturated monomers. This is
because in radical polymerization free radicals are the
unique chemical structure of the growing species and
double bonds are the only chemical groups involved in polymerization (1), (2).

Ring-opening polymerizations involve a variety of
the ionic growing species. Moreover, some of the heterocyclic monomers may react ambidently and, therefore, produce chemically isomeric structures of active centers.
Polymerization of lactones or polymerization of substituted α-oxides, both with two possible ways of ring-opening,
are the typical examples.

Thus, the actual chemical structures have to be determined first and then their proportions and contributions in the chain growth have to be established. The
further step is the determination of the rate constants
of the elementary reactions involving all of these species
that have to be correlated.

Correlations are made first for a given monomer propagating with various growing species, then reactivities
of monomers belonging to the same class of chemical
compounds are determined and eventually correlation
between monomers with different heteroatoms can be given.

0097–6156/85/0286–0117$06.00/0
© 1985 American Chemical Society

   This above discussed correlations require that identical chemical mechanisms are compared. For instance, the
cationic polymerizations of heterocyclics are known to
proceed by S_N1, S_N2, and A_C2 mechanisms. Besides, there
are two different S_N2 mechanisms, and both can involve
the same monomer, namely proceeding with onium ions or
with actived monomer (3).
   This paper describes problems outlined above, methods
of determination of the chemical structures in both anionic and cationic ring-opening polymerizations, equilibria
between different active species, the corresponding mechanisms of propagation and related rate constants of
propagation on these species, and finally the available
correlations.

Determination of the chemical structures of the growing
species

Anionic polymerization. For some heterocyclic monomers
the unique chemical structure of the growing species
follows unequivocally from the monomer structure. However,
in many cases isomeric structures have to be taken into
account. For instance, for symmetrical monomers, like
thietane, the carbanion but not the thiolate anion was
proposed (4). Unsymmetrically substituted monomers can
provide active species by α- or β- ring scission. Unusual
structure of activated monomer was proposed for NCA and
lactams. These structures can not be distinguished by
spectrophotometric methods, and application of ^{1}H- or
^{13}C-NMR looks more promising.
   We have recently elaborated a method based on the
anion capping with $ClP(O)(OC_6H_5)_2$, followed by determination of the structure of the parent anion in $^{31}P\{^1H\}$-NMR,
and comparing chemical shifts with these of the indepent-
ly studied model compounds (5).
   Some examples are given in Table 1; below is the
general scheme:

$$RX^{\ominus}, Mt^{\oplus} + ClP(O)(OC_6H_5)_2 \longrightarrow RX\overset{O}{\underset{\|}{-}}P(OC_6H_5)_2 + MtCl \qquad (1)$$

where X=heteroatom and Mt=e.g. Na , K

   Thus, trapping provides a possible way of determining the isomeric structures during polymerization,
measuring their proportions and their rates of intercon-
version.
   The corresponding structures of the growing species
have been established by comparing the observed chemical
shifts of the trapped products with these of the model
compounds. Thus, trapped $CH_3O^{\ominus}$, $CH_3CH(C_6H_5)O^{\ominus}$,
$C_6H_5CH_2CH_2O^{\ominus}$, and $CH_3-CH(CH_3)S^{\ominus}$ give the following
chemical shifts: -11.5, -13.1, -12.7 (in THF) and +19.0
ppm δ (in C_6H_6). The carboxylate and silanolate models

both lead to the same chemical shift equal to -26.3 ppm δ (THF) and characteristic for $[(C_6H_5O)_2(O)P]_2O$ (5).

These and related methods allowed us recently to re-evaluate the structure of active centers in anionic polymerization of simple, unsubstituted lactones, β-propiolactone. The rationale was put forward in terms of stereo - electronic factors to explain why β-propiolactone propagates on carboxylate and ε-caprolactone on alcoholate anions. This is shown in scheme below :

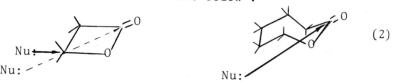

(2)

The broken arrow in the pictures above indicates the hampered direction of attack for the approaching nucleophile.

Table I. Monomers, structures of the growing species and trapped anions, according to $^{31}P\{^1H\}$-NMR (5).

Monomer	Product of trapping	$\delta^{31}P$ ppm	Growing species
$\overset{\ulcorner}{C}H_2CH_2\overset{\urcorner}{O}$	$...-CH_2OP(O)(OC_6H_5)_2$	-11.5	$...-CH_2O^{\ominus}$
$\overset{\phi}{\underset{\ulcorner}{C}H_2\overset{\urcorner}{C}HO}$	$...-CH_2\overset{\phi}{C}HOP(O)(OC_6H_5)_2$ $...-\overset{}{\underset{\phi}{C}HCH_2}OP(O)(OC_6H_5)_2$	-13.0 -12.6	$...-CH_2\overset{\phi}{C}HO^{\ominus}$ $...-\underset{\phi}{C}HCH_2O^{\ominus}$
$\overset{\ulcorner}{C}H_2CH_2O\overset{O}{\overset{\|}{\underset{\urcorner}{C}}}$	$...-CH_2\overset{O}{\overset{\|}{C}}OP(O)(OC_6H_5)_2$ pyrophosphate	-26.0	$...-CH_2C\overset{O}{\underset{O}{\diagdown}}{}^{\ominus}$
$\overset{\ulcorner}{(}CH_2\overset{}{)}{}_5O\overset{O}{\overset{\|}{\underset{\urcorner}{C}}}$	$...-(CH_2)_5OP(O)(OC_6H_5)_2$	-11.5	$...-CH_2O^{\ominus}$
$\overset{\ulcorner}{[}(CH_3)_2SiO\overset{}{]}{}_3$	$...-\underset{CH_3}{\overset{CH_3}{Si}}OP(O)(OC_6H_5)_2$ pyrophosphate	-26.3	$...-\underset{CH_3}{\overset{CH_3}{Si}}O^{\ominus}$
$\underset{\ulcorner}{CH_2}\overset{CH_3}{\underset{\urcorner}{C}HS}$	$...-CH_2\underset{CH_3}{C}HSP(O)(OC_6H_5)_2$	+18.9	$...-CH_2\underset{CH_3}{C}HS^{\ominus}$

Quantitative determination of the concentration of macroanions in the anionic polymerization of heterocyclics is based on the same approach of end-capping with P-containing compounds.

<u>Cationic polymerization.</u> Similarly, trapping in the cationic polymerization with R_3P, where R=alkyl or aryl, allows one to determine the structure of the parent cations, shown below for onium ions (<u>6</u>), (<u>7</u>):

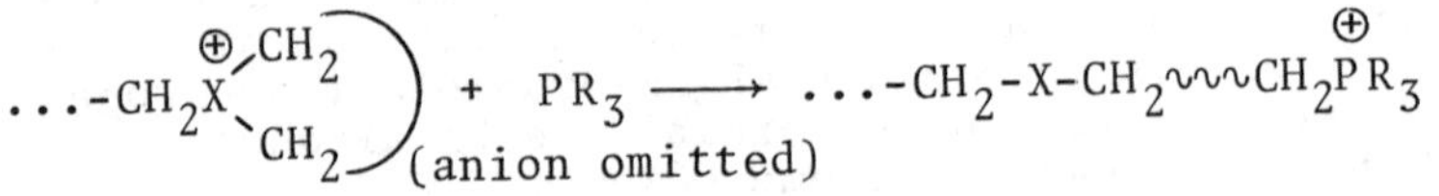

$$\ldots-CH_2\overset{\oplus}{X}\overset{CH_2}{\underset{CH_2}{\diagdown}}\Big) + PR_3 \longrightarrow \ldots-CH_2-X-CH_2\sim\sim CH_2\overset{\oplus}{P}R_3 \quad (3)$$
$$\text{(anion omitted)}$$

These methods are related to the better known trapping of radicals, trapping of carbanions elaborated by Szwarc (<u>8</u>) and, more recently, end-capping with phenolates in cationic polymerization (<u>9,10,11</u>).

The phosphine ion-trapping, in contrast to the methods using UV spectrophotometry for further identification, provides information about the fine structure of the growing species. More information can follow from multiplicities of the ^{31}P-NMR spectra.

Some examples of the application of the trapping of cations with phosphines are given below :

Table II.   Monomers, and structures of the related
             growing species and quaternary phosphonium
             salts (trapped cations) (<u>6</u>).

Monomer	Product of trapping	$\delta^{31}P$ ppm	Growing species
 (ring, 4, O)	$\ldots-O(CH_2)_3-\overset{\oplus}{P}(C_6H_5)_3$	23.8	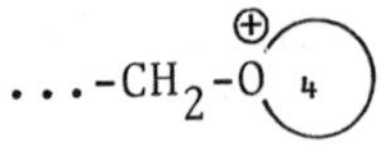 $\ldots-CH_2-\overset{\oplus}{O}$ (ring, 4)
 (ring, 5, O)	$\ldots-O(CH_2)_4-\overset{\oplus}{P}(C_6H_5)_3$	23.4	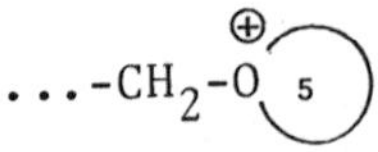 $\ldots-CH_2-\overset{\oplus}{O}$ (ring, 5)
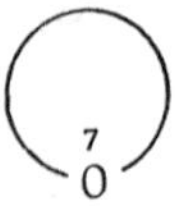 (ring, 7, O)	$\ldots-O(CH_2)_6-\overset{\oplus}{P}(C_6H_5)_3$	23.0	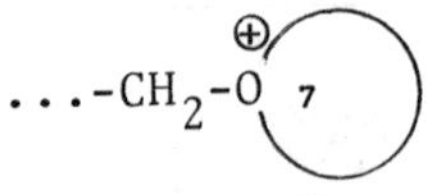 $\ldots-CH_2-\overset{\oplus}{O}$ (ring, 7)
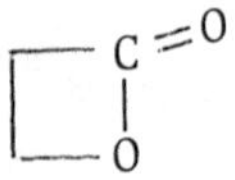 (ring, C=O, O)	$\ldots-CH_2OCCH_2CH_2\overset{\oplus}{P}(C_6H_5)_3$ (with O)	23.4	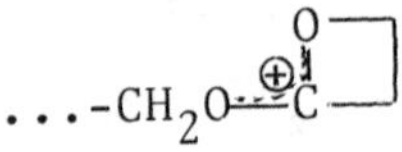 $\ldots-CH_2O\overset{\oplus}{-}C$ (ring, O)
 (ring, 5, O O)	$\ldots-CH_2OCH_2-\overset{\oplus}{P}(C_6H_5)_3$	16.7	$"\ldots-CH_2-\overset{\oplus}{O}CH_2"$

It has previously been shown in our laboratory that in the terpolymerization of oxetane, THF and oxepane the active end group of all three growing species could be simultaneously observed (<u>6</u>).

Particularly important is the established structure of the growing species in the polymerization of β-propiolactone, which differs from the accepted earlier acylium cation. The tertiary oxonium ion structure, observed by us also for ε-caprolactone, has been confirmed by other methods (12).

^{1}H and ^{13}C-NMR have also been successfully used in determination of structures of onium ions as the growing species. This has already been reviewed by us (13).

Isomerism of the ionic active centers

The chemical structures described above of active ionic centers have been considered as the unique ones. Thus, although they may exist in several physical forms as various ion-pairs, their aggregates or "free" ions, but there is only one chemical structure they propagate on. We recently observed however systems in which ionic growing species differing in chemical structure coexist in these systems and participate in the chain growth.

In anionic polymerization, β-propiolactone initiated with potassium alcoholate, gives, in initiation, both alcoholate and carboxylate anions. Alcoholate ions in every next step convert partially into carboxylate whereas carboxylate reproduce themselves quantitatively. Thus, after a few steps only carboxylate anions are left (14). Related situation was observed in the polymerization of styrene oxide (15). Here, however, it is only due to the structure of the initiator used. Thus, when in the initiation step both secondary and primary alcoholate anions are formed, due to the low steric requirements, in the next step apparently only the attack on the least substituted carbon atom takes place and already in the second step exclusively secondary alcoholate anions are present.

In these two systems eventually macromolecules are formed by one kind of active species winning early enough in competition with the other species. However, it has been observed in this laboratory, particularly in the cationic polymerization, especially in the polymerization of cyclic acetals (16) and orthoesters (17), that two or more chemically different kinds of active species may coexist throughout the whole polymerization process, their proportions may depend (cyclic acetals) on the monomer conversion. Thus, in the polymerization of cyclic acetals the carbenium-oxonium equilibria have to be taken into account (16):

$$...-CH_2OCH_2 \overset{\oplus}{} + O \rightleftharpoons ...-CH_2OCH_2O \overset{\oplus}{} \qquad (4)$$

where O is a monomer molecule or another macromolecule.

More recently one of us with Szymański observed that the
active species holding monomer molecule can isomerize and
the following equilibrium was directly observed by [1]H-
and ^{13}C-NMR in model compounds (18):

$$(5)$$

As indicated by the direction of the arrows the isomeric
7-membered oxonium ion dominates in the polymerization
of the 5-membered 1,3-dioxolane whereas in the polymerization of the 7-membered 1,3-dioxepane cationated monomer
dominates. This is apparently due to the differences in
strain of the involved rings. Kinetic analysis of the polymerization of these two monomers has shown that the
isomeric (enlarged) oxonium ions can be treated as the
kinetically dormant species; propagation and depropagation on these species proceed with almost identical rates.
This explains why for the same starting concentration of
initiator, as observed by Plesch (19), 1,3-dioxepane polymerizes over 100 times faster than 1,3-dioxolane. This is
because the proportion of the productively active species
is higher for the former than for the latter monomer.

Covalent growing species

Closely related to the ionic polymerization of heterocyclic monomers is, what we can call, pseudoionic polymerization (or sometimes, perhaps, cryptoionic). We use the
preffix pseudo- in the same meaning as it   was first
used in the vinyl cationic polymerization. It means that
propagation actually proceeds on the covalent species
that could have been in equilibrium with their ionic
counterparts. Several systems falling to this category
have recently been described for both anionic and cationic polymerization of heterocyclics. In the anionic processes derivatives of Zn or Al alkyls or alcoholates are
believed to function this way. However, for none of these
systems the absence of ionic contribution was shown. Two
catalytic systems are of particular interest, namely the
-Zn-O-Al< systems (20) and >Al-alkyl modified by bulky
porphyrin derivatives (21). Both are discussed in this
volume and both have been clearly shown to produce living
systems. The former with ε-caprolactone and the latter

with ethylene oxide, propylene oxide and β-propiolactone (22).
    As indicated above, in the anionic processes only
"either- or" situation was observed, i.e. when covalent
species are present no ions in equilibrium were detected.

Covalent active species in the cationic polymerization.
In the cationic polymerization several systems were
studied, in which both covalent and ionic growth have
been simultaneously studied. For the first time the covalent growth was described for oxazolines by Saegusa (23).
    In the polymerization of THF the presence of cova-
lent species was assumed by Smith and Hubin (24) and
shortly after the covalent species were directly observed
in our laboratory (^{1}H-NMR) (25) as well as by Saegusa
(^{19}F-NMR) (26) and Pruckmayr (^{13}C-NMR) (27). ^{1}H-NMR
clearly showed the existence of two distinct species,
covalent and ionic, with their characteristic chemical
shifts identical to model compounds.
    In the polymerization of heterocyclic monomers, the
covalent species in equilibrium with their ionic counterparts were observed directly, thus the corresponding
equilibrium constant could be determined for polymeri -
zing systems. There are two reaction pathways possible
for the ionization reaction :

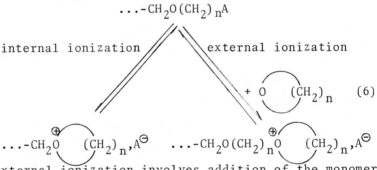

$$...-CH_2O(CH_2)_nA$$

internal ionization   external ionization

$$+ O \quad (CH_2)_n \quad (6)$$

$$...-CH_2O \quad (CH_2)_n, A^{\ominus} \quad ...-CH_2O(CH_2)_nO \quad (CH_2)_n, A^{\ominus}$$

The external ionization involves addition of the monomer
molecule to the covalent active species and, thus, means
the covalent propagation.
    The contribution of each of the two mechanisms shown
in scheme (6) and operating simultaneously may be
estimated on the basis of the dependence of the |ion|/
|ester| ratio on conversion. For unimolecular internal
reaction this proportion should be independent of monomer concentration (thus conversion) while for the bimolecular, external ionization the proportion of ions should
decrease with conversion. It was shown that for the
most thoroughly studied system, i.e. polymerization of
THF, the internal ionization dominates (28).
    More recent results indicate that in the polymeriza-
tion of the 7-membered cyclic ether: oxepane (Ox), both
intra- and intermolecular ionizations have to be

considered. Thus in CH_3NO_2 solvent at $25°$ $k_{ii}=2.3 \cdot 10^{-4}$ s^{-1} and $k_{ei}=1.35 \cdot 10^{-4}$ $mol^{-1} \cdot l \cdot s^{-1}$ meaning that both processes proceed with the same rates for $|Ox|=1.7$ $mol \cdot l^{-1}$ (effective monomer concentration). For the discussed earlier THF case the effective monomer concentration would be above 100 $mol \cdot l^{-1}$ i.e. much above the concentration which may be achieved even in bulk (29).

Reactivities of covalent active species

In the polymerization of heterocyclic compounds rate constants of propagation on covalent species were determined for several systems and compared with the corresponding rate constants of ionic growth. In the polymerization of THF, $k_{pc}=5 \cdot 10^{-4}$ $mol^{-1} \cdot l \cdot s^{-1}$ in CH_3NO_2 at $25°$; the similar value $3 \cdot 10^{-4}$ $mol^{-1} \cdot l \cdot s^{-1}$ was measured in the Ox polymerization. Although the values of the rate constants of covalent propagation are close to each other, the contribution of covalent growth is considerably different because the corresponding ionic rate constants are different: $k_{pi}=2.4 \cdot 10^{-2}$ $mol^{-1} \cdot l \cdot s^{-1}$ for THF and $1.3 \cdot 10^{-4}$ $mol^{-1} \cdot l \cdot s^{-1}$ for Ox. The observed relations are due to the low steric requirements of covalent growth and the much larger role of steric hindrance for ionic growth, as discussed by us in Ref. 13:

$$\bigcirc O \longrightarrow \underset{H \quad H}{\overset{\overset{\displaystyle CH_2}{|}}{C}} - OSO_2CF_3 \quad ; \qquad ; \qquad (7)$$

Macroion-pairs and macroions

Below, in Table III, some typical data on dissociation of the macro- ion-pairs for both anionic and cationic ring--opening polymerization are given.
    There is a number of similarities in behaviour of macroions derived from various monomers. Thus, macroion--pairs of living poly(ethylene oxide) and polycaprolactone in THF solvent with $K^{\oplus}$ cations, both have very low K_D. Dissociation of living poly-β-propiolactone, with carboxylate growing anion and crowned $K^{\oplus}$ counterion, in which electrostatic intereaction within the ion-pair is much weaker than in alcoholate ion-pairs, resembles this of the tertiary oxonium ions. For both systems in CH_2Cl_2 solvent K_D is approximately equal at 10^{-5} $mol \cdot l^{-1}$, i.e. 10^5 times larger than for alcoholate ion-pairs in THF solvent.

Table III. Dissociation constants of some macroion-pairs in the anionic and cationic ring-opening polymerization

Monomer	Growing species	Counter-ion	Solvent	K_D, 25° mol·l⁻¹	Ref.
CH_2CH_2O	$\ldots -CH_2CH_2O^{\ominus}$	$K^{\oplus}$	THF	$1.8 \cdot 10^{-10}$	30
$(CH_2)_5OCO$	$\ldots -CH_2CH_2O^{\ominus}$	$K^{\oplus}$	THF	$4 \cdot 10^{-10}$ (20°)	31
CH_2CH_2O	$\ldots -CH_2CH_2O^{\ominus}$	$Cs^{\oplus}$	THF	$2.7 \cdot 10^{-10}$	30
CH_2CH_2O	$\ldots -CH_2CH_2O^{\ominus}$	$K^{\oplus}/|222|$	THF	$2.0 \cdot 10^{-7}$	30
$CH_2CH(CH_3)S$	$\ldots -CH_2CH(CH_3)S^{\ominus}$	$K^{\oplus}$	THF	$2.1 \cdot 10^{-8}$	32
CH_2CH_2OCO	$\ldots -CH_2CH_2C\overset{O}{\underset{O}{<}}$	$K^{\oplus}/DB18C6$	CH_2Cl_2	$5 \cdot 10^{-5}$	33
$(CH_2)_4O$	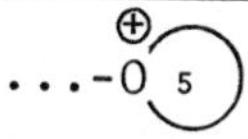 $\ldots -O^{\oplus}_{5}$	$AsF_6^{\ominus}$	CH_2Cl_2 / CH_3NO_2	$3 \cdot 10^{-5}$ (0°) / $2 \cdot 10^{-3}$	34
$(CH_2)_6O$	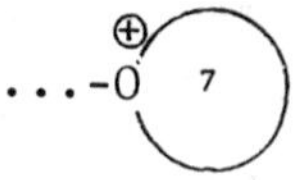 $\ldots -O^{\oplus}_{7}$	$SbF_6^{\ominus}$	CH_2Cl_2 / $C_6H_5NO_2$	$3.1 \cdot 10^{-5}$ (0°) / $1.6 \cdot 10^{-3}$	35
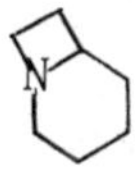 (bicyclic amine)	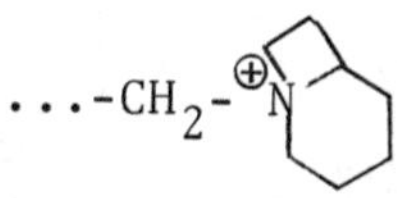 $\ldots -CH_2-{}^{\oplus}N$ (bicyclic)	$I^{\ominus}$	$C_6H_5NO_2$	$3 \cdot 10^{-2}$	36

The dissociation constants discussed above were determined from the conductometric data according to Fuoss. The large majority of the cationic processes are well described by a simple scheme of ion-pair dissociation; the K_D determined for both the low molecular models and the high polymer fitted with the ion-pair at the end give similar results. The high nucleophilicity of monomer, strongly solvating the cation, and large size of anions decrease the interaction within the ion-pair in both thermodynamic and kinetic sense.

In the anionic polymerization the situation is different. The negative charges are highly concentrated at the chain end at least for alcoholate and thiolate anions, alkali metal cations usually used as counterions have

smaller size, and strongly interact with anions. Therefore, the electrostatic attraction within an ion pair is
stronger and K_D extremly low. Besides, these ion-pairs
are lees susceptible to solvation and strongly self-asso-
ciate into aggregates. Thus, analysis of the fine structure of ion pairs on the bases of Fuoss equation as well
as its applicability in the analysis of dissociation is
less straightforward than for the solvated (or solvent
separated) ion-pairs. The latter do not change the degree
of solvation in dissociation :

$$\ldots -X^{\ominus}(Mt\cdot nS)^{\oplus} \xrightleftharpoons{K_d} \ldots -X^{\ominus} + (Mt\cdot nS)^{\oplus} \tag{6}$$

whereas the former may require at least two discrete
steps for dissociation, namely the preliminary solvation
and then dissociation of the thus solvated ion-pair.
Fuoss (37) has recently stressed that the determination
of the distance between ions for such a multistep process
may require an approach differing from the application of
a simple dependence of K_D on the dielectric constants or
reciprocal of the absolute temperature (i.e. the Fuoss
equation).

    Aggregation of ion-pairs has been demonstrated in
the polymerization of ethylne oxide ($\ldots -CH_2O^{\ominus}K^{\oplus}$ in THF
solvent); apparently cyclic trimers of ion pairs domina-
te, formed with the equilibrium constants equal approx.
to $10^{6} \sim 10^{7}$ $l^{2}\cdot mol^{-2}$. This value was determined from the
analysis of the kinetics of polymerization (30). Polyme-
rization of ε-caprolactone with $Na^{\oplus}$ as counterion in THF
solvent also shows the 1/3 dependence of the rate of polymerization on the total concentration of active species
(38) whereas with $K^{\oplus}$ counterion pairs do not aggregate
in THF (31). However in the polymerization of less polar
dimethyl siloxane trimer (D3) ($Na^{\oplus}$ cation) ion-pairs
efficiently aggregate in THF (39). The observed concentration dependences strongly indicate the formation
aggregates but there are no other more direct proofs of
their existance. According to Kazanski (40), all of the
attempts to determine the state of association from the
viscosity measurements have to be considered as unsuccessful after closer examination of the conditions of measurements and related theoretical fundamentals-particularly
when the 3/4 law derived for concentrated solutions (or
polymer melts) is being applied to the dilute solutions
in which polymerization proceeds.

    Increasing the solvent polarity in both anionic and
cationic systems increases significantly K_D. Thus, K_D of
alcoholate macroion-pairs from poly(ethylene oxide) with
$K^{\oplus}$ in DMSO solvent is equal to $4.7\cdot10^{-2}$ $mol\cdot l^{-1}$ and K_D in
the polymerization of THF in CH_3NO_2 solvent equals 10^{3}
$mol\cdot l^{-1}$.

    The same effect is observed when in the anionic poly-
merization larger cations are introduced. Particularly
when crowned or cryptated cations are used.

Macroions and macroion-pairs in propagation

There are a few systems for which, using K_D to establish proportions of macroion-pairs and macroions, the rate constants of propagation on these species were determined. In the cationic polymerization of THF, OXP, and more recently conidine, it has been shown that $k_p^{\mp} = k_p^{+}$ (36). This was explained by assuming weak interactions of counterions within the ion-pairs, due to dissipitation of the positive charge in the onium ions, as well as by the stereochemical course of the propagation step (bordeline $S_N 2$) in which the monomer approach hardly requires the pulling apart of the anion.

Table IV. Rate constants of propagation in anionic polymerization of heterocyclic compounds

Monomer polymerization conditions	Active species	$k_p^{\mp}$	k_p^{-}	Ref.
		$mol^{-1} \cdot l \cdot s^{-1}$		
 THF, 20°C	$\cdots -CH_2CH_2O^{\ominus}K^{\oplus}$	$4.8 \cdot 10^{-2}$	-	32
	$\cdots -CH_2CH_2O^{\ominus}Cs^{\oplus}$	$1.22 \cdot 10^{-1}$	-	32
	$\cdots -CH_2CH_2O^{\ominus}K^{\oplus}$ 222	$2.5 \cdot 10^{-2}$	1.67	32
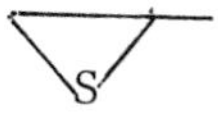 THF, -30°C	$\cdots -CH_2CH(CH_3)S^{\ominus}Na^{\oplus}$	$2.5 \cdot 10^{-3}$	3.8	32
	$\cdots -CH_2CH(CH_3)S^{\ominus}Cs^{\oplus}$	$2.3 \cdot 10^{-1}$	-	32
	$\cdots -CH_2CH(CH_3)S^{\ominus}Na^{\oplus}|222|$	11.9	5.6	32
$\{Si(CH_3)_2O\}_3$ Benzene, 20°C	$\cdots -Si(CH_3)_2O^{\ominus}Li^{\oplus}|211|$	1.4	-	32
$(CH_2)_2OCO$ CH_2Cl_2, 25°C	$\cdots -CH_2CH_2COO^{\ominus}K^{\oplus}DB18C6$	$7.0 \cdot 10^{-4}$	$1.6 \cdot 10^{-1}$	33
$(CH_2)_5OCO$ THF, 20°C	$\cdots -C(O)(CH_2)_5O^{\ominus}K^{\oplus}$	4.7	-	31

In the anionic polymerization there are three monomers only that have been studied in more detail, namely ethylene oxide, propylene sulfide, and β-propiolactone. Some

preliminary data on ε-caprolactone have become available
more recently.
  Polymerizations of ethylene oxide and propylene sulfide were reviewed several times by the authors of the
original results, namely the Paris and the Moscow groups
(32), (40). One of us with Kazanski reviewed recently the
recent data, including also polymerization of lactones
(30). In the polymerization of ε-caprolactone with $K^{\oplus}$
counterion in THF propagation proceeds exclusively on the
ion-pairs (31). These ion-pairs practically do not dissociate and do not aggregate at the polymerization condit-
ions (temp. from 0 to 20°, THF, $|\varepsilon CL|_0 = 0.5$ mol·l^{-1}). The
comparison of the rate constants of propagation on the
alcoholate ion pairs with $K^{\oplus}$ counterions in the homopolymerization of ε-caprolactone ($k_p^{\mp}$ (20°)=4.7 mol^{-1}·l·s^{-1}
(31) with that of oxirane ($k_p^{\mp}$ (20°)=4.8·10^{-2} mol^{-1}·l·s^{-1}
(32) reflects the much higher reactivity of the former
monomer.
Presumably this is because the higher ring strain of
oxirane, in comparison with that of ε-caprolactone, is
overweighed by the higher reactivity of the ester group
in εCL in comparison with the reactivity of the ether
linkage.

Solvation phenomena

Heterocyclic monomers and polymers present in their polymerization strongly interact with the growing species.
This is manifested in facts already described in this
paper.
Cationic polymerization. In the cationic polymerization
of cyclic ethers, sulfides, or amines in CH_2Cl_2 or even
in nitrosolvent, monomers and resulting polymers are the
most nucleophilic components of the system. Therefore,
explaining equal reactivities of macroions and macroion-
-pairs in the cationic polymerization of heterocyclic
monomers, we assumed that both ion-pairs and ions are solvated by monomers themselves. This decreases the electrostatic interaction within the ion-pairs. However, more
detailed analysis of $\Delta H_p^{\neq}$ and $\Delta S_p^{\neq}$ (activation parameters
of propagation) revealed that these monomers (at least THF
and oxepane) do not polymerize merely in clusters of monomer and polymer (8), but that solvent molecules are
also present in the immediate vicinity of the active
species (34), (41). This conclusion was based on the fact
that $\Delta H_p^{\neq}$ and $\Delta S_p^{\neq}$ measured in various solvent differ tremendously; e.g. $\Delta H_p^{\neq}$ for THF in THF solvent equals 14.0
kcal·mol^{-1} whereas in THF/CC1$_4$ mixture equals 8.6
kcal·mol^{-1}. Due to the compensation by the adjusting changes of $\Delta S_p^{\neq}$ the corresponding rate constants measured in
these solvent did not change more than two- three times.

Anionic polymerization. In anionic polymerization ethylene oxide, propylene sulfide or their corresponding

polymers are able to solvate cations. This state of solvation should differ with temperature and since solvation is exothermic, the lower the temperature the higher the contribution of solvation to the energetics of reactions. Thus, discussing any correlation between structure and reactivity not only the electronic and structural ele - ments of the active species and monomers but also the solvation phenomena should be taken into an account.

Below, in Table V the solvation power of ethylene oxide, propylene oxide, THF, and poly(ethylene oxide), $\overline{M}_n=6000$, are compared.

Table V. Equilibrium constants of complexation of $Na^{\oplus}$ by ethers and polyether solvents at 25° (30).

Ligand	n	K_n, $1 \cdot mol^{-1}$
ethylene oxide	1	0.41
propylene oxide	1	0.36
THF	1	0.69
poly(ethylene oxide) 6000	6	3000

The equilibrium constants listed in Table V, measured by using ^{23}Na and ^{133}Cs-NMR indicate that in the polymerization of ethylene oxide the polymer formed should strongly and selectively solvate $Na^{\oplus}$ counterion. This is also true for $K^{\oplus}$ and $Cs^{\oplus}$ cations; the corresponding K_n for poly(ethylene oxide) are equal to 500 and 200 $1 \cdot mol^{-1}$.

Solvation of cations by poly(ethylene oxide) chain is highly cooperative, showing the phenomenon "nothing or everything", i.e. the cation is either not solvated or fully solvated using its complete coordination ability:

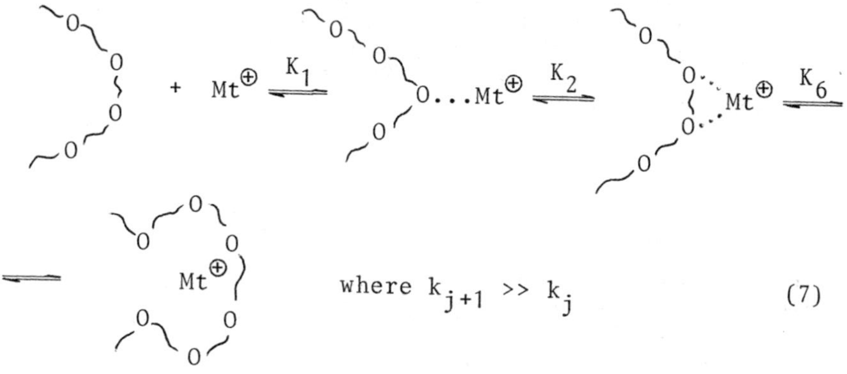

$$\text{where } k_{j+1} \gg k_j \qquad (7)$$

The high cooperative tendency is due to the much higher loss of entropy in the first stage of cooperation than

in the subsequent stages. The quantitative treatment
revealed that macromolecule joined at one segment will
fill up the free vacances with a probability higher than
0.9 (427, (43).

This high tendency of poly(ethylene oxide) to solvate cations and from the polymeric shell around the counterion leads to autoacceleration in polymerization (polymerization faster on solvated ion-pairs), and increase
of conductivity with monomer conversion. Moreover, polymerization is not sensitive to the "external" solvating
agents, e.g. crown ethers.

Solvation by monomer itself was postulated by two of
us in the polymerization of β-propiolactone. This is
highly polar and powerfully solvating compound. It could
be assumed that due to its dipolar feature β-propiolactone could solvate not only cation but also the growing
anion.

The anionic polymerization of β-propiolactone in
CH_2Cl_2 led to the dependence of the ratio of reactivities
of macroions and macroion pairs on the solvating power of
the medium (33). For higher proportions in the mixture
of the more powerfully ion solvating component (β-pro-
piolactone) the ratio of $k_p^-/k_p^{\mp}$ decreased (at 35°C $k_p^-/k_p^{\mp}$=
=210 and 150 for the systems with $|\beta PL|_o$=1 and 3 mol·l^{-1}
respecively). Moreover, $k_p^{\mp}$ in this system practically
does not depend on the composition of the mixture and the
decrease of the ratio is due to the decrease of k_p^- (33)).
We found also that the ratio of $k_p^-/k_p^{\mp}$ decreases also when
the temperature is lowered, e.g. in a system with $|\beta PL|_o$=
=3 mol·l^{-1} from 150 at 35°C to 5.6 at -20°C.

Although the inversion point (above: $k_p^->k_p^{\mp}$, below
$k_p^-<k_p^{\mp}$) has not been reached, because at lower temperature,
when the rate constants almost merge, polymerization
becomes very low, but the observed dependence of $k_p^-/k_p^{\mp}$
(or reverse) on temperature indicates that simple comparison of "reactivities" of ions and ion-pairs measured
at one single temperature can be just misleading. One can
even wonder whether the dependence $k_p^{\mp}$ (crowned or crypd.)
>k_p^- observed at -30° for propylene sulfide would not
change to the usually considered as "normal" i.e. $k_p^{\mp}>k_p^{\mp}$
at higher temperature.

Correlation of the structures and reactivities

Cationic polymerization. The available and reliable k_p
values in the cationic ring-opening polymerization,equal
for ions and ion-pairs (at 25°C), are listed below in
Table VI.

Table VI. Rate constants and basicities of different heterocyclics

Monomer k_p	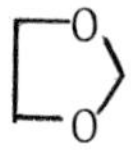	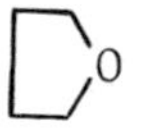		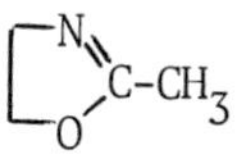	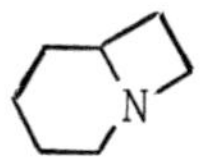
mole$^{-1} \cdot l \cdot s^{-1}$	$\sim 10^4$	$4 \cdot 10^{-2}$	$\sim 10^{-8}$	$5 \cdot 10^{-5}$	$7 \cdot 10^{-3}$
pK_a	-6.5	-2.0	1.2	3.4	11

Thus, the data above indicate that the higher the monomer basicity the lower its homopropagation rate constant. The proper correlation analyses of the reactivities of the series of given species (i.e. monomers or active centers) requires that one compares additions of various monomers to the same active species as well as the addition of the same monomer to the different active species.

Kinetics of addition of various heterocyclic monomers to the same active species was studied recently in our laboratory. The 1-methyltetrahydrofuranium cation was used as a model of ionic species while ethyltriflate was employed as a model of the covalent ones.

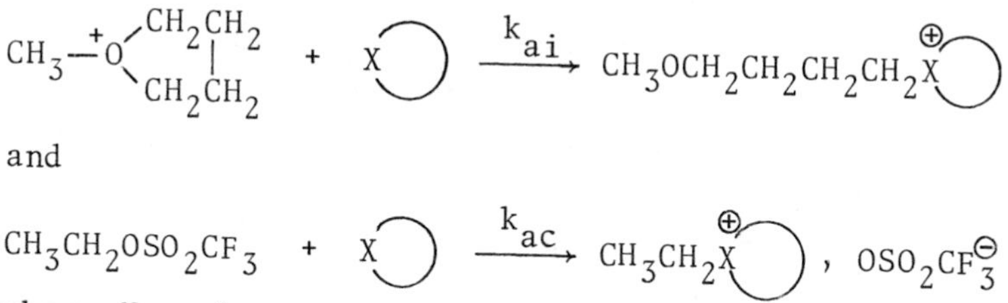

$$CH_3 - \overset{+}{O} \overset{CH_2 CH_2}{\underset{CH_2 CH_2}{|}} + X \bigcirc \xrightarrow{k_{ai}} CH_3OCH_2CH_2CH_2CH_2\overset{\oplus}{X}\bigcirc$$

and   (8)

$$CH_3CH_2OSO_2CF_3 + X\bigcirc \xrightarrow{k_{ac}} CH_3CH_2\overset{\oplus}{X}\bigcirc , \; OSO_2CF_3^{\ominus}$$

where X; a heteroatom or a group involving heteroatom(s).

The values of the corresponding rate constants determined in this way are given in Table VII, complete with values of these constants determined previously.

These data indicate that the rate constants of reactions between various monomers and model active species, both: ionic and covalent are the function of monomer basicity; on the other hand no correlation between ring strain and reactivity was found (44).

In order to correlate the reactivities of various onium ion (active species) with their structures, the reaction between ions, modelling different active species, and the highly nucleophilic monomer conidine was studied. The corresponding rate constants are given in Table VIII.

Table VII.   Rate constants for the addition of heterocyclic monomers to the model active species $(C_6H_5NO_2, 35°)$ $\underline{(44)}$, shown in scheme 8.

Monomer	$\dfrac{k_{ai}}{mol^{-1}\cdot l\cdot s^{-1}}$	$\dfrac{k_{ac}}{mol^{-1}\cdot l\cdot s^{-1}}$	pK_a
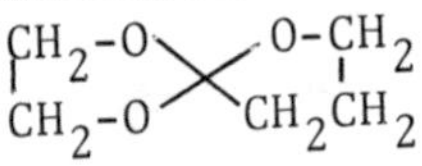 $\begin{array}{l}CH_2-O\diagdown\quad\diagup O-CH_2\\CH_2-O\diagup\diagdown CH_2CH_2\end{array}$	$4.5\cdot10^{-3}$	$4\cdot10^{-5}$	-7.3
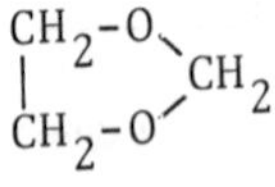 $\begin{array}{l}CH_2-O\diagdown\\\quad\quad\quad CH_2\\CH_2-O\diagup\end{array}$	10^{-4}	$6\cdot10^{-5}$	-6.5
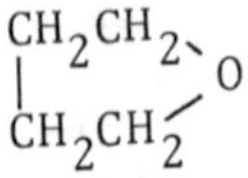 $\begin{array}{l}CH_2CH_2\diagdown\\\quad\quad\quad O\\CH_2CH_2\diagup\end{array}$	$4\cdot10^{-2}$	$2\cdot10^{-4}$	-2.0
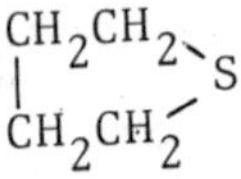 $\begin{array}{l}CH_2CH_2\diagdown\\\quad\quad\quad S\\CH_2CH_2\diagup\end{array}$	20	$4.4\cdot10^{-2}$	1.2
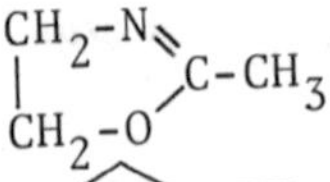 $\begin{array}{l}CH_2-N\diagdown\\\quad\quad\quad C-CH_3\\CH_2-O\diagup\end{array}$	120	$4\cdot10^{-1}$	3.4
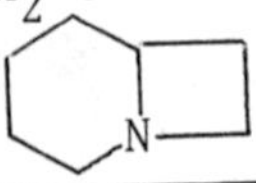	500	170	11

Table VIII.   Rate constants of addition of conidine to various onium ions, differing in heteroatoms $(C_6H_5NO_2$ solvent, $35°)$ $\underline{(44)}$.

Active centre	$\dfrac{k}{mol^{-1}\cdot l\cdot s^{-1}}$
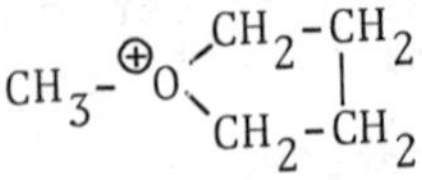 $CH_3-\overset{\oplus}{O}\diagdown\begin{smallmatrix}CH_2-CH_2\\\ \ \ \ \ \ \ \ \\CH_2-CH_2\end{smallmatrix}$, $SbF_6^{\ominus}$	$5\cdot10^2$
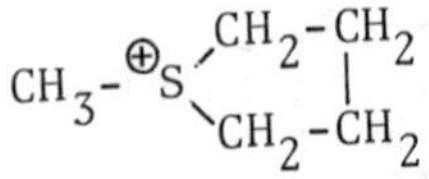 $CH_3-\overset{\oplus}{S}\diagdown\begin{smallmatrix}CH_2-CH_2\\\ \ \ \ \ \ \ \ \\CH_2-CH_2\end{smallmatrix}$, $CF_3SO_3^{\ominus}$	$1\cdot10^{-3}$
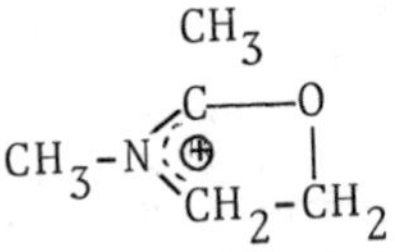 $CH_3-N\overset{\oplus}{\diagdown}\begin{smallmatrix}CH_3\\C-O\\\ \ \ \ \ \ \\CH_2-CH_2\end{smallmatrix}$, $CF_3SO_3^{\ominus}$	$9\cdot10^{-2}$
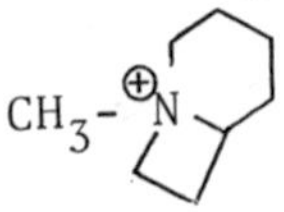 $CH_3-\overset{\oplus}{N}$, $CF_3SO_3^{\ominus}$	$7\cdot10^{-3}$

The data shown in Table VIII indicate that the higher is the basicity of the parent monomer the lower is the reactivity of active species   from this monomer toward the standard monomer.

Thus the observed order of reactivities in homopropagation is parallel to the order of reactivities in reaction of standard monomer with different active species and reverse to that observed for reaction of different monomers with standard active species.

This is a clear demonstration, that in passing from the ground state to the transition state the bond-breaking is more advanced than the bond-making.

With further shift into the direction of still more advanced breaking of the bond within active species this borderline S_N2 mechanism could eventually convert into the S_N1 mechanism. This should be promoted by the presence of the stabilizing group located closely to the carbenium ion (like in cyclic acetals polymerization) and/or high ring strain (like in the three membered rings). Indeed, contribution of S_N1 mechanism in both cases has been postulated for polymerization of 1,3-dioxolane and isobutylene oxide but there is still no clear-cut evidence for its operation.

Cationic polymerization of heterocyclic monomers can proceed not only by the S_N2 mechanism involving onium ions located at the chain end, and analysed in this paper, but also by another S_N2 mechanism involving activated monomer, adding to the neutral chain ends. For the latter no correlation is yet available and can differ from these described for onium ions in this section.

Literature Cited

1. Bamford, C.H.; Barb, W.G.; Jenkins, A.D.; Onyon, P.F. "The Kinetics of Vinyl Polymerization by Radical Mechanism"; Butterworths: London, 1958.
2. Bagdasarian, K.S. "Theory of Radical Polymerization" (in Russian); AN SU: Moscow 1959.
3. Penczek, S.; Kubisa, P.; Matyjaszewski, K.; Szymański, R. In "Cationic Polymerization"; Goethals, E.J., Ed.; Academic: in press.
4. Morton, M.; Kammereck, R.F.; Fetters, L.J. Br.Polym.J. 1971, 3, 120.
5. Duda, A.; Sosnowski, S.; Słomkowski, S.; Penczek,S. Makromol.Chem., submitted.
6. Brzezińska, K.; Chwiałkowska, W.; Kubisa, P.; Matyjaszewski, K.; Penczek, S. Makromol.Chem. 1977, 178, 2491.
7. Matyjaszewski, K.; Penczek, S. Makromol.Chem. 1981, 182, 1735.
8. Szwarc, M. Adv.Polym.Sci. 1983, 49, 1.
9. Saegusa, T.; Matsumoto, S. J.Polym.Sci. 1968, 6, 1559.
10. Barzykina, R.G.; Komratov, G.N.; Korovina, G.V.; Entelis, S.G. Vysokomol.Soed. 1974, 16, 906.

11. Sawamoto, M.; Furukawa, A.; Higashimura, T. <u>Macromolecules</u> 1983, 16, 518.
12. Hofman, A.; Szymański, R.; Słomkowski, S.; Penczek,S. <u>Makromol.Chem.</u> in press.
13. Penczek, S.; Kubisa, P.; Matyjaszewski, K. <u>Adv.Polym. Sci.</u> 1980, 37, 1.
14. Hofman, A.; Słomkowski, S.; Penczek, S. <u>Makromol.Chem.</u> 1984, 185, 91.
15. Jedliński, Z.; Kasperczyk, J.; Dworak, A.; Matuszewska, B. <u>Makromol.Chem.</u> 1982, 183, 587.
16. Szymański, R.; Kubisa, P.; Penczek, S. <u>Macromolecules</u> 1983, 16, 1000.
17. Matyjaszewski, K. <u>J.Polym.Sci.,Polym.Chem.Ed.</u> 1984, 22, 29.
18. Szymański, R.; Penczek, S. <u>Makromol.Chem.</u> 1982, 183, 1587.
19. Plesch, P.H.; Westermann, P.H. <u>J.Polym.Sci.</u> 1968,C16, 3837.
20. Ouhadi, T.; Hamitou, A.; Jerome, R.; Teyssie, P. <u>Macromolecules</u> 1976, 9, 927.
21. Aida, T.; Inoue, S. <u>Macromolecules</u> 1981, 14, 1162.
22. Yasuda, T.; Aida, T.; Inoue, S. <u>Macromolecules</u> 1983 16, 1792.
23. Saegusa, T. <u>Makromol.Chem.</u> 1974, 175, 1199.
24. Smith, S.; Hubin, A.J. <u>J.Macromol.Sci.-Chem.</u> 1973, A7, 1399.
25. Matyjaszewski, K.; Penczek, S. <u>J.Polym.Sci.,Polym. Chem.Ed.</u> 1974, 12, 1905.
26. Kobayashi, S.; Danda, H.; Saegusa, T. <u>Macromolecules</u> 1974, 7, 415.
27. Pruckmayr, G.; Wu, T.K. <u>Macromolecules</u> 1975, 8, 954.
28. Buyle, A.M.; Matyjaszewski, K.; Penczek, S. <u>Macromolecules</u> 1977, 10, 269.
29. Baran, T.; Brzezińska, K.; Matyjaszewski, K.; Penczek, S. <u>Makromol.Chem.</u> 1983, 184, 2497.
30. Kazanskii, K.S.; Penczek, S. <u>Vysokomol.Soed.</u> 1983,25 1347.
31. Sosnowski, S.; Słomkowski, S.; Penczek, S. <u>J.Macromol. Sci.-Chem.</u>in press.
32. Boileau, S. ACS SYMPOSIUM SERIES No. 166, Mc Grath, J.E. Editor,American Chemical Society: Washington, D.C., 1981; p. 283.
33. Słomkowski, S.; Penczek, S. <u>Macromolecules</u> 1980, 13, 229.
34. Matyjaszewski, K.; Słomkowski, S.; Penczek, S. <u>J.Polym.Sci.,Polym.Chem.Ed.</u> 1979, 17, 2413.
35. Brzezińska, K.; Matyjaszewski, K.; Penczek, S. <u>Makromol.Chem.</u> 1978, 179, 2387.
36. Matyjaszewski, K. <u>Makromol.Chem.</u> 1984, 185, 51.
37. Fuoss, R.M. <u>J.Amer.Chem.Soc.</u> 1978, 100, 5576.
38. Sosnowski, S.; Słomkowski, S.; Penczek, S.; Reibel, L. <u>Makromol.Chem.</u> 1983, 184, 2159.
39. Chojnowski, J.; Mazurek, M. <u>Makromol.Chem.</u> 1975, 176, 2999.

40. Kazanskii, K.S. Pure & Appl.Chem. 1981, 53, 1645.
41. Penczek, S. Macromolecules 1979, 12, 1010.
42. Arkhipovitch, G.N.; Dubravskii, S.A.; Kazanskii, K.S.; Shupik, A.N. Vysokomol.Soedin. 1981, 23, 1653.
43. Dimov, D.K.; Panayotov, I.M.; Lazarov, V.N.; Tsvetanov, J.Polym.Sci.,Polymer Chem.Ed. 1982, 20, 1389.
44. Matyjaszewski, K.; Penczek, S. Macromolecules, submitted.

RECEIVED November 15, 1984

Control of Ring-Opening Polymerization with Metalloporphyrin Catalysts
Mechanistic Aspects

SHOHEI INOUE and TAKUZO AIDA

Department of Synthetic Chemistry, Faculty of Engineering, University of Tokyo, Hongo, Bunkyo-ku, Tokyo 113, Japan

Aluminum porphyrins are initiators (catalysts) used particularly in ring-opening polymerization to yield products with well-defined molecular weight and with narrow distribution. The reaction can be extended to the polymerization of epoxide and β-lactone, and the copolymerization of epoxide with carbon dioxide and with cyclic acid anhydride. The formation of a copolymer with narrow molecular weight distribution from epoxide and cyclic acid anhydride is of interest because of the two different propagating species involved: an alkoxide and a carboxylate. An aluminum porphyrin coupled with a quaternary ammonium or phosphonium salt is a good catalyst system for the alternating copolymerization of epoxide and carbon dioxide or epoxide and cyclic acid anhydride. The latter reaction is the first example of a catalytic process occurring on both sides of a metalloporphyrin plane.

Aluminum porphyrins are initiators (catalysts) used particularly in ring-opening polymerization to yield products of well-defined molecular weight and with narrow distribution. The reaction has been extended from the polymerization of epoxide (1) to that of β-lactone (2), and also to the copolymerization of epoxide and carbon dioxide (3) or epoxide and cyclic acid anhydride (4). Because of the living nature of the polymerization, block copolymerization from different epoxides, for example, has been accomplished with high efficiency (5). Every aluminum atom in the metalloporphyrin carries one growing polymer molecule in the polymerization of epoxide and β-lactone. This fact and the strong effect of the ring current of porphyrin on NMR spectrum are of great advantage for the investigation of the structure of the growing species and the reactivity.

0097–6156/85/0286–0137$06.00/0

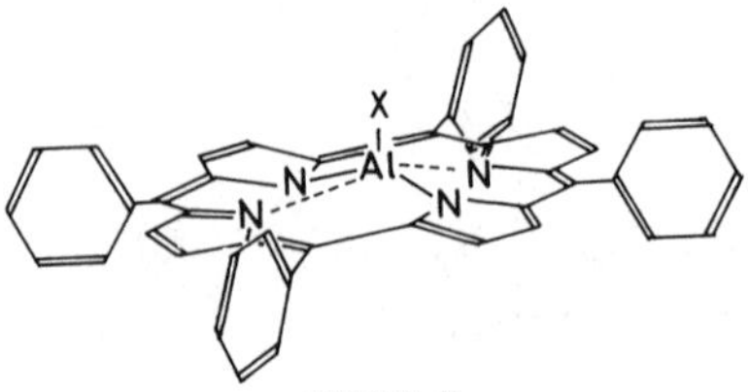

TPPAl-X

1 a X=Cl
b X=OR
c $X=O_2CR$
d X=OAr
e $X=C_2H_5$

Structure and Reactivity of Growing Species in the Polymerization of Epoxide

The growing species in the polymerization of epoxide initiated with tetraphenylporphinatoaluminum chloride (TPPAlCl, 1a) is a porphinatoaluminum alkoxide (1b) (6). The 1H-NMR spectrum of the reaction mixture in the polymerization of propylene oxide shows a doublet signal at an unusually high magnetic field (-2.0 ppm). This signal is due to the methyl group at the growing end attached to the metalloporphyrin (Figure 1).

$$\text{(Al-Cl} + x\ \underset{\text{O}}{CH_2\text{-}\overset{CH_3}{CH}} \longrightarrow \text{(Al-O-}\overset{CH_3}{CH}\text{-}CH_2\text{-}(\text{O-}\overset{CH_3}{CH}\text{-}CH_2)_{x-1}\text{-Cl} \qquad (1)$$

TPPAlCl

Polymerization of ethylene oxide proceeds similarly to give TPPA1-O-CH2-CH2- as the growing end; a characteristic signal is shown at -1.4 ppm (Figure 2). On the other hand, tert-butylethylene oxide reacts with TPPAlCl to give the corresponding alkoxide, TPPAl-O-CH {C(CH3)3}-CH2-Cl (-1.55 ppm); further propagation reaction proceeds with difficulty. By taking advantage of this fact, the reactivity of porphinatoaluminum alkoxide as the growing species can be evaluated: In the reaction of tert-butylethylene oxide with an equimolar mixture of the living ends from ethylene oxide and from propylene oxide, the porphinatoaluminum alkoxide corresponding to tert-butylethylene oxide increased at the expense of the living end from ethylene oxide, whereas the living end from propylene oxide remained almost intact. Thus, TPPAl-O-CH2-CH2- is concluded to be much more reactive towards epoxide than TPPAl-O-CH(CH3)-CH2-.

Structure of Growing Species in the Polymerization of β-Lactone

In the ring-opening reaction of β-lactone, two different modes of cleavage are possible. One is the cleavage at the alkyl-oxygen bond to give a porphinatoaluminum carboxylate (Equation 2a), and the other is acyl-oxygen bond scission to form a porphinatoaluminum alkoxide carrying an acyl chloride group (2b).

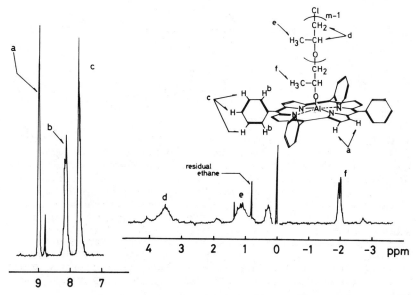

Figure 1. 1H-NMR spectrum of a living oligomeric propylene oxide prepared with TPPAlCl. (Reproduced from Ref. 6. Copyright 1981 American Chemical Society.)

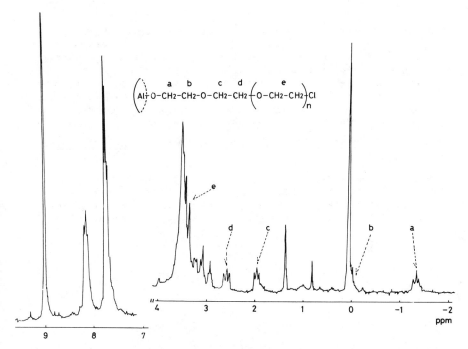

Figure 2. 1H-NMR spectrum of a living oligomeric ethylene oxide prepared with TPPAlCl. (Reproduced from Ref. 6. Copyright 1981 American Chemical Society.)

$$\text{(a)} \quad \underset{\substack{| \\ O-C=O}}{\overset{\substack{R \\ |}}{CH-CH_2}} \quad \overset{Al-Cl}{\longrightarrow} \quad Al-O-\underset{\substack{\| \\ O}}{\overset{}{C}}-CH_2-\overset{\substack{R \\ |}}{CH}-Cl \qquad (2a)$$

$$\text{(b)}$$

$$\qquad\qquad\qquad\qquad Al-O-\overset{\substack{R \\ |}}{CH}-CH_2-\underset{\substack{\| \\ O}}{\overset{}{C}}-Cl \qquad (2b)$$

$$R = H \text{ or } CH_3$$

In the IR spectrum of the equimolar reaction product of β-propiolactone and TPPAlCl, a strong absorption bond due to the carboxylate group was observed at 1600 cm (Figure 3). However, the absorption due to the acyl chloride group at 1800 cm-1 was not seen. The 1H-NMR spectrum of this reaction product is characterized by a triplet signal at -0.7 ppm, from protons of the methylene group of a porphinatoaluminum carboxylate (2a, R=H) (Figure 4). If the ring opening of β-propiolactone took place at the acyl-oxygen bond (2b, R=H), a triplet signal from the TPPAl-O-CH2-CH2- group should appear at -1.4 ppm, similarly to the species from ethylene oxide. The 1H-NMR spectrum of the equimolar reaction product of TPPAlEt (1e) and 3-chloro-propionic acid was identical to the spectrum of the TPPAlCl-propiolactone system.

$$Al-Et + H-O-\underset{\substack{\| \\ O}}{\overset{}{C}}-CH_2-CH_2-Cl \longrightarrow Al-O-\underset{\substack{\| \\ O}}{\overset{}{C}}-CH_2-CH_2-Cl + EtH \qquad (3)$$

Thus, the ring opening of β-propiolactone takes place almost exclusively at the alkyl-oxygen bond to form a porphinatoaluminum carboxylate (Equation 2a) (7). The same conclusion was obtained for the oligomerization of β-propiolactone and β-butyrolactone as a result of 1H-NMR and 13C-NMR spectroscopy. As expected, the equimolar reaction product of TPPAlEt (1e) and carboxylic acid (cf. Equation 3) is a good initiator for the polymerization of β-lactone and yields a polymer with narrow molecular weight distribution.

Copolymerization of Epoxide with CO2, and Epoxide with Cyclic Acid Anhydride

TPPAlOR (1b) was effective for the copolymerization of epoxide with CO2 (8-9) or with cyclic acid anhydride and produced copolymers with ester and ether linkages. These copolymers have a narrow molecular weight distribution, but do not yield alternating structures.

$$\underset{\substack{\diagdown O \diagup}}{\overset{\substack{CH_3 \\ |}}{CH_2-CH}} + CO_2 \longrightarrow \left(\!\!O-\overset{\substack{CH_3 \\ |}}{CH}-CH_2\!\!\right)_{\!x}\!\!\left(\!\!O-\underset{\substack{\| \\ O}}{\overset{}{C}}-O-\overset{\substack{CH_3 \\ |}}{CH}-CH_2\!\!\right)_{\!1-x} \qquad (4)$$

$$\underset{\substack{\diagdown O \diagup}}{\overset{\substack{CH_3 \\ |}}{CH_2-CH}} + \underset{\substack{\diagdown O \diagup}}{\overset{}{O=C \quad C=O}} \rightarrow \left(\!\!O-\overset{\substack{CH_3 \\ |}}{CH}-CH_2\!\!\right)_{\!x}\!\!\left(\!\!O-\underset{\substack{\| \\ O}}{\overset{}{C}} \quad \underset{\substack{\| \\ O}}{\overset{}{C}}-O-\overset{\substack{CH_3 \\ |}}{CH}-CH_2\!\!\right)_{\!1-x} \qquad (5)$$

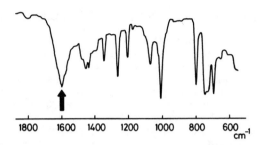

Figure 3. IR spectrum of the equimolar reaction product of TPPAlCl and β-propiolactone. (Reproduced from Ref. 7. Copyright 1983 American Chemical Society.)

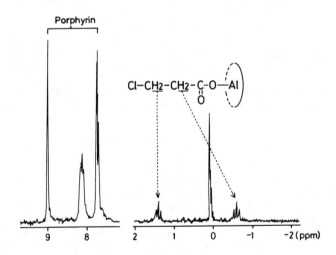

Figure 4. 1H-NMR spectrum of the equimolar reaction product of TPPAlCl and β-propiolactone. (Reproduced from Ref. 7. Copyright 1983 American Chemical Society.)

This result is particularly interesting because the propagation is presumed to involve two different types of reactions occurring on the same aluminum atom (i.e., the reaction of aluminum alkoxide with CO_2 or cyclic acid anhydride and the reaction of aluminum carboxylate with epoxide).

Aluminum Porphyrin Coupled with Ammonium or Phosphonium Salt

In order to enhance the reactivity of aluminum porphyrins (<u>1</u>), especially towards CO_2 in the copolymerization with epoxide (Equation 4), the effect of addition of an amine or phosphine as a possible sixth ligand to the aluminum porphyrin was examined. The enhancement in reactivity by the addition of a tertiary amine such as <u>N</u>-methylimidazole was actually observed for the epoxide-CO_2 reaction. The product, however, was a cyclic carbonate (<u>10</u>), not a linear copolymer. On the other hand, the addition of triphenylphosphine was very effective in the formation of an alternating copolymer from epoxide and CO_2, or from epoxide and cyclic acid anhydride. Because preliminary studies indicated that triphenylphosphine was converted to a quaternary salt in the reaction, the effect of a quaternary phosphonium or ammonium salt separately prepared was examined. As a result of this investigation, the system containing an aluminum porphyrin and phosphonium or ammonium salt was found to be a novel, effective catalyst for these alternating copolymerization reactions and to yield products with narrow molecular weight distribution.

For example, the alternating copolymerization of propylene oxide and phthalic anhydride (Equation 5, x = 0) with the TPPAlCl-ethyltriphenyl-phosphonium bromide (EtPh3PBr) system proceeds at room temperature (Figure 5) much more readily than with other catalysts (<u>11</u>). The molecular weight of the product increased linearly with conversion, and the narrow molecular weight distribution was retained (Figure 6). TPPAlCl-tetraalkylammonium halide and TPPAlO2CR-tetraalkylammonium carboxylate (R4NO2CR) systems are also effective. Tetraalkylammonium or phosphonium salt alone is ineffective under similar conditions.

Catalytic Reaction on Both Sides of a Metalloporphyrin Plane

Of particular interest is the fact that in the alternating copolymerization of epoxide and phthalic anhydride with the TPPAlCl-EtPh3PBr system every aluminum atom of the catalyst carries two growing polymer molecules. As seen in Table I, the observed molecular weight of the copolymer, determined by vapor pressure osmometry (VPO), is about one-half the molecular weight calculated on the assumption that every aluminum atom carries one polymer molecule.

In this connection, it is important to note that with the TPPAlO2CR and R4NO2CR system, the catalyst system for this copolymerization, a novel porphinatoaluminum complex with two carboxylate axial ligands is formed.

$$\overset{\frown}{\underset{\smile}{}}Al-O_2CR \ + \ RCO_2^- \overset{+}{N}R_4' \longrightarrow [RCO_2 - \overset{\frown}{\underset{\smile}{}}Al-O_2CR']^- \overset{+}{N}R_4 \qquad (6)$$

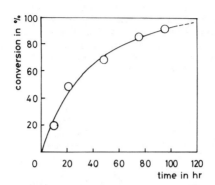

Figure 5. Alternating copolymerization of propylene oxide (PO) and phthalic anhydride (PA) with the TPPAlCl-Et3PhPBr system. [PO] = [PA]/ [Cat] = 25, in CH2Cl2 at room temperature.

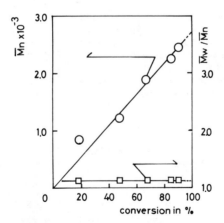

Figure 6. The influence of conversion on molecular weight in the alternating copolymerization of propylene oxide and phthalic anhydride catalyzed by the TPPAlCl-Et3PhPBr system.

The formation of a similar porphinatoaluminum complex was observed in the copolymerization of propylene oxide and phthalic anhydride with the TPPAlO2CCMe3-Et4NO2CMe system. In the 1H-NMR spectrum of the reaction mixture, we observed that the signals assigned to the phthalate group attached to porphinatoaluminum were shifted to a higher magnetic field (Figure 7a). Their intensity values are estimated to be twice that of the signal due to the porphyrin ligand. Thus, Equation 7 is suggested as a possible mechanism for the copolymerization proceeding on both sides of a metalloporphyrin plane, where both X and Y groups, corresponding to Me3CCO2- and MeCO2- groups, respectively, in this reaction, can initiate the reaction.

$$Al-X \quad + \quad R_4\overset{+-}{N}Y \quad + \quad \underset{O}{\overset{C-C}{\diagdown\diagup}} \quad + \quad \text{(phthalic anhydride)}$$

$$\longrightarrow \quad Y-C-C-O \sim\sim\sim CO_2 \cdots \begin{pmatrix} Al \end{pmatrix} \cdots O_2C \sim\sim\sim O-C-C-X \tag{7}$$

$$(R_4N)^+$$

   More evidence in favor of polymer chain growth on both sides of a metalloporphyrin plane was obtained in the copolymerization of propylene oxide and phthalic anhydride. The catalyst was the combination of EtPh3PBr and TPPAl-(-O-CHMe-CH2-)-Cl (TPPAlPPO), which can be obtained by the polymerization of propylene oxide with TPPAlCl (Equation 1). If the copolymerization proceeds on both sides of a metalloporphyrin plane, a block copolymer, polyetherpolyester, is expected to be formed on one side, and a polyester is expected on the other side. In fact, GPC of the reaction product (Figure 8) showed two narrow peaks and clearly indicated the formation of polymers with different chain lengths. Thus, this reaction provides the first example of a catalytic reaction occurring on both sides of a metalloporphyrin plane.

Table I. Alternating Copolymerization of Epoxide and Phthalic Anhydride with TPPAlCl-EtPh3PBr System

Epoxide[a]		$\overline{M}_n \times 10^{-3}$	$\overline{M}_w/\overline{M}_n$	$\overline{M}_{calc}$	N/Al [b]
R	R'	VPO	GPC	for $N/Al=1$	$(\overline{M}_{calc}/\overline{M}_n)$
CH3	H	2.97	1.09	5.15	1.74
CH3	CH3(cis)	3.25	1.11	5.50	1.69
CH3	CH3(trans)	3.39	1.09	5.50	1.62
(CH2)4		4.00	1.17	6.15	1.54
Ph	H	2.71	1.14	6.70	2.47
PhOCH2	H	3.60	1.08	7.45	2.07

a) RHC-CHR'   b) N: Number of polymer molecules
        $\diagdown\diagup$
         O

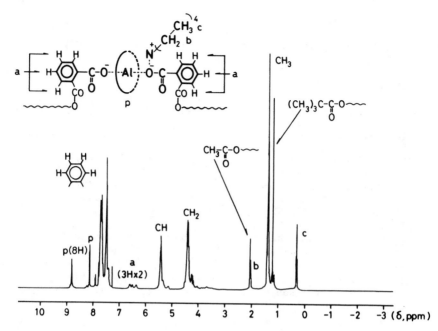

Figure 7. 1H-NMR spectrum of the copolymerization of propylene oxide and phthalic anhydride with the TPPAlO2CCMe3-Et4NO2CMe system (in CDCl3 at room temperature).

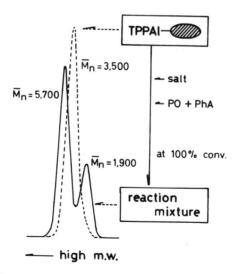

Figure 8. GPC profile of the copolymerization of propylene oxide and phthalic anhydride with the TPPAlPPO-EtPh3PBr system. Key: ---, the product; and ---, TPPAlPPO.

Acknowledgment

The author is grateful to Messrs. T. Yasuda, M. Ishikawa, and
K. Sanuki for their collaboration.

Literature Cited

1. Aida, T.; Mizuta, R.; Yoshida, Y.; Inoue, S. Makromol. Chem.,
   1981, 182, 1073.
2. Yasuda, T.; Aida, T.; Inoue, S. Makromol. Chem., Rapid Commun.,
   1982, 3, 585.
3. Aida, T.; Inoue, S. Macromolecules, 1982, 15, 682.
4. Aida, T.; Inoue, S. Polymer Preprints Japan, 1983, 32, 217-218.
5. Aida, T.; Inoue, S. Macromolecules, 1981, 14, 1162.
6. Aida, T.; Inoue, S. Macromolecules, 1981, 14, 1166.
7. Yasuda, T.; Aida, T.; Inoue, S. Macromolecules, 1983, 16, 1792.
8. Aida, T.; Inoue, S. Macromolecules, 1982, 15, 682.
9. Inoue, S.; Yamazaki, N., Eds. "Organic and Bioorganic Chemistry
   of Carbon Dioxide"; Kodansha : Tokyo, 1981; p. 167.
10. Aida, T.; Inoue, S. J. Am. Chem. Soc., 1983, 105, 1304.
11. Ishii, Y.; Sakai, S. In "Ring-Opening Polymerization";
    Frisch, K. C.; Reegen, S. L., Eds.; Marcel Dekker: New York,
    1969; p. 91.

RECEIVED October 4, 1984

Anionic Ring-Opening Polymerization of Octamethylcyclotetrasiloxane in the Presence of 1,3-Bis(aminopropyl)-1,1,3,3-tetramethyldisiloxane

P. M. SORMANI, R. J. MINTON[1], and JAMES E. MCGRATH

Department of Chemistry and Polymer Materials and Interfaces Laboratory, Virginia Polytechnic Institute and State University, Blacksburg, VA 24061

Polyorganosiloxanes are probably the most widely used and studied class of "semi-inorganic" polymers. There are a variety of interesting and useful properties exhibited by these materials that make them worthy of study. For example, they exhibit high lubricity, low glass transition temperatures, good thermal stability, high gas permeability, unique surface properties, and low toxicity ($\underline{1}$).

Cyclic organosiloxanes and silanol oligomers may be readily prepared by the hydrolysis of chlorosilanes, according to Scheme 1 ($\underline{1}$). The predominant cyclics are those corresponding to x=4 or 5, while the strained cyclic trimer is present only in small quantities.

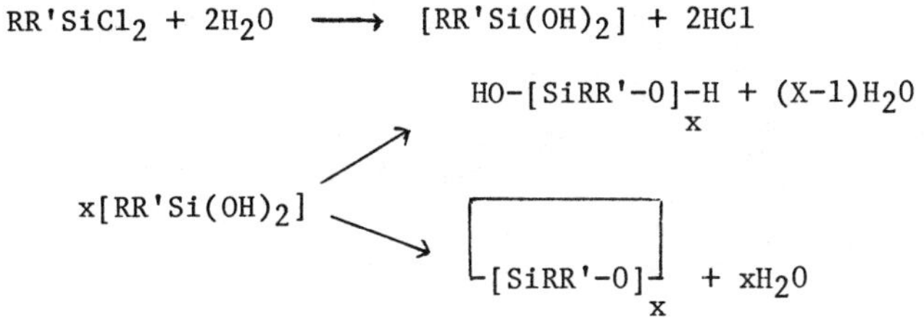

$$RR'SiCl_2 + 2H_2O \longrightarrow [RR'Si(OH)_2] + 2HCl$$

$$HO-[SiRR'-O]_x-H + (X-1)H_2O$$

$$x[RR'Si(OH)_2]$$

$$\left[SiRR'-O\right]_x + xH_2O$$

Scheme 1. Preparation of cyclic organosiloxanes and chlorosilanes.($\underline{1}$)

Polydimethylsiloxane oligomers may be easily prepared by the acid or base catalyzed ring opening polymerization of the cyclic tetramer, octamethylcyclotetrasiloxane. The molecular weight of the polymer prepared may be controlled by the addition of a linear disiloxane as an endblocker ($\underline{2},\underline{3},\underline{4}$). When the disiloxane is hexamethyldisiloxane, this is the well-studied case of the

[1]Current address: Thoratec Laboratories, 2023 Eighth Street, Berkeley, CA 94710

preparation of silicone oil. However, it is the case of functional
disiloxanes that has been of interest in our laboratories for quite
some time (3,5). Table I shows a list of the various functional
disiloxanes that have been used to prepare functionally terminated
siloxane oligomers (6). Scheme II shows a general outline for the
preparation of these oligomers. It should be noted that in the
absence of any endblocker, a high molecular weight silicone gum is
formed.

Table I

End blockers used to prepare functionally-terminated polysiloxane
oligomers

$$[H_2N \ (-CH_2)_3 \underset{\underset{CH_3}{|}}{\overset{\overset{CH_3}{|}}{Si}}]-O_2$$ α, ω aminopropyl 1,3 tetramethyldisiloxane

$$(CH_3)_2N-[\underset{\underset{CH_3}{|}}{\overset{\overset{CH_3}{|}}{Si}-O}]_x-\underset{\underset{CH_3}{|}}{\overset{\overset{CH_3}{|}}{Si}}-N(CH_3)_2$$ low molecular weight silylamine end
blocker

$$[HN\overset{\frown}{\underset{\smile}{}}N-(CH_2)_2NHC\overset{\overset{O}{\|}}{}-(CH_2)_3-\underset{\underset{CH_3}{|}}{\overset{\overset{CH_3}{|}}{Si}}-O]_2$$ piperazine-terminated disiloxane

$$[HOC\overset{\overset{O}{\|}}{}-(CH_2)_3-\underset{\underset{CH_3}{|}}{\overset{\overset{CH_3}{|}}{Si}-}]-O_2$$ α, ω carboxypropyl 1,3 tetramethyldisiloxane

$$[CH_2\overset{O}{\frown}CH-CH_2-O(CH_2)_3-\underset{\underset{CH_3}{|}}{\overset{\overset{CH_3}{|}}{Si}}]-O_2$$ α, ω glycidoxypropyl 1,3 tetramethyldisiloxane

These ring-opening polymerizations are referred to as
equilibration reactions. Since a variety of interchange reactions
can take place, a quantitative conversion of the tetramer to high
polymer is not achieved and there is, at thermodynamic equilibrium,
a mixture of linear and cyclic species present. Scheme III shows
examples of the types of redistribution reactions thought to be
occurring in these systems. It is generally convenient to use "D"
to refer to a difunctional siloxane unit and "M" to refer to a
monofunctional siloxane unit. Thus, D_4 represents the cyclic
siloxane tetramer and MM represents the linear hexamethyldisiloxane.

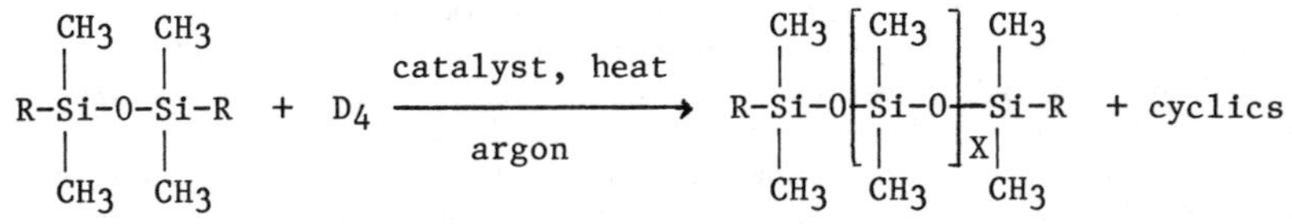

$$\begin{array}{c}
\underset{\underset{CH_3}{|}}{\overset{\overset{CH_3}{|}}{R-Si-O-Si-R}}\underset{\underset{CH_3}{|}}{\overset{\overset{CH_3}{|}}{}} + D_4 \quad \xrightarrow[\text{argon}]{\text{catalyst, heat}} \quad \underset{\underset{CH_3}{|}}{\overset{\overset{CH_3}{|}}{R-Si-O}}\left[\underset{\underset{CH_3}{|}}{\overset{\overset{CH_3}{|}}{Si-O}}\right]_{x}\underset{\underset{CH_3}{|}}{\overset{\overset{CH_3}{|}}{Si-R}} + \text{cyclics}
\end{array}$$

Scheme II.   Preparation of functional siloxane oligomers.   R = any
functional group listed in Table I.

This terminology normally applies only to dimethylsiloxy units.   It
should be noted that with the exception of using organolithium
catalysts in the anionic polymerization of the D_3 cyclic,
significant amounts of redistribution cannot be avoided (7,8).

(1) $-D_x^- + D_4 \longrightarrow -D_{(x+4)}^-$

(2) $-D_x^- + MM \longrightarrow MD_xM$

(3) $MD_xM + MM \longrightarrow MD_{(x-5)}M + MD_5M$

(4) $MD_xM + MD_yM \longrightarrow MD_{(x+w)}M + MD_{(y-w)}M$

Scheme III. Redistribution reactions occurring during a siloxane
equilibration.

There are a variety of catalysts that can be used in the
preparation of polysiloxane oligomers by equilibration reactions.
The choice of catalyst depends upon the temperature of the
equilibration as well as the type of functional disiloxane that is
used.   For example, in preparing an aminopropyl terminated siloxane
oligomer, a basic catalyst is used, rather than an acidic catalyst
which would react with the amine end groups.   The discussion here
will be limited to basic catalysts.
Bases such as hydroxides, alcoholates, phenolates and
siloxanolates of the alkali metals, quaternary ammonium and
phosphonium bases and the corresponding siloxanolates and fluorides,
and organoalkali metal compounds have all been found to catalyze the
polymerization of cyclic siloxanes (1,2,9,10).   It is believed that
all catalysts generate the siloxanolate anion in situ, and it is
this species which breaks the silicon-oxygen bond in either the
linear or cyclic siloxanes present.   However, the reactivities of
the disiloxane and the various cyclic siloxanes differ.   The rate of
reaction increases in the order MM < MDM < MD$_2$M < D$_4$ < D$_3$, where MM
again represents hexamethyldisiloxane, a non-functional endblocker.
Catalysts based on the quaternary ammonium and phosphonium
bases are referred to as transient catalysts, since they decompose
above certain temperatures to products which are not catalytically
active toward siloxanes.   An example of this is the tetramethyl-
ammonium siloxanolate catalyst, prepared by the reaction of tetra-
methylammonium hydroxide with D$_4$ (11).   This catalyst polymerizes D$_4$
at temperatures up to perhaps 120°C.   Above this temperature, the
catalyst fairly rapidly decomposes to trimethyl amine and methoxy-
terminated siloxane.

However, catalysts such as the potassium siloxanolate catalyst are not transient. Non-transient catalysts must be neutralized or removed by some other method in order to give a thermally stable polymer. If the catalyst is not removed, it will cause depolymerization at high temperatures. For example, a silicone gum prepared by reacting D_4 with 0.01% KOH has been reported to lose over 99% of its weight at 250°C in 24 hours (11). Non-transient catalysts can often be used at much higher temperatures than the transient catalysts, leading, of course, to faster rates of reaction.

The kinetics of these processes have been of interest to a number of workers (12). However, there has been no investigation of these equilibration reaction kinetics using functional endblockers. Studies have been done on the kinetics of formation of silicone gum or the reaction of D_4 with hexamethyldisiloxane. For example, Grubb and Osthoff studied the kinetics of the KOH catalyzed polymerization of D_4 (12).

It was found that the ring-opening polymerization of D_4 by KOH proceeds according to first order kinetics, with a square root dependence on the catalyst concentration. The square root dependence on the catalyst concentration is believed to be due to the existence of an equilibrium between an active ion pair and a an inactive associated form $(-SiOM)_2$. Rate constants were determined at different catalyst levels and temperatures. An activation energy of about 18 kcal/mole was determined by an Arrhenius plot, in agreement with other workers in the field (9).

It is important to consider the effect of solvent on the rate of polymerization as well as on the amount of cyclics that are present. The rate of polymerization can be greatly enhanced by the action of dipolar aprotic solvents such as DMSO. This has been demonstrated by Cooper (13). However, the presence of a solvent will also increase the amount of cyclics present in a fully equilibrated sample. This can be understood in a qualitative way by considering that the siloxanolate species can attack not only silicon-oxygen bonds in the cyclics present, but also a phenomenon known as back-biting can occur. Back-biting refers to the attack of the siloxanolate anion on a silicon-oxygen bond along the same chain at least four repeat units away. Scheme IV gives an illustration of this as well as other types of reactions occurring. When the siloxanes present are diluted by the presence of a solvent, the siloxanolate anion will be less likely to encounter a cyclic to attack, and so back-biting will become more prevalent. There is, in fact, a critical concentration, above which only cyclic molecules will exist. Generally therefore, it is more desirable to perform these equilibration reactions in bulk, and so limit the formation of cyclic species as much as possible.

There have been a variety of analytical techniques used to study these equilibration reactions. For example, gel permeation chromatography (GPC), or size exclusion chromatography (SEC), and gas-liquid chromatography (GLC) have been useful techniques (5,14). While GPC is useful for monitoring the overall molecular weight distribution of the polysiloxane, there are some limitations. For example, aminopropyl-terminated polysiloxane oligomers cannot be run on styragel based SEC columns due to adsorption of the oligomer on the column. Either a silanized silica gel-based column must be used, or the oligomer must be derivatized. In addition to these problems, while GPC can be used to monitor the disappearance of D_4 as the equilibration proceeds, it does not address the question of whether or not the functional disiloxane has been quantitatively consumed.

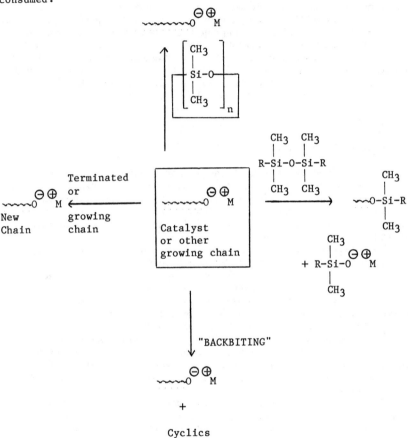

Scheme IV.   Possible reactions occurring during a siloxane equilibration.

Gas-liquid chromatography has the potential to discriminate between the different oligomeric species formed. We have found that capillary GC may be used to measure the concentration of α,ω aminopropyl 1,3 tetramethyldisiloxane in equilibration reactions. Although the presence of catalyst could potentially lead to the generation of additional cyclics at the elevated temperatures necessary for GC, there was no effect on the amount of disiloxane present. Of course, to measure the D_4 concentration, the samples must be free of catalyst.

High-Performance Liquid Chromatography (HPLC) has been used for the analysis of oligomers (15,16). A clear advantage of HPLC is that the analysis can be done at room temperature, thus eliminating the possibility of generating additional reaction products such as cyclics. HPLC has therefore been of use in measuring the amount of D_4 present as the equilibration reaction proceeds, as well as in observing the appearance of new oligomeric species.

This paper will discuss investigations of the polymerization of D_4 in the presence of α,ω aminopropyl 1,3 tetramethyldisiloxane with potassium siloxanolate catalyst. The effects of temperature and catalyst concentration on the rate of disappearance of both starting materials will be discussed. Demonstration of the utility of both non-aqueous reversed-phase HPLC and capillary GC for the investigation of these reactions will be presented.

Experimental

A. Catalyst Preparation

The catalyst used in this work was prepared either in bulk or using toluene as an azeotroping agent. Octamethylcyclotetrasiloxane, D_4, and potassium hydroxide (KOH) were used as received. The KOH was crushed into a fine powder and added to enough D_4 to make a mixture with a molar ratio of D_4 to KOH of 10/1 in the case of the bulk catalyst. This corresponds to 2 wt% KOH.

The mixture was then put into a flask equipped with an argon inlet, an overhead stirrer, an attached Dean-Stark trap, with a condenser attached to the Dean-Stark trap. Argon was bubbled through the mixture with stirring, and the mixture was heated to 120°C. The high temperature and argon stream were necessary to eliminate water present in the base as well as any water formed during the reaction. As the KOH reacted with the D_4, the mixture gradually became more viscous. After all the KOH had dissolved, the mixture was still transparent and colorless. In general, within 12 hours, the mixture was a milky white, viscous material. After approximately 24 hours, the mixture was clear and colorless and able to be removed by pipet.

Titrations of the catalysts were performed on a Fischer automatic titrator. Isopropanol (100 mls.) and 20 mls. water were used as the solvent media. Alcoholic HCl (0.0995 N.) was used as the titrant. The calculated amount of KOH present was between 1.9 and 2.7%, which compared favorably with the theoretical value of 2.0 wt%.

The procedure for the preparation of catalyst using toluene as an azeotroping agent is similar. The same relative amounts of potassium hydroxide and D_4 were used and enough toluene was added to make a 50% solution. The reaction mixture was heated for about 12 hours at 95°C and then at 120°C for an additional 12 hours.

In each case, the catalyst was stored under argon in vials sealed with teflon tape and placed in a dessicator until use. To remove catalyst for use in reactions, the vial was warmed slightly, if necessary, to reduce the viscosity, and the desired amount of catalyst removed by pipette. Argon was then flushed through the catalyst remaining in the vial, which was then resealed and returned to the dessicator.

B. Equilibration Reactions

Octamethylcyclotetrasiloxane and the α, ω aminopropyl 1,3
tetramethyldisiloxane were put into a three neck flask, fitted with
a reflux condenser, an argon inlet, and a magnetic stirring bar.
One neck of the flask was covered with a septum for the removal of
samples by a syringe equipped with a stopcock. Potassium
siloxanolate catalyst was pipetted into the flask. The flask was
then immediately fitted with the argon inlet and heated by a
silicone oil bath. Samples were removed at various times and put
into sample vials which were capped with septums. These were stored
in a refrigerator until analysis by HPLC. Some samples were
analyzed immediately by capillary gas chromatography.

C. High Performance Liquid Chromatography

Quantitative analysis of D_4 in equilibration samples was carried out
using a Waters Model 450 solvent delivery system. The columns
employed were ODS columns, either Dupont Zorbax columns obtained
from Fischer, or the Regis Little Giant. The Little Giant is a 5
cm. long, 10 mm. i.d. column packed with 3 micron particle size
ODSII packing. The Dupont Zorbax columns were 25 cm. long with a 10
mm. i.d. and a 10 micron particle size ODS packing. A differential
refractometer and a fixed wavelength infrared detector were used.
    Samples were made up in one ml. volumetric flasks in toluene.
Typical concentrations were in the range of 10-17%. The mobile
phase was composed of 35% acetone and 65% acetonitirle in the case
of the Dupont columns with a flow rate of 1.5 ml./min. The Little
Giant column used a mobile phase of 20% acetone and 80% acetonitirle
at a flow rate of 0.8 ml./min. The change in mobile phase and flow
rate was necessary to restore sufficient resolution for quantitative
analysis while still maintaining fast analysis times. All solvents
were HPLC grade solvents and were used as received with no further
purification.
    A calibration curve was prepared for D_4 by plotting peak height
in millimeters vs. micrograms injected. A 20 microliter sample loop
was used with the DuPont columns to ensure reproducible sample size.
A 35 microliter sample loop was used with the short column. Larger
sample volumes could be used in this case due to a lower operating
pressure. A typical calibration curve for D_4 is shown in Figure 1.
D_4 calibration curves were prepared using both the refractive index
and the infrared detector. The μg of D_4 corresponding to the
measured D_4 peak height of an equilibration sample can be read from
the calibration curve. Knowing the μl injected and the total sample
weight, the D_4 concentration can be determined.

D. Capillary Gas Chromatography

Capillary gas chromatography was used to measure the amount of
aminopropyl disiloxane present in the samples. An 11 m. column with
an internal diameter of 0.2 mm. coated with a dimethylsiloxane
stationary phase was used. A splitter injector was employed with a

split ratio of 100/1, at a temperature of 310°C. Samples were dissolved in methylene chloride and one microliter of solution injected. A flame ionization detector was employed, at a temperature of 275°C. Temperature programming was necessary to give good resolution and reasonable analysis times. The program followed was as follows: 80°C to 170°C at 5°C/min., 170°C to 225°C at 30°C/min. The carrier gas flow rate (He) was 1.7 ml/min at 80°C. Peak areas were calculated using a Perkin Elmer 3600 data station. Tetradecane (C_{14}) was used as an internal standard, and added from a stock solution to disiloxane or equilibration sample solutions. Shown in Figure 2 is a typical calibration curve, where the disiloxane/C_{14} area ratio is plotted against the disiloxane/C_{14} weight ratio. Knowing the experimentally determined disiloxane/C_{14} area ratio, and the weight of added C_{14} and equilibration sample, the amount of disiloxane present in any sample may be determined. The amount of disiloxane found is approximately plus or minus 1%.

Results and Discussion

A set of control experiments was first performed, where no aminopropyldisiloxane "end blocker" was used in the reaction. Reactions were done at 82°C, 111°C, 117°C and 140°C using sufficient catalyst to make 0.02 wt% KOH (0.034 mole/liter). Samples were removed at convenient intervals by syringe. In the case of the reaction done at 140°C, after only 15 minutes the reaction mixture was too viscous to be removed by syringe. At 111°C it took 120 minutes for the mixture to become too viscous for removal by syringe, while at 82°C it took 8.5 hours to reach this point.

The observed increase in viscosity is expected in the absence of endblocker and corresponds to the appearance of high molecular weight species. This presents a difficulty in the kinetic analysis of this data. Shown in Figure 3 is a plot of ln [D_4] vs. time for all four reaction temperatures. If the reaction is first order with respect to D_4 concentration, plotting ln [D_4] vs. time should give straight lines where the slope = -rate constant, k. However, the concentration of D_4 should be monitored to at least 75% conversion to differentiate between first and second order kinetics. In fact, in this concentration range, a second order plot - [D_4] vs. [D_4] time - also gives straight lines. However, it is known that this reaction is first order in D_4 concentration (11). In fact, a second order plot of Grubb and Osthoff's data (11) gives a plot that is indeed linear in the same concentration range as we have studied, but that then curves at higher concentrations. Therefore, the similarity between our first and second order kinetic plots is expected in the concentration range studied.

Assuming first order kinetics, a plot of ln k vs. 1/T (°K) was made (Figure 4), and the activation energy calculated to be 18.1 kcal/mole, in good agreement with previous values calculated by other workers (9). Reactions which used either the bulk catalyst or the catalyst prepared using toluene as an azeotroping agent gave rate constants which fell on the same line in the Arrhenius plot, indicating that the efficiency of each method of catalyst preparation is roughly the same.

The next set of experiments involved the determination of the

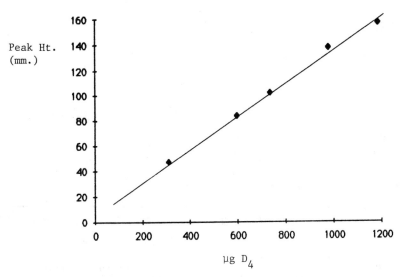

Figure 1. HPLC calibration curve for D$_4$, infrared detector.

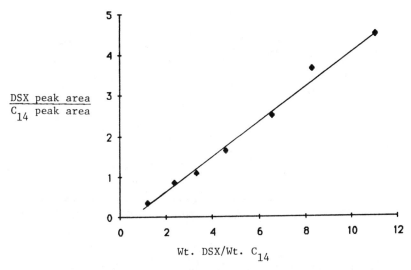

Figure 2. Capillary GC calibration curve for aminopropyl disiloxane (DSX).

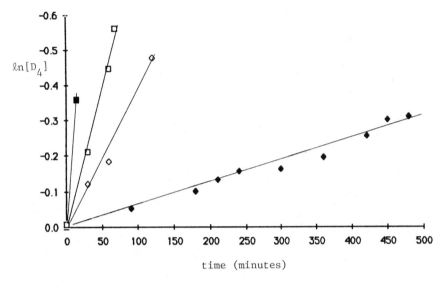

Figure 3.   Disappearance of D_4 at various temperatures. ■ = 140°C,
            □ = 117°C, ◇ = 111°C, ◆ = 82°C.

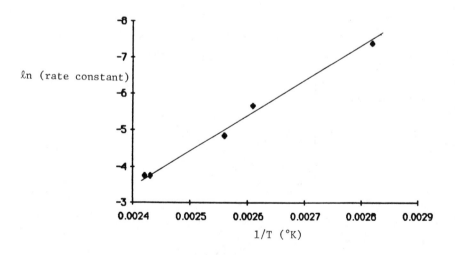

Figure 4.   Arrhenius plot for the reaction of D_4 with
            potassium siloxanolate catalyst at 0.02 Wt. %
            KOH.

rate of disappearance of D_4 in the presence of disiloxane. Shown in Figure 5 is a plot of D_4 concentration vs. time at 0.02 wt% KOH. In each case, the D_4 concentration decreases from 75 wt% to under 10 wt% in approximately 30 minutes. The ratio of D_4 to disolxane in this case should give an oligomer with $\langle Mn \rangle$ = 1000. Similar results were obtained at 0.12 wt% KOH and 91°C, 111°C, and 131°C, shown in Figure 6. No huge increase in viscosity was observed in any of these cases, in contrast to the reactions done in the absence of endblocker. In these reactions, the aminopropyl disiloxane is functioning analogously to a chain transfer agent to control the molecular weight. There is no large buildup in viscosity because the growing chains react with the aminopropyl disiloxane and are terminated. It is also interesting to note that in the absence of disiloxane at 140°C the amount of D_4 has decreased from 99 wt% to about 65 wt% at 20 minutes, which is much more than the amount that is remaining at the same time in the presence of disiloxane. One possible explanation of this is that since the reaction mixture is much less viscous in the presence of disiloxane, the D_4 can more easily react with the siloxanolate species. In other words, since this is a diffusion-controlled process, the lower viscosity has a dramatic effect on the rate of reaction of D_4 with potassium siloxanolate catalyst.

Lastly, the rate of reaction of the aminopropyl disiloxane was investigated. On the basis of electronegativity differences, it would be expected that the aminopropyl disiloxane would react more slowly than D_4 with the potassium siloxanolate catalyst. This was indeed found to be the case. At levels of 0.02 wt% KOH, the reaction was rather slow. However, at 0.12 wt% KOH, the reaction proceeded at convenient rates. Shown in Figure 7 is a plot of disiloxane concentration vs. time at 0.12 wt% KOH and temperatures of 90°C, 105°C, 129°C, 140°C. For example, at a temperature of 129°C, after approximately 6.5 hours, the disiloxane concentration has decreased from 19 wt% to 2.7 wt%. (In these reactions, the initial D_4 concentration was 80 wt%. This ratio of D_4 to disiloxane should yield a 1200 $\langle Mn \rangle$ oligomer.) This corresponds to an 85% decrease in the amount of disiloxane present. In contrast, at 131°C and the same catalyst level, the amount of D_4 present has decreased by over 90% after only 60 minutes.

Conclusions

It has been found that the attack of potassium siloxanolate catalyst on octamethylcyclotetrasiloxane is greatly accelerated in the presence of α, ω aminopropyl 1,3 tetramethyldisiloxane. The disiloxane functions analogously to a chain transfer agent and serves to prevent a large increase in viscosity, leading to a faster rate of reaction of D_4 with catalyst.

The rate of reaction of aminopropyl disiloxane with potassium siloxanolate catalyst was significantly slower than the rate of reaction of D_4. Temperatures above 100°C appear to most efficiently incorporate the functional disiloxane at the catalyst levels studied.

The studies described herein have been most useful (6,17,18,19) in establishing reaction conditions for the synthesis of well

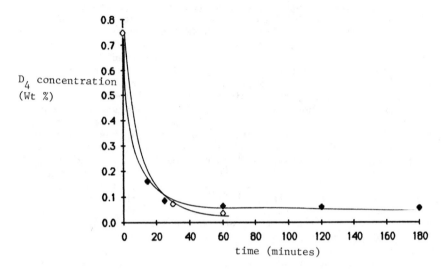

Figure 5.   Disappearance of D_4 in the presence of
aminopropyl disiloxane at 115°C (◊) and
140°C (♦), 0.02 Wt% KOH.

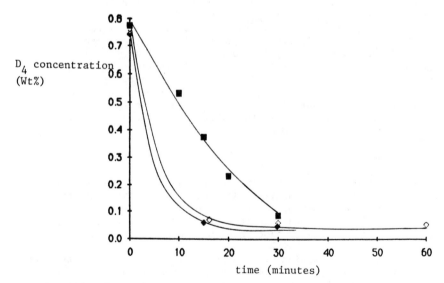

Figure 6.   Disappearance of D_4 in the presence of aminopropyl
disiloxane with 0.12 wt% KOH at temperatures of 91°C
(■), 111°C (◊), and 131°C (♦).

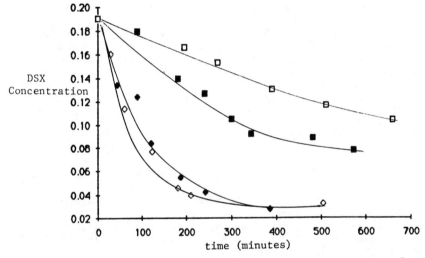

Figure 7. Influence of temperature on the rate of reaction of aminopropyl disiloxane, at 0.12 wt% KOH. □ = 90°C, ■ = 105°C, ◆ = 129°C, ◇ = 140°C.

defined amino alkyl functional oligomers, which in turn have been employed to produce novel segmented copolymers. [17-19]

Acknowledgments

The authors would like to thank Mr. M. Ogden for assistance with the capillary GC work and the Army Research Office for supporting this research under Grant DAAG-29-85-G-0019. They also thank the Exxon Educational Foundation for partial support.

Literature Cited

1. Wright, P. V., in Ring-Opening Polymerizations, K. J. Ivin and T. Saegaser, Editors, Elesevier Press (1984).
2. Noll, W., Chemistry and Technology of Siloxanes, Academic Press, New York (1968).
3. McGrath, J. E., Riffle, J. S., Banthia, A. K., Yilgor, I., Wilkes, G. L., in "Initiation of Polymerization", ACS Symposium Series No. 212, F. E. Bailey Jr., Editor, 1983, p. 145-172.
4. Kantor, S. W., Grubb, W. T., Osthoff, R. C., J. Amer. Chem. Soc. 76, 5190 (1954).
5. Riffle, J. S., Yilgor, I., Banthia, A. K., Wilkes, G. L., McGrath, J. E., in Epoxy Resins II, R. S. Bauer, Editor, ACS Symposium Series No. 221 (1983).
6. Yilgor, I., Riffle, J. S., McGrath, J. E., in "Reactive Oligomers", F. Harris, Editor, ACS Symposium Series, in press (1985).
7. Saam, J. C., Gordon, D. J., and Lindsay, S., Macromolecules 3, 4 (1970).
8. Bostick, E. E., in Block Polymers, edited by S. L. Aggarwal, Plenum Press (1970).
9. Voronkov, M. G., Mileshkevich, V. P., Yuzhelevskii, Y. A., The Siloxane Bond, Consultants Bureau, New York and London (1978).
10. Stark, F. O., Falender, J. R., Wright, A. P., in Comprehesive Organometallic Chemistry, Sir Geoffrey Wilkinson, FRS, Editor, Pergamon Press (1983).
11. Gilbert, A. R., Kantor, S. W., J. Polym. Sci. 11, 35 (1959).
12. Grubb, W. T., Osthoff, R. C., J. Am. Chem. Soc. 77, 1405 (1955).
13. Cooper, G. D., J. Polym. Sci. A-1, 4, 603 (1966).
14. Carmichael, J. B., Heffel, J., J. Phys. Chem., 69, 2213 (1964).
15. Andrews, G., Macromolecules, 14 (5), 1603 (1981).
16. Andrews, G., Macromolecules, 15 (6), 1580 (1982).
17. Johnson, B. C., Ph.D Thesis, VPI and SU, June 1984, and forthcoming publications.
18. Tran, C., Ph.D Thesis, VPI and SU, November 1984, and forthcoming publications.
19. Yorkgitis, E., Tran, C., Eiss, N. S., Yu, T. Y., Yilgor, I., Wilkes, G. L., McGrath, J. E., Siloxane Modifiers for Epoxy Resins, in "Rubber-Modified Thermoset Resins," C. K. Reiw and J. K. Gillham Editors, Adv. Chem. Series No. 208 (1984).

RECEIVED April 1, 1985

An Improved Process for ε-Caprolactone-Containing Block Polymers

H. L HSEIH and I. W. WANG

Research and Development, Phillips Petroleum Company, Bartlesville, OK 74004

A new process, which involves the conversion of polymeric oxyl-lithium to oxyl-aluminum chain end as modified ring-opening site for lactones, has proven to be very effective in eliminating trans-esterifications. Therefore, uniform block polymers containing styrene, butadiene, and ε-caprolactone can be prepared with well defined structure.

Among its many attributes, poly(ε-caprolactone) (PCL) is particularly unique in its capability to blend with different commercial polymers over a wide composition range. Incorporating PCL segment with other rubbery or glassy blocks will allow us to form novel multiphase block polymers. Such specialty polymers are often used as "oil-in-oil" type of emulsifiers in polymer blends, whose mechanical properties can be improved if better homogeneity is achieved by the emulsifier additives (1,2).

As to its other characteristics, styrene-butadiene-caprolactone (S-B-CL) triblock terpolymer with high butadiene content behaves much like a thermoplastic elastomer, with raw tensile strength equal to and ozone resistance better than S-B-S type copolymer (3). The impact-resistant resin by blending 25 parts of S-B-CL triblock with 75 parts of styrene/acrylonitrile (SAN) copolymer resembles ABS type material in such properties as tensile strength, flexural modulus, oil resistance, and transparency (4).

The melt condensation of acid and hydroxyl functional group normally requires exact stoichiometry, elevated temperature, and a long reaction cycle. Such a route would not be possible to utilize to produce block polymers from lactones and other vinyl monomers. However, a rather facile route leading to polyester formation can be realized by the ring-opening polymerization of lactones as seen from the scheme:

0097–6156/85/0286–0161$06.00/0

$$n \quad \overset{O}{\underset{(CH_2)_m}{\overset{\|}{C-O}}} \quad \longrightarrow \quad \overset{O}{\underset{}{\overset{\|}{-[C-(CH_2)_m-O]_n}}}$$

The anionic ring-opening mechanism which involves acyl-oxygen cleavage with subsquent propagation through an alkoxide anion raises the possibility of block polymerizing lactone molecules with other vinyl monomers. For example, the alkyllithium-initiated polystyryl and polydienyl anions, or their corresponding oxyl-lithium anions, have been employed as cross-initiator for ε-caprolactone block polymerization (3,5). Idealistically, well defined block polymers of ε-caprolactone, styrene, and butadiene can be made through such a method.

However, in the ring-opening polymerization of lactones under anionic conditions it is difficult to eliminate the unzipping and ester scrambling interferences, which are two major side reactions concurrent with the polyester formation (5). The depolymerization due to unzipping, or back-biting, phenomenon can be visualized through the intramolecular version of ester interchange:

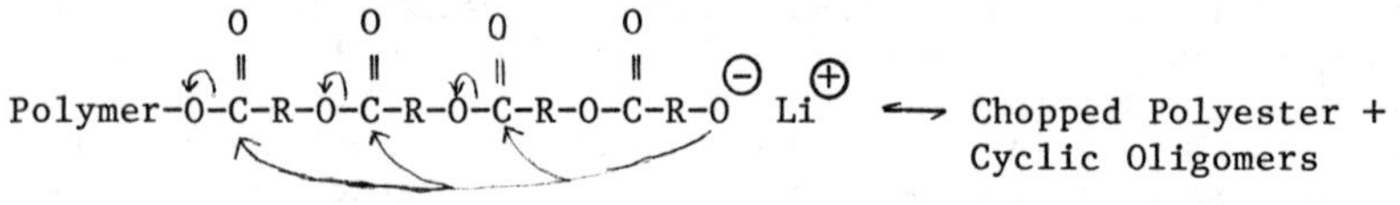

$$\text{Polymer-O-C-R-O-C-R-O-C-R-O-C-R-O}^{\ominus} \text{ Li}^{\oplus} \quad \longleftrightarrow \quad \text{Chopped Polyester +}$$
$$\text{Cyclic Oligomers}$$

Back-biting

In addition, the intermolecular transesterification will account for the scrambling:

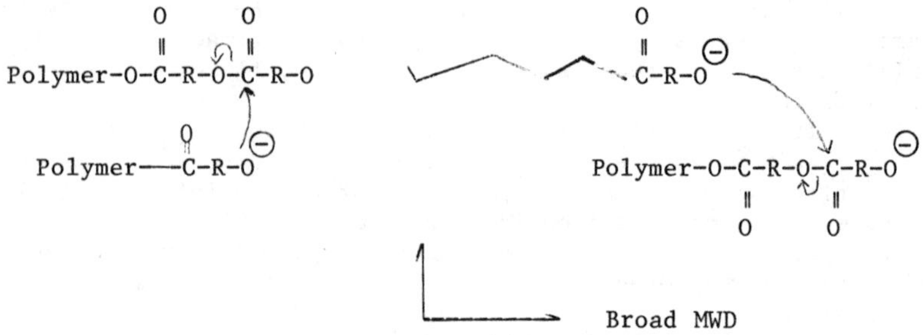

$$\text{Polymer-O-C-R-O-C-R-O}$$
$$\text{Polymer——C-R-O}^{\ominus}$$

$$\cdots\text{C-R-O}^{\ominus}$$
$$\text{Polymer-O-C-R-O-C-R-O}^{\ominus}$$

Broad MWD

Scrambling

As a result, in the anionic polymerization of lactones, low polymer yield, uncontrollable molecular weight, broad distribution, coupling linkage, and cyclic ester oligomers contamination have been frequently encountered (3,5,6). In this paper, a new improved process was revealed to eliminate such transesterifications has been developed. Block copolymers with styrene and butadiene have been prepared and characterized by different techniques.

Experimental

Materials. The solvent, cyclohexane was Phillips polymerization
grade and was dried by counter current scrubbing with purified N_2,
followed by two consecutive passages through Alcoa F-1 activated
alumina. Tetrahydrofuran (THF) was stored over activated alumina
after distillation from powdered calcium hydride (CaH_2) under N_2
blanket.

Butadiene (B) monomer was flashed at 46°C from its dimer and
inhibitor, then condensed by passing through activated alumina into
the container at -15°C under N_2 pressure. Styrene (S), was either
dried through activated alumina, or refluxed over CaH_2 then vacuum
distilled. ε-Caprolactone (CL) was distilled from CaH_2 at reduced
pressure. Ethylene oxide (EO) was purchased from Eastman Chemical
Co., diluted in dried cyclohexane before use and used without
further purification. n-Butyllithium (n-BuLi) in heptane was
obtained from Alfa Products; and diethylaluminum chloride (DEAC) was
from Texas Alkyls. They were diluted in purified cyclohexane before
use.

Polymerization and Characterization. Block polymerizations via
sequential monomer additions were performed, using well known bottle
polymerization techniques (3). All polymer samples were isolated by
coagulation in agitated isopropanol. The precipitated polymer was
filtered, washed and then vacuum dried at 50°C overnight before
final conversion was calculated.

Different characterization techniques were performed to
establish the qualities of the block polymers containing
ε-caprolactone. They included calibrated infrared (IR)
determination of CL content, in-house gel permeation chromatography
(GPC), GPC with univeral calibration curve, GPC equipped with low
angle laser light scattering (LALLS/GPC), and Rheovibron measurement
of the transition temperatures.

Results and Discussion

In order to cope with the ester exchange interferences occuring in
the anionic ring-opening polymerization of lactones, reducing the
nucleophilicity of alkoxide chain end would be an appropriate
approach to solve the problems. A new process was thus conceived to
convert the relatively stronger basic oxyl-lithium chain end to a
oxyl-aluminum moiety as the new propagating site for ε-caprolactone
polymerization. Several literature references have stated that the
1:1 insertion of lactone molecule into the metal heteratom bond of
certain organometallic compounds do take place (7,8,9). More
recently, Ph. Teyssie and his co-workers also showed that a
bimetallic oxoalkoxide catalyst with structure simply represented
as:

$$(RO)_2Al-O-Zn-O-Al(OR)_2$$

is a capable initiator to prepare CL homopolymer and its
corresponding CL-B-CL block copolymer (10). The preparation of such
bimetallic compounds and the overall block copolymerization

procedures, however, are rather complicated and require multiple
independent steps.

Similar to their approach and extended from Hsieh's previous
paper, (3), a simple but much improved route was devised. Stepwise
addition of an alkylaluminum halide, which follows that of an
alkylene oxide to the alkyllithium-initiated "living" prepolymer,
provides the desired oxyl-aluminum chain end as a modified ring-
opening site for cyclic esters. The improved process is illustrated
in Figure 1.

Block Copolymerization of Butadiene and ε-Caprolactone. The
improved process was first conducted as shown in Table I where 70
parts of butadiene and 30 parts of ε-caprolactone were sequentially
added in Step 1 and 4, respectively. After the addition of ε-caprolactone (Step 4), the solutions were short stopped at various times.
Final conversions, caprolactone contents and GPC data were obtained
(Table I, Figures 2 and 3). The % CL found in each sample is
consistant with the final conversion. These diblock copolymer
samples have relatively narrow MWD as illustrated by the calculated
values (Table I) and GPC curves (Figure 2). The time versus
conversion curves in Step 4 for both 50°C and 70°C are shown in
Figure 3. The stability of these block polymers made by the
improved process is apparent. For comparison purpose, the results
using the original process where the propagating site is
oxyl-lithium (3) are also included in Figure 3. One can see the
extremely rapid polymerization of ε-caprolactone followed by the
undesirable side reactions. Indeed, the rather dramatic effect of
reaction time on MWD of block polymers of this type was reported by
Hsieh earlier (3).

B-CL Diblock Copolymers. By varying the amounts of n-BuLi charged,
a series of B-Cl diblock copolymers with different molecular weight
but fixed composition of 70-30 weight percent were prepared and
listed in Table II with their characterization results. The molar
ratio of n-BuLi: EO: DEAC was, however, always kept constant at
1:3 : 1. High quality block copolymer was prepared via the new
process.

For samples 64-1 to 64-4 of B-CL (70-30 wt %) diblock, absolute
MWD was further determined by using GPC equipped with low angle
laser light scattering along with the Mark-Houwink relationships.
The results are summarized in Table III. Narrow, monomodal
distributions were seen, with M_w/M_n <1.1. The M_n correlations with
intrinsic viscosity in toluene were plotted in Figure 4. Since
viscosity average molecular weight $M_n = (M^a_w)^{1/a}$ depends upon a, an
interactive procedure was used to arrive at final values of a. The
following Mark-Houwink equations were thus derived from least
squares fit:

$$B-CL \ (70-30 \ wt \ \%):$$
$$[\eta] = 8.549 \times 10^{-4} \ M_n \ 0.636 \ (THF)$$
$$[\eta] = 8.859 \times 10^{-4} \ M_n \ 0.632 \ (Toluene)$$

The correlations are essentially identical in both THF and toluene
solvents.

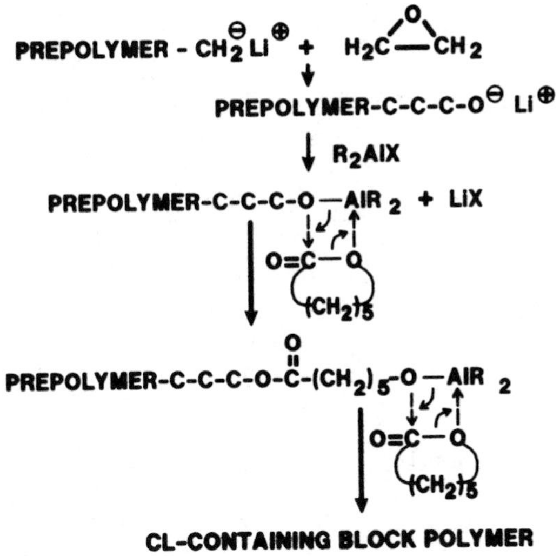

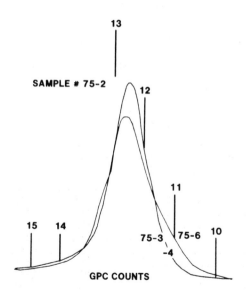

Figure 1. Simplified scheme of the improved process for caprolactone ring-opening polymerization.

Figure 2. GPC of B-Cl diblock copolymers via the improved process.

TABLE I. The Improved Process

Agent

1st Step
Cyclohexane, g   780
THF, g   0.05
Butadiene, g   70
n-BuLi, m mole   1.68
Temp. °C   50
Time, hr   1.5

2nd Step
Ethylene oxide, m mole   4.2
Temp. °C   50
Time, hr   0.25

3rd Step
THF, g   5.0
Diethylaluminumchloride, m mole   1.68
Temp., °C   70
Time, hr   0.25

4th Step
ε-Caprolactone, g   30
Temp., °C   50
Time, hr.   variable

Sample	Block Pzn. Time of CL at 4th Step, hr.	Final[a] Conv. %	% CL[b] found	MWD	Mw x 10^{-3}[c] —————————— Mn x 10^{-3}
75-1	0.25	73.5	2.6	1.17	54/46
75-2	0.5	76	5.3	1.17	54/46
75-3	1.0	83	11	1.17	58/49
75-4	2.0	91	22	1.22	63/52
75-5	3.0	99	28	1.25	66/52
75-6	4.0	100	29	1.26	65/52
75-7	5.0	100	30	1.27	69/54

a.   Isopropanol coagulation
b.   IR analysis
c.   GPC with universal calibration curve

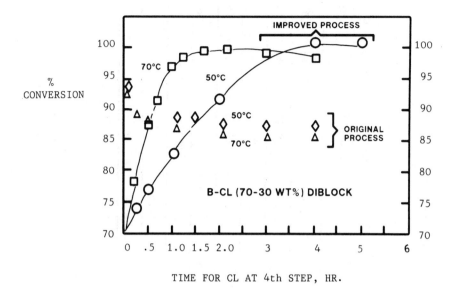

Figure 3. Comparison between original and improved process for B-Cl diblock copolymerization.

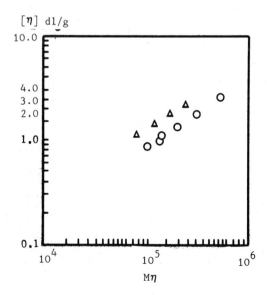

Figure 4. Molecular weight correlation with intrinsic viscosity in toluene.

Table II. B-CL (70-30 wt%) Diblock Copolymers

Sample	CL Block Pzn Time, hr[a]	% Final Conv.[b]	Wt.% CL found in B-CL[c]	Wt.% Acetone Extractable	MWD	Mwx10⁻³d Mnx10⁻³	Transition Temp., °C Tg	Tm
13-5	6	100	30	0	1.37	101/74	-86	57
13-8	4	99	29	0	1.41	92/65	-86	56
16-2	3	99	30	0	1.26	64/51	-86	56
16-4	4	100	30	0	1.29	69/53	-86	56
64-1	8	98	-	-	1.31	283/216	-	-
64-2	8	99	-	-	1.36	200/147	-	-
64-3	8	100	-	-	1.30	143/110	-	-
64-4	8	100	-	-	1.27	100/78	-	-

a. At 50°C.
b. Alcohol coagulation.
c. IR analysis.
d. GPC/universal calibration curve.
e. Rheovibron measurement at 11 Hz.

Table III. LALLS/GPC Data for Caprolactone Copolymers and Terpolymers

Sample	$[\eta]$Tol dl/g	$[\eta]$THF dl/g	In Thousands			
			M_z	M_w	M_n	$M_\eta{}^a$
			B-CL 70-30			
64-1	2.21	2.23	265	241	225	237
64-2	1.17	1.79	191	168	156	165
64-3	1.41	—	134	121	112	119
64-4	1.10	1.10	84	79	74	78
76-2	1.78	1.86	407	318	155	300
76-3	1.29	1.38	262	209	149	200
76-4	0.98	—	147	129	112	126
76-5	0.87	0.92	117	105	94	103
73-4	1.09	—	147	135	126	134

a. $M_\eta = (M^a)_w{}^{1/a}$

S/B-CL and S-B-CL Block Terpolymers. Synthesis of S/B-CL diblock and S-B-CL triblock terpolymers (Figure 5) was also carried out utilizing the improved process to incorporate caprolactone units with random styrene/butadiene (S/B) block and styrene-butadiene (S-B) diblock copolymer, respectively. As judged from the analytic data of the resultant samples (Table IV), the conversion from oxyl-lithium to oxyl-aluminum chain end significantly improves the ring-opening polymerization of ε-caprolactone. Again, the viscosity average molecular weight-intrinsic viscosity relationships for S/B-CL (30/40-30 wt %) were similarly established by using LALLS/GPC data (Table III).

S/B-CL (30/40-30 wt %):

$$[\eta] = 5.393 \times 10^{-4} \, M_n \, 0.644 \, \text{(THF)}$$
$$[\eta] = 5.313 \times 10^{-4} \, M_n \, 0.642 \, \text{(Toluene)}$$

$[\eta]$ in THF is about 3% higher than in toluene, reflecting relatively better solvent power of THF for the terpolymer containing styrene units.

Tg and Tm Measurement. Glass temperature (Tg and melt temperature (Tm) of some dicumyl peroxide cured block polymers of ε-caprolactone, styrene and butadiene were determined on the Rheovibron at 11 Hz. Some of the results are listed in the last two columns of Table II and IV.

The transition temperatures were essentially the same for a similar type of polymers regardless of molecular weight range in this study. The lower transition (-86°C) in the B-CL of 70-30 wt% diblock is associated with Tg of the polybutadiene block and the upper transition (56°C) with the crystalline melting temperature of polycaprolactone segment. No polycaprolactone Tg (about -60°C for its amorphous part) was observed in the B-CL diblock, evidently because of its lower content and proximity to the polybutadiene Tg. It can be detected at -59°C, however, in the S/B-Cl of 33/44-23 wt% diblock terpolymer. In the latter, the Tg of the S/B block (-20°) was essentially the same as the random S/B copolymer control which is coded as S/B-CL: 33/44-0 wt% in Table IV.

The transition behavior of S-B-CL of 30-40-30 wt% triblock terpolymer in Sample 54-3 (Figure 6) did show the polystyrene Tg at 93°C which is essentially the same as in the S-B diblock control coded as S-B-CL: 30-40-0 wt%. Its low transition temperature at -68° is quite broad and likely results from the lack of resolution of the Tg's of polybutadiene and polycaprolactone segments.

Conclusions

1. Converting the "living" prepolymer from oxyl-lithium to oxyl-aluminum chain end can successfully eliminate the undesired trans-esterification in the ε-caprolactone ring-opening polymerization; and its more uniform block polymers with styrene and butadiene were thus prepared.

    2. Conversion, IR analysis, acetone extraction, GPC and

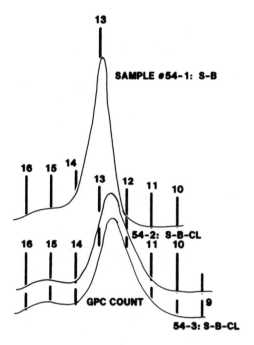

Figure 5.   GPC of S-B-Cl compared to S-B.

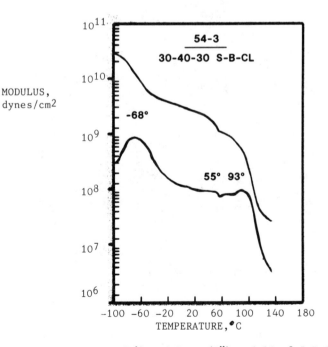

Figure 6.   Storage (E') and loss (E") moduli of S-B-Cl terpolymer.

Table IV. Terpolymers of ε-Caprolactone, Styrene, and Butadiene

Sample	Wt.% Composition S/B-CL	% Final[a] Conv.	Wt.% CL[b] Found	MWD	$Mwx10^{-3}$ $Mnx10^{-3}$[c]	Tg °C	Tm
23-1	33/44-0	100	0	1.06	50/47	-22	--
23-2	33/44-23	91	23[e]	1.33	63/47	-59,-20	56
23-3	33/44-23	91	24[f]	1.35	77/57	-58,-21	56
73-4	30/40/30	99	33	1.34	145/108	--	--
76-3	30/40-30	100	33	1.64	222/135	--	--
76-4	30/40-30	99	34	1.35	129/96	--	--
76-5	30/40-30	99	35	1.38	108/79	--	--
	S-B-CL						
54-1	30-40-0	0	0	1.22	51/42	-82,+97	55
54-3	30-40-30	32	32	1.39	68/49	-68,+93	

a. Alcohol coagulation
b. IR analysis
c. GPC/universal calibration curve
d. Rheovibron measurement at 11 Hz
e. 2.2% acetone extractable
f. 0.7% acetone extractable

Rheovibron measurements confirmed the integrity of the block structure; and GPC equipped with LALLS established the Mark–Houwink relationships for some of the block polymers.

3. PCL-containing multi-armed block polymers could also be prepared by replacing alkylene oxide with either diepoxides, lactones (1/2 molar ratio to n-BuLi), or CO_2 as coupling agent for alkyllithium-initiated living prepolymer prior to the addition of alkylaluminum halide and lactone monomer.

Acknowledgment

The authors would like to thank C. J. Stacy and K. W. Rollmann for LALLS/GPC and Rheovibron measurements and related characterization data.

Literature Cited

1. Paul, D. R. in "Polymer Blends", Vol. 2; Paul, D. R.; Newman, S. Ed; Academic Press, 1978; Chap. 12.
2. Paul, D. R.; Barlow, J.W. J. Macromol. Sci.-Rev. Macromol. Chem. 1980, C18(1), 109.
3. Hsieh, H. L. J. Appl. Poly. Sci. 1978, 22, 1119.
4. Clark, E.; Childers, C. W. J. Appl. Poly. Sci., 1978, 22, 1081.
5. Yamashita, Y. in "Anionic Polymerization"; J. E. McGrath Ed.; ACS SYMPOSIUM SERIES No. 166, American Chemical Society: Washington, D.C., 1981; p. 199.
6. Ito, K.; Hashizuka, Y.; Yamashita, Y. Macromolecules 1977, 10, 821.
7. Itoh, K.; Sakai, S.; Ishii, Y. J. Org. Chem. 1966, 31, 3948.
8. Itoh, K.; et. al J. Organometal. Chem. 1967, 10, 451.
9. Noltes, J. G.; Verbeek, F.; Overmars, H. G. J.; Boersma, J. J. Organomet. Chem. 1970, 25, 33.
10. Heuschen, J.; Jerome, R.; Teyssie, Ph. Macromolecules, 1981, 14, 242.

RECEIVED April 8, 1985

Organolithium Polymerization of ε-Caprolactone

MAURICE MORTON and MEIYAN WU[1]

Institute of Polymer Science, The University of Akron, Akron, OH 44325

Butyllithium and lithium t-butoxide were used as
initiators to polymerize ε-caprolactone in H_4-furan
and benzene solvents, at 25°C. Rapid reactions (<6
min to 100% conversion) and high molecular weights
(>50,000) were obtained. The relative reaction rates
were: t-BuOLi/benzene>n-BuLi/benzene>t-BuOLi/H_4-furan
>s-BuLi/H_4-furan. Although the initial polymers had
a narrow MWD characteristic of "living" polymers, all
systems exhibited a rapid intramolecular and intermolecular ester interchange reaction of the active
chain ends, leading to a broadening of molecular
weight distribution with time.

Previous studies (1-6) of the anionic ring-opening polymerization
of ε-caprolactone by organoalkali initiators have shown that this
system is subject to a ring-chain equilibrium, and that the more
electropositive alkali metal cations lead to a faster attainment of
this equilibrium. Yamashita and co-workers (3-5) proposed that
linear polymer is first formed and breaks down to cyclic oligomers
by a "back-biting" reaction of the active chain ends, the rate of
such a depropagation reaction depending on the alkali metal present,
e.g., being very fast (<1 min.) with K^+, but much slower with Li^+.
They also suggested that the cyclic dimer was the most thermodynamically stable, but that these cyclic oligomers can be subject to
further polymerization. These studies have led to the general
impression that it is not possible to obtain high polymers by anionic
polymerization of ε-caprolactone. The work reported herein describes
the preliminary results obtained by organolithium ring-opening
polymerization of this lactone with different initiators and solvents
and indicates under what conditions high molecular weights can be
obtained.

[1]Current address: Institute of Chemistry, Academia Sinica, Beijing, People's Republic of China

0097–6156/85/0286–0175$06.00/0
© 1985 American Chemical Society

Experimental

ε-Caprolactone was dried with calcium hydride for several days, then distilled twice over calcium hydride under argon at 0.1 mm. The fraction of BP 64°C was collected and stored under argon.

Lithium t-butoxide was prepared by the reaction of equimolar quantities of n-butyllithium and t-butanol in n-hexane, in vacuo.

H_4-furan was dried successively over calcium hydride and sodium, in vacuo, and stored over sodium naphthalene.

Benzene was treated with conc. sulfuric acid, and dried, in vacuo, over calcium hydride, followed by sodium dispersion.

Polymerizations were carried out in a 100 ml. flask equipped with a side arm sealed with a serum cap, and attached to a high vacuum line. Solvent was first distilled into the flask, then the initiator was injected through the serum cap by means of a syringe, under a blanket of argon. The monomer was then injected in the same way. All polymerizations were carried out at 25°C. Polymerizations were terminated and the polymer stabilized by addition of aqueous HCl.

Results and Discussion

The following systems were used in this study:

a) Lithium t-butoxide in H_4-furan and benzene
b) s-Butyllithium in H_4-furan
c) n-Butyllithium in benzene

Stoichiometry and molecular weight distribution. All of the above systems gave very rapid reactions, yielding 100% conversion within 15 mins. or less. The polymers were isolated by evaporation of the solvent, under vacuum, and their molecular weights determined by intrinsic viscosity in benzene, using the equation (1):

$$[\eta] = 9.94 \times 10^{-5} M_w^{0.82} \tag{1}$$

Their MWD was determined by GPC. Tables I to IV show the types of molecular weights and distributions obtained when the polymerizations were permitted to proceed to 100% conversion. The following conclusions can be drawn from the data:

1. The molecular weights are high (>50,000) and consistently higher than predicted for a "living" polymer. This may be due to a persistent side reaction of the initiator with impurities.

2. The MWD is broader than expected for a "living" polymer, especially in the case of the H_4-furan systems. This can be expected on the basis of the reaction between active chain ends and ester groups in the polymer chain, i.e., intramolecular and intermolecular ester interchange. This reaction also leads to marked degradation when the active chain ends are kept for longer period of time (see Table I).

Table I. Lithium t-Butoxide in H_4-Furan at 25°C (100% Conversion)

Time (min.)	[Monomer]	[t-BuOLi] ($\times 10^2$)	M_S* ($\times 10^{-4}$)	M_n** ($\times 10^{-4}$)	M_W/M_n
3	2.26	1.15	2.24	3.63	1.58
360	2.26	1.15	2.24	∿1.3	--
3	2.26	0.80	3.22	3.8	1.55

*M_S = stoichiometric mol. wt. = $\dfrac{[Monomer]}{[t\text{-}BuOLi]} \times 114$

**M_n was calculated from the M_W value obtained from Equation (1) and the M_W/M_n values obtained from GPC.

Table II. Lithium t-Butoxide in Benzene at 25°C

[Monomer] = 1, [t-BuOLi] = 0.5 $\times 10^{-2}$ Conversion = 100%

Time	M_S ($\times 10^{-4}$)	M_n ($\times 10^{-4}$)	M_W/M_n
6 min	2.3	4.2	1.46
6 min	2.3	4.0	1.50

Table III. s-Butyllithium in H_4-Furan at 25°C (100% Conversion)

Time	[Monomer]	[s-BuLi] ($\times 10^2$)	M_S ($\times 10^{-4}$)	M_n ($\times 10^{-4}$)	M_W/M_n
6 min.	2.26	1.6	1.60	1.9	1.60
6 min.	2.26	0.8	3.22	3.5	1.59

Table IV. n-Butyllithium in Benzene at 25°C

[Monomer] = 1,   [n-BuLi] = 0.5 $\times 10^{-2}$ (100% Conversion)

Time	M_S ($\times 10^{-4}$)	M_n ($\times 10^{-4}$)	M_W/M_n
6 min.	2.3	4.1	1.41
6 min.	2.3	3.8	1.49

Mechanism of Polymerization

Proton NMR analysis of the polymers indicated that the mechanism of polymerization is similar for both types of initiator, i.e., butyllithium and lithium alkoxide. The propagating chain is a lithium alkoxide in both cases, as evidenced by the presence of terminal OH groups in the polymers.

Change in Molecular Weight with Conversion

Since the data in Tables I to IV were based on polymers at 100% conversion, it was thought of interest to observe the changes in molecular weight which occur with increasing conversion. The GPC traces shown in Figures 1 through 3 illustrate these changes during the very early times of reaction. Thus Figure 1 shows that the conversion of monomer (left-hand trace) to polymer (right-hand trace) occurs without any intermediate oligomer formation. Such oligomers were found when the active polymer solution was allowed to stand for longer periods of time, apparently because of degradative reactions.
      Figures 2 and 3 both show the same phenomenon, i.e., that the MWD broadens even before 100% conversion is reached, presumably due to the degradative ester-interchange reactions mentioned above. However, even at complete conversion, the MWD is still relatively narrow, i.e., $M_w/M_n < 1.5$.

Reaction Rates

Since the data in Tables I to IV show that 100% conversion was attained in as little as 3 minutes, it was decided to determine the actual rates for the butyllithium and lithium t-butoxide in benzene and H_4-furan. A first-order plot is shown in Figure 4 and the first-order rate constants are listed in Table V. Of the two initiators, the t-BuOLi apparently leads to the faster rate in benzene, but both

Table V.   First-Order Propagation Rate Constants at 25°C

Initiator	Solvent	k (sec^{-1}) x 10^3
s-BuLi	H_4-furan	3.2
t-BuOLi	H_4-furan	4.0
n-BuLi	Benzene	9.0
t-BuOLi	Benzene	19.0

initiators give much slower rates in H_4-furan. There is apparently a 6-fold difference between the fastest and slowest systems.

Effect of Terminating Agent

The susceptibility of the polyester chain to hydrolysis, etc., makes it sensitive to the type of termination used at the end of the reaction. Figure 5 shows that the use of methanol (or water) can

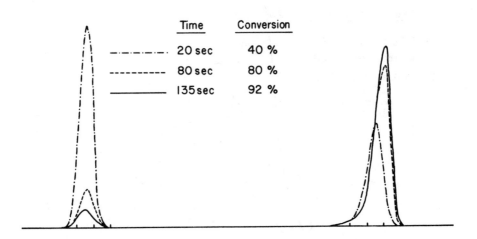

Figure 1. Gel Permeation Chromatograms (GPC) of ε-Caprolactone
Polymerization by Lithium t-Butoxide in Benzene at 25°C.
([M] = 1   [t-C₄H₉OLi] = 0.5 x 10⁻²)

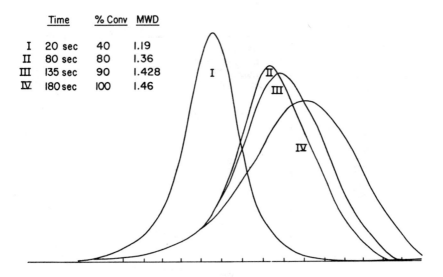

Figure 2. Effect of Conversion on MWD in t-C₄H₉OLi-Initiated
Polymerization of ε-Caprolactone in Benzene at 25°C.
([M] = 1   [t-C₄H₉Li] = 0.5 x 10⁻²)

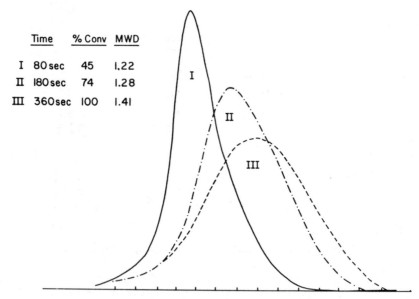

Time	% Conv	MWD
I 80 sec	45	1.22
II 180 sec	74	1.28
III 360 sec	100	1.41

Figure 3.   Effect of Conversion on MWD in n-C$_4$H$_9$Li-Initiated
Polymerization of ε-Caprolactone in Benzene at 25°C.
([M] = 1   [n-C$_4$H$_9$Li = 0.5 x 10^{-2}]

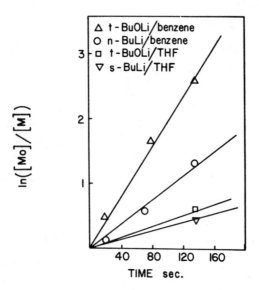

Figure 4.   First-order Rate Plots of ε-Caprolactone Polymerization
([M] = 1   [Initiator] = 0.5 x 10^{-2})

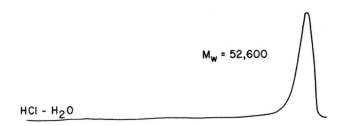

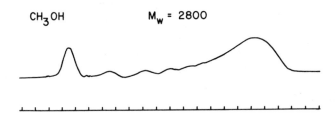

Figure 5.   Effect of Terminating Agent on GPC of Poly-ε-caprolactone (t-C₄H₉Li in benzene)

result in a 20-fold reduction in the molecular weight with substantial broadening of the MWD, while the use of aqueous hydrochloric acid leads to a stable chain end.

Conclusions

The use of lithium t-butoxide or butyllithium to polymerize ε-caprolactone in benzene leads to very rapid polymerization at room temperature to high mol. wt. (100% conversion in 3 min.). The MWD is quite narrow $(M_w/M_n \sim 1.2)$ at the start but broadens with conversion, presumably due to the interaction of the alkoxide chain ends and the ester units in the chains. The rate is much slower in H_4-furan with a concomitant increase in intramolecular and intermolecular exchange reactions.

Acknowledgment

This work was supported in part by National Science Foundation Grant DMR 78-09024.

Literature Cited

1. Perret, R.; Skoulios, A. Makromol. Chem. 1972, 152, 291.
2. Deffieux, A.; Boileau, S. Macromolecules 1976, 2, 369.
3. Ito, K.; Hishazuka, Y.; Yamashita, Y. Macromolecules 1977, 10, 821.
4. Ito, K.; Yamashita, Y. Macromolecules 1978, 11, 68.
5. Yamashita, Y. Polym. Prepr. 1980, 21(1), 51.
6. Sigwalt, P. Angew. Makromol. Chem. 1981, 94, 161.

RECEIVED March 28, 1985

Cationic Heterocyclic Polymerization

E. FRANTA and L. REIBEL

Centre de Recherches sur les Macromolecules (CNRS), 6 rue Boussingault, 67083 Strasbourg
Cédex, France

The preparation of well defined polymers and copolymers with low polydispersity in weight,
composition and structure has been achieved within
the last 20 years thanks to the advent of the living
anionic technique (1,2), the number of relevant
monomers being however limited. The extension of
this approach to cationically polymerizable monomers
has been attempted and some success achieved in the
last few years concerning olefinic monomers, one
giving rise to the so called "quasi living" polymers
(3) and another one leading to a truly living system
when one uses a particular iodine-based initiation
in relation with some stabilized monomers (4). Other
interesting candidates are the heterocyclic monomers.
It has been shown that some cyclic imines can
produce living systems (5) and that tetrahydrofuran
polymerization propagates through a tertiary oxonium
cation which is stable provided that the counter-ion
is suitably chosen (6); the same is true for cyclic
acetals such as dioxolane and dioxepane when proper
initiators are chosen (7).

For THF polymerization, in order to obtain living systems,
initiation has to be performed by appropriate initiators such as
triflic derivatives (8) or oxocarbenium salts (9). The use of
organic halides in conjunction with various silver salts has also
been advocated (10-12), but conflicting results have been reported
concerning the "efficiency" and the mechanism of initiation. In
order to use this silver salt based method to prepare copolymers,
one has to know with accuracy how the halides carried by a polymer
will react with the silver salt used if one is to achieve controlled
grafting, i.e. determine beforehand the number and length of the
grafts and to avoid the formation of homopolymer; therefore we have
studied various organic halides that we have considered as model-
compounds.

0097–6156/85/0286–0183$06.00/0
© 1985 American Chemical Society

Experimental

All of the experiments have been carried out under high vacuum. Materials have been purified according to standard procedures (12); kinetic measurements by dilatometry have been described elsewhere (9) as well as the polymerization procedure.

    ^{1}H NMR was used to characterize the initiator groups attached on oligopoly-THF formed as well as UV spectroscopy whenever possible; a vapor phase chromatograph coupled to a mass spectrometer was used to separate and identify low molecular weight compounds formed during initiation. Molecular weight have been determined by gel permeation chromatography or osmometry.

Mechanism and Kinetics

Two different steps have to be distinguished:

- the first one is relative to the metathetic reaction between the organic halide and the silver salt:

$$RX + AgSbF_6 \rightarrow R^+SbF_6^- + AgX$$

the carbenium formed is very reactive and usually cannot be isolated, though this has been possible in a few cases: when chlorodiphenylmethane (13) or 9-bromofluorene (14) are the halides.

- the second step is related to the initiation itself. Several competitive ways are operative.

a) Initiation by Addition.

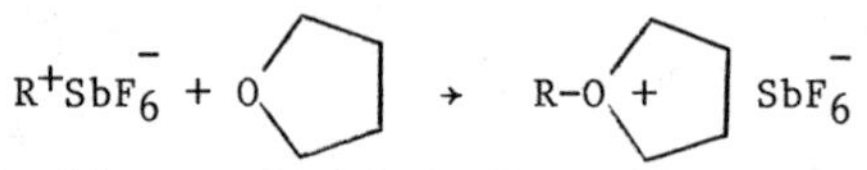

the tertiary cyclic oxonium formed in this process is an efficient initiator for the polymerization of THF leading to a living system. This is the kind of initiation that is desirable for our purpose since it produces narrow molecular weight distribution for the poly-THF obtained, easy control of the number average degree of polymerization and if the organic group is a polymer, we will obtain blocking or grafting – depending where the halide is located – in the absence of homo-poly-THF.

b) Initiation by Proton Elimination.

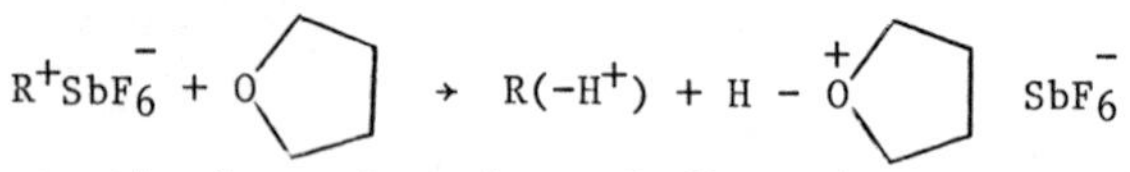

This leads to the formation of an olefin and to a secondary cyclic oxonium which is a slow initiator for THF polymerization. Besides if R is a polymer, one obtains a mixture of homopolymers and no grafting.

c)   Initiation by Hydride Abstraction.

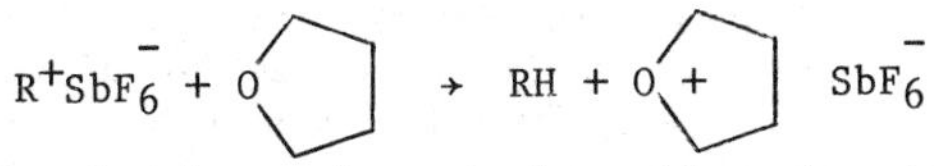

$$R^+SbF_6^- + O\hspace{-1em}\diagup\diagdown \;\; \rightarrow \;\; RH + O\hspace{-1em}\diagup\diagdown{}^+ \;\; SbF_6^-$$

This mechanism has been shown to be active when trityl salts are used (15) but is actually more complicated than proposed (13). It has not been observed for other initiators so far.
After the solution of the organic halide and of the silver salt have been mixed, the first reaction to take place is the metathetic one, with its own kinetics which varies very much depending on the halide and on the organic group. Then initiation per se takes place with various possible mechanisms as we have seen above and also with various kinetics: fast in the first case, (initiation by addition) slow in the two other cases. It is therefore meaningless and deceptive to try to characterize an initiator by simply observing the quantity of poly-THF formed after a certain time: involving two subsequent reactions this says nothing about the fraction of organic halide having reacted and about the mechanism. This global estimation has been sometimes used in the literature but in our opinion should be avoided. We have, therefore, studied the reaction of various organic halides with a silver salt (silver hexafluoroantimonate mostly but also silver triflate when stated): dilatometric measurements of the kinetics of polymerization provide us with information relative to the overall concentration in active sites involved in the propagation. ^{1}H NMR determination conducted on polymers after purification enable us to determine the end groups and to relate them to the structure of the initiator, provided that the molecular weight of the samples are sufficently low to allow for a suitable accuracy. UV spectroscopy was used for the same purpose when the organic halide carried a chromophore. These measurements enable us to determine the fraction of initiator which has initiated the THF polymerization by addition. Whenever necessary we have analyzed the reaction mixture to determine the presence of low molecular weights originating from the initiation process: a vapor phase chromatograph coupled to a mass spectrometer enabled us to separate and identify the low molecular weight compounds and thus to confirm our hypothesis concerning the mechanism. Molecular weight of the samples were determined usually by membrane osmometry and gel permeation chromatography was used to determine the molecular weight distribution. The values obtained were then compared to the theoretical ones, supposing a living system i.e.: each initial organic halide gives rise to one macromolecule on which it is attached at the end and no transfer to monomer and no termination take place.

Results and Discussion

Influence of the Leaving Halide and the Organic Group.   In order to determine the influence of the leaving group we have studied the series of allyl iodide, bromide and chloride: allyl iodide within minutes produces a fast and nearly quantitative (≈90%) initiation;

less than quantitative initiation is probably due in this case to
difficulties in handling the fairly unstable starting material;
quantitative silver iodide precipitation is immediate. Allyl
bromide is much slower: only 50% of the silver bromide has
precipitated after 120 minutes and the same proportion of active
sites is created during this period as compared to the initially
present allyl bromide. Allyl chloride is still much slower: after
two hours a small proportion of silver chloride has precipitated and
less than 10% of the initially present allyl chloride has produced
an active site.

^{1}H NMR determination on low molecular weight material enabled us to
determine that for allyl iodide, all of the poly-THF chains carried
an allyl group: this is not surprising since proton elimination is
not possible in this case. Benzyl bromide and benzyl chloride were
also studied and the difference in activity because of the leaving
group was observed as well; benzyl bromide has quantitatively
reacted after about 30 minutes: that is, each molecule of organic
halide produced one macromolecule and is attached to it. The
chloride being not so good a leaving group, we observed that after
10 hours, only 8% of the halide had produced an active site; the
benzyl group is attached to the polymer which again is not
surprising since proton elimination is not possible.

Terminal iodide, bromide and chloride were also studied in detail;
one observes vast differences in the rates of silver halide
precipitation: Although complete within minutes for the iodide, it
takes less than an hour for the bromide and several days for the
chloride. This is clearly the trend that one expects because of the
nature of the leaving group, but the proportion of active sites
versus time behaves similarly for the 3 halides (Figure 1); the
chloride remains the slowest but the iodide and the bromide behave
identically, within experimental errors i.e. both are very slow
initiators: after 2 hours only about 12% of the initially present
halide have given rise to an active site.

Mass spectroscopy enabled us to characterize various compounds
like isobutene, two of its dimers ($(CH_3)_3C-CH=C(CH_3)_2$ and
$(CH_3)_3C-CH_2--C=CH_2$) and some trimers. We can conclude that
$$\quad\quad\quad\quad\quad\quad |$$
$$\quad\quad\quad\quad\quad\quad CH_3$$
initiation by addition onto the tertiarybutyl cation is practically
non existent since it could not be detected, but that initiation
took place by proton elimination producing isobutene; this latter,
in an acidic medium, gives rise to some dimers and trimers
terminated by a double bond after elimination (16).

We have also used 2-iodopropane and 3-iodopentane as models for
polybutadiene carrying some secondary iodide obtained after reacting
HI onto polybutadiene under controlled conditions to avoid
cyclization. The results are presented on Fig. 2. Under our
conditions 2-iodopropane is a fast, quantitative initiator proceding
by addition (16) through proton elimination is theoretically
possible; 3-iodopropane, a better model for the polymer shows a
drastically different behavior: only 25% of the initiator proceeds
by addition, the rest i.e. 75% proceeds by elimination and indeed we
determined the presence in the solution of various compounds such as
1-pentene and 2-pentene, and several of their dimers and trimers

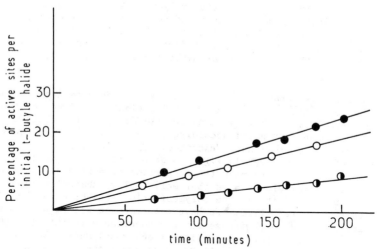

Figure 1.   Efficiency of different t-butyl halides inn the polymerization of THF: t-butyl chloride (◑) t-butyl bromide
(●), t-butyl iodide (o); $|M_o|$ = 12.3 M,1.5 x 10^{-3}M $<|I_o|$
<5.7 x 10^{-3} M, t = 25°C.

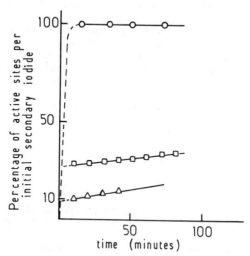

Figure 2.   Efficiency of different secondary iodides in the polymerization of THF: 2-iodopropane(o), 3-iodopentane(□),
hydroiodated polybutadiene (...-CH_2-CH=CH-$(CH_2)_2$-CHI-
$(CH_2)_3$-CH=CH-.....) (Δ); $|M_o|$ = 12.3 M, 1.5 x 10^{-3}M $<|I_o|$
<5.7 x 10^{-3}M, t = 25°C.

(16). Using polybutadiene partially hydroiodated led to an even worse efficiency of the secondary iodide: only 10% reacted by addition, the rest by elimination (16). We could produce this way graft copolymers made of a polybutadiene backbone carrying a number of poly-THF grafts but could not avoid the formation of homo-poly-THF.

α-ω-dichloropolyisobutene could be an interesting case leading to triblock formation, the central one being elastomeric and the two outside ones crystalline. We used an α-ω-dichloropolyisobutene (M_n = 1900) sample prepared according to the "inifer" technique (17) and treated it as previously described. The dilatometric measurements are reported in Figure 3 on a first order plot: the slope enables us to determine the proportion of active sites versus time: the initial slope shows that only 1.3% of the chloride has produced an active site in the polymerization of THF; after 100 hours only 2.5% are active. We have not tried to determine which mechanism operates in this case but can conclude that this technique cannot be used to prepare the desired block copolymers. We used also silver triflate and came to the same conclusion: only a few per cent are active in both cases we have been able to separate up to 90% of the polyisobutene homopolymer.

We have also tried to graft poly-THF onto polyvinyl chloride (PVC) used as a would-be backbone by reacting a PVC sample with a silver salt (silver hexafluoroantimonate in THF). Polymerization proceeds slowly and we have observed that the number of active sites varies greatly with the origin of the PVC and that oligomers play a paramount role (18); UV spectrometry showed that PVC undergoes extensive dehydrochlorination and that it can be recovered as a homopolymer: no observable grafting takes place. Since secondary chlorides under our conditions do not react with the silver salt used and since tertiary chlorides exist in too small amount in the PVC samples used, we think that the metathetic reaction involves some allylic chloride groups present – the number of which depends on the origin of the PVC – leading to the formation of a carbenium ion:

$$...-CH-CH_2-CH-CH=CH-CH_2-... + AgSbF_6$$
$$\qquad |\qquad\quad |$$
$$\qquad Cl\qquad\ Cl$$

$$\rightarrow\quad ...-CH-CH_2-CH^+-CH=CH-CH_2-... + AgCl$$
$$\qquad\qquad |$$
$$\qquad\qquad Cl\qquad SbF_6^-$$

The allylic chlorine atoms are carried predominantly on the low molecular PVC chains which explains the higher reactivity of this fraction.

The carbenium formed reacts on THF and leads to β elimination of a proton:

$$\longrightarrow \ldots -CH_2-CH-CH=CH-CH=CH-CH_2-\ldots \ + \ O+$$

(with substituent Cl below the first CH, and H below the $O+$)

This creates a new allylic chloride that can react the same way with silver salt producing a carbenium ion that undergoes β elimination and generates another allylic chloride group as well as a sequence of conugated double bonds (18) (Figure 4). This process can continue until no allylic chloride can be formed anymore because of a chain defect. The protons eliminated are the true initiators for the THF polymerization and this explains why only homo-poly-THF is formed.

We observe that the substituted allylic chloride groups are thus not suitable for grafting as they can readily undergo proton elimination. Silver triflate also provoked extensive PVC dehydrochlorination but this was not studied in detail (19).

Cyclic Acetal Polymerization

Cyclic acetals are known to produce long lived species provided proper initiators are used (7). Nevertheless, their polymerization is more complex than that of THF: backbiting causes the formation of small cycles and transacetalization is important (8): their active sites are tertiary cyclic oxonium groups - like THF - but in addition are also in equilibrium with alkoxycarbenium species. The enhanced reactivity of the latter could lead to interesting synthetic applications. We have intestigated the polymerization of 1,3-dioxolane (DX^L) and 1,3-dioxepane (DXP) in order to find the proper conditions giving as close as possible to living systems and to apply them to the preparation of block and graft copolymers.

Carbocations and Triflic Anhydride as Initiators for 1,3-dioxolane and 1,3-dioxepane.

First we have used the same kind of initiators that were successful for THF polymerization.

Organic halides were reacted with silver hexafluoroantimonate in CH$_2$Cl$_2$ solution in the presence of DXL or DXP ($\approx$5 M at low temperature (- 55°C)); the polymers have been recovered and characterized as described above. As organic halides we used chlorodiphenylmethane, 2-iodopropane and benzylbromide; the first one exhibited mostly hydride shift and the two others presented incomplete initiation; this along with extensive transfer to monomer made it impossible to control the molecular weight (20). Triflic anhydride, already advocated by PENCZEK et al. (7) was then used to prepare dicationically active polyacetals. The temperature and initial monomer concentration were changed and this enabled us to define conditions where the experimental and theoretical degree of polymerization agree (e.g. 40,000 instead of 50,000) and where cyclic oligomers constitute less than 5% of the total amount of polymers obtained. When the initial monomer concentration was much lower than 1 M, a large fraction of the oligomers is obtained; when it is about 3 M, viscosity becomes so high that stirring becomes

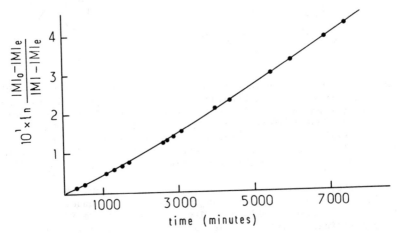

Figure 3.   Kinetics of THF polymerization initiated by α,ω-dichloro-
polyisobutene, $AgSbF_6$; initial and equilibrium
concentration of monomer $|M_0| = 12.3$ M and $|M_e| = 3.1$ M
resp., $|I_0| = 2.83 \times 10^{-3}$ M, $t = 25°C$.

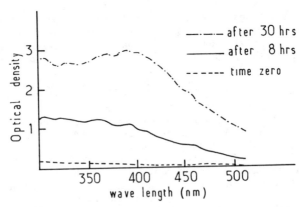

Figure 4.   Ultra-violet spectra of PVC when reacted with $AgSbF_6$ in
Thf at $25°C$; PVC: 1.9 g in 102 ml, $AgSbF_6$: 0.95g, 1 cm
quartz cell.

impossible and the molecular weight distribution broadens. Our best results were obtained at -40°C with initial monomer concentrations between 1 and 1.5 mol. 1^{-1} (20).

We have illustrated the original nature of the active sites, namely; the reactive alkoxycarbenium type ions through block and graft copolymers formation. The first stage of the polymerization is relative to poly-1,3-dioxepane initiated with triflic anhydride under the conditions described above. Then after 1 to 2 hours, the second monomer namely; 1,2-dimethoxyethylene (DME) was added and left for 12 hours at the same temperature; then the polymers were recovered and analyzed. Gel permeation chromatography (Figure 5) shows a monomodal distribution: the curve is shifted to higher molecular weight as compared to the initial poly-DXP. Membrane osmometry indicates that the number average molecular weight agrees well with the theoretical one (e.g. 77,000 and 73,000. respectively) assuming a living system. Fractionation using methylene chloride as a solvent and iso-octane as a non-solvent has been carried out: each fraction was treated with a selective solvent and characterized by ^{1}H NMR: no extractible homopolymer is present and each fraction contains both DXP and DME. This exemplifies the possibility to prepare a triblock copolymer via direct initiation of a vinylic monomer using a diliving polyacetal (21).

Grafting onto Experiments

In order to make use of the dual nature of the growing sites of living polyacetals we have carried out grafting experiments by reacting the latter onto phenyl containing backbones such as polystyrene. One thus expects a Friedel and Crafts type reaction to proceed which will produce the desired graft copolymers.

The active poly-DXL was prepared by stoichiometric reaction of silver hexafluoroantimonate onto benzoyl chloride ($\simeq 2.10^{-2}$M) in methylene chloride at -15°C in the presence of DXL ($\simeq 3$M) for 2 hours. After an aliquot was taken for characterization of the poly-DXL, the remainder of the solution was reacted with a solution of polystyrene in CH_2Cl_2 for 18 hours. After termination with the polymers MeONa were recovered and characterized.

Results are presented in Table I.

Table I - Reaction of Active Polydioxolane onto Polystyrene

Polystyrene M_n	$\dfrac{\|O^+\|}{\|PS\|}$	Nature of the soluble part	Weight Fraction of the insoluble part % (*)	Nature of the insoluble part
160,000	63	some ill defined graft copolymer	60	polystyrene
10,000	5.5	graft copolymer + homo-poly-DXL	No	—

(*): as compared to the initial polystyrene introduced

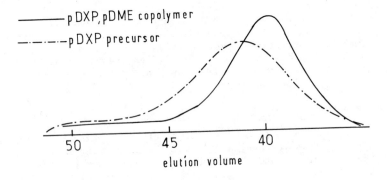

Figure 5.   Gel permeation chromatograms of the p DXP, p DME
copolymer and its p DXP precursor.

When the number of active sites per polystyrene chain is high
($= \dfrac{|O^+|}{|PS|} = 63$) we observe that an insoluble fraction is present
which represents 60% of the initially present polystyrene: it is
composed chiefly of polystyrene along with trace amounts of
poly-DXL. Some soluble but ill defined graft copolymer is obtained
as well.

In contrast, when the ratio is only 5.5, no gel formation is
observed: all of the polymer formed is soluble in refluxing
2-methoxyethanol (solvent for poly-DX1 and a non-solvent for
polystyrene) which shows that all of the polystyrene must be grafted
with poly-DXL. Refluxing ethanol was used to separate
homo-poly-DXL. Grafting efficiency for polystyrene is 100% but only
20% for the poly-DXL present.

These results corroborate the findings concerning the
particular nature of the poly-DXL active sites and the presence of
highly reactive species such as alkoxycarbenium ions.

Literature Cited

1. Szwarc, M.; Adv. in Polym. Sci. 49, (1983).
2. Rempp, P.; Franta, E.; Herz, J. ACS Symp. Series 166 (1981)
   J. E. McGrath Ed.
3. Kennedy J. P. et al.; J. Macromol. Sci. (Chem.) A-18, 1189
   (1982/83) 1381.
4. Higashimura, T.; Deng, Y. X.; Sawamoto, M. Polym. J. 15(5)
   (1983).
5. Goethals, E. J. J. Polym. Sci. (Polym. Symp.) 56, 271 (1976).
6. Dreyfuss, P.; Dreyfuss, M. P. Polym. 6, 93 (1965).
7. Chwialkowska, W./ Kubisa, P.; Penczek, S. Makromol. Chem. 183,
   753 (1982).
8. Penczek, S.; Kubisa, P.; Matyjaszewski, K. Adv. in Polym. Sci.
   37, (1980).
9. Franta, E.; Reibel, L.; Lehmann, J.; Penczek, S. J. Polym.
   Sci. (Symp.) 56, 139 (1976).
10. Dreyfuss, P.; Kennedy, J. P. J. Polym. Sci. (Symp.) 56, 129
    (1976).
11. Burgess, F. J.; Cunliffe, A. V.; Richards, D. H.; Thompson, D.
    Polym. 19, 334 (1978).
12. Afshar-Taromi, F.; Scheer, M.; Rempp, P.; Franta, E. Makromol.
    Chem. 179, 849 (1978).
13. Reibel, L.; Scheer, M.; Rempp, P.; Franta E. Preprints IUPAC
    Microsymposium Karlovy Vary (CS) M 15 (1980).
14. Smid, J.; Scheer, M.; Franta, E. Unpublished results.
15. Dreyfuss, M. P.; Westfahl, J. C.; Dreyfuss, P. Macromol. 1,
    437 (1968).
16. Schweickert, J. C. Thesis, Strasbourg (1980).
17. We wish to thank Dr. O Nuyken and Dr. S. Pask for providing us
    with several well characterized samples.
18. Franta, E.; Zilliox, J. G. IUPAC 6th Int. Symp. on "Cationic

Polymerization and Related Processes" Ghent (B) (1983), Abstr. 157.

19. Zilliox, J. G.; Franta, E. Unpublished results.
20. Reibel, L.; Durand, C.; Franta, E. Can J. Chem. (submitted for publication)
21. Penczek, S.; Kubisa, P.; Reibel, L.; Franta, E. to be published.

RECEIVED March 27, 1985

Thermally or Photochemically Induced Cationic Polymerization

J. V. CRIVELLO

Corporate Research and Development Center, General Electric Company, Schenectady, NY 12301

Diaryliodonium salts are a new class of versatile initiators of cationic polymerization which are characterized by their exceptional latency. On irradiation with UV light, these compounds are very efficient photoinitiators whose reactivity and absorption characteristics can be tailored at will by structural modification as well as through the use of photosensitizers. When diaryliodonium salts are combined with catalytic amounts of Cu(II) compounds, they can be used to thermally initiate cationic polymerization at temperatures above 80°C. The further addition of reducing agents produces initiator systems in which cationic polymerization occurs spontaneously on mixing at 25°C. Examples of polymerizations carried out using these photochemical and thermal initiator systems are given along with the proposed mechanism of the reactions involved.

In recent years, research at this laboratory has centered about the development of new initiators for cationic polymerization. Among the most interesting and useful of these initiators are diaryliodonium salts whose structure is shown below.

$$Ar_2I^+ \; X^-$$

where $X^- = BF_4^-$, PF_6^-, AsF_6^-, SbF_6^-, etc.

These compounds are stable, colorless, crystalline, ionic salts which are readily soluble in organic solvents but nearly insoluble in water. Especially useful is their excellent solubility in a wide variety of cationically polymerizable monomers. Unlike carbenium salts such as trityl and tropylium salts and trialkyloxonium salts which spontaneously initiate cationic polymerization on contact with susceptible monomers, solutions of diaryliodonium salts in these same monomers are stable and show no tendency to polymerize even when heated to temperatures up to 150°C.

0097–6156/85/0286–0195$06.00/0

Photoinitiated Cationic Polymerization

Although diaryliodonium salts are stable toward thermolysis, they exhibit a surprising degree of photosensitivity. On irradiation with 254 nm light, these compounds undergo a facile irreversible photolysis as shown in Equation 1 to give an aryl radical, aryliodoinium cation-radical pair (1).

$$Ar_2I^+ \ X^- \ \xrightarrow{h\nu} \ [Ar_2I^{\cdot+} \ X^-]^* \ \longrightarrow \ ArI^{\cdot+} \ X^- \ + \ Ar\cdot \qquad (1)$$

The quantum yield for the above process has been estimated to be approximately 0.2 on the basis of the aryliodide formed. Work by Pappas and Gatechair(2) and by Timpe and his coworkers(3) indicates that the quantum yield for this reaction may be as high as 0.7 based on the amount of protonic acid which is formed. If the above photolysis is carried out in the presence of a monomer, spontaneous cationic polymerization is observed. The species responsible for initiating cationic polymerization is the aryliodinium cation-radical which may undergo direct electrophilic attack on the monomer (Equation 2). Alternatively, this cation-radical can react with other species present in the reaction mixture to generate Brønsted acids which may subsequently initiate polymerization (Equation 3).

$$ArI^{\cdot+} \ X^- \begin{cases} \xrightarrow{\text{Attack on M}} \text{Polymer} & (2) \\[2em] \xrightarrow{\text{HX Formation}} \xrightarrow[\text{HX on M}]{\text{Attack of}} \text{Polymer} & (3) \end{cases}$$

The observation that only a very small portion of the polymer chains which were produced using diaryliodonium salts contain end groups which are derived from initiator fragments suggests that the process shown in Equation 3 in which Brønsted acids are formed is dominant.

The rate of photolysis of diaryliodonium salts and hence the number of initiating species generated per given irradiation time and light intensity is related to the structure of the cation which is the light absorbing species. A bathochromic shift in the absorption bands is observed when electron releasing substituents are introduced into the ortho and para positions of the aromatic rings. Conversely, the absorption bands are shifted to shorter wavelengths when electron withdrawing substituents are placed at these positions. Using these general guidelines, it is possible to design photoinitiators whose absorption characteristics lie in virtually any desired portion of the ultraviolet spectrum. Although the anion of a diaryliodonium salt plays no role in its photochemistry, it is the dominant factor in the subsequent polymer chemistry since it determines the reactivity of both the initiating and propagating species as well as controlling which termination processes occur. Among the most useful diaryliodonium salt photoinitiators are those which bear the very weakly nucleophilic anions such as BF_4^-, PF_6^-, AsF_6^-, and

SbF_6^-. These photoinitiators are capable of polymerizing almost every known type of cationically polymerizable monomer. Due to the very weakly nucleophilic character of these anions, termination is very slow if not absent and in certain cases such as in the polymerization of tetrahydrofuran, living cationic polymerizations are observed with such initiators.

The photolysis of diaryliodonium salts can be carried out in the long wavelength UV and in the visible region of the spectrum although they do not absorb at these wavelengths provided that photosensitizers are employed(4,5). Diarylketones, condensed ring aromatic hydrocarbons and phenothiazines are excellent photosensitizers for use in the UV, while the acridinium and benzothiazolium dyes, acridine orange and setoflavin-T are active photosensitizers for the short wavelength visible region.

A mechanism involving electron transfer has been implicated in photosensitization and is depicted in Equations 4-7.

$$P \xrightarrow{\;h\nu\;} P^* \tag{4}$$

$$P^* + Ar_2I^+ X^- \longrightarrow [P\cdots Ar_2 I^+ X^-]^* \tag{5}$$

$$[P\cdots Ar_2I^+ X^-]^* \longrightarrow P^{\cdot+} X^- + Ar_2I\cdot \tag{6}$$

$$n\,M \xrightarrow{\;P^{\cdot+} X^-\;} -(M)_n- \tag{7}$$

The key feature of this mechanism is that the excited photosensitizer, P^*, is oxidized by the diaryliodonium salt which is correspondingly reduced. This mechanism is substantiated by first, the direct experimental observation of photosensitizer cation-radical species by UV and ESR spectroscopy(6) and second, by a direct correlation between the activity of a photosensitizer and the reduction potential of its excited state relative to the diaryliodonium salt(3,5). It is interesting to note that the cation-radical, $P^{\cdot+}$, derived from the photosensitizer rather than from the photoinitiator is responsible for initiating polymerization in this instance.

Thermally Initiated Cationic Polymerization

Initiators Activated by Elevated Temperatures. Although, as mentioned earlier, diaryliodonium salts possess considerable thermal stability and, therefore, cannot be used directly in thermally activated polymerizations, we sought to find some way in which the thermal latency of these initiators could be broken. This appeared to be possible on the basis of a recent general reaction discovered in our laboratory(7,8). Diaryliodonium salts undergo facile reaction with nucleophiles whereby the nucleophile is arylated as depicted in Equation 8.

$$Ar_2I^+ X^- + Nu \xrightarrow[\Delta]{Cu(II)} ArNu^+ X^- + ArI \tag{8}$$

This reaction proceeds smoothly at temperatures from 100-125°C in the presence of a catalytic amount of a copper compound to give high yields of the arylated product. Furthermore, as shown in Scheme 1,

the reaction is applicable to a wide variety of substrates. Even
compounds as poorly nucleophilic as diphenylsulfide are quantitatively arylated in one hour at 125°C.

Scheme 1

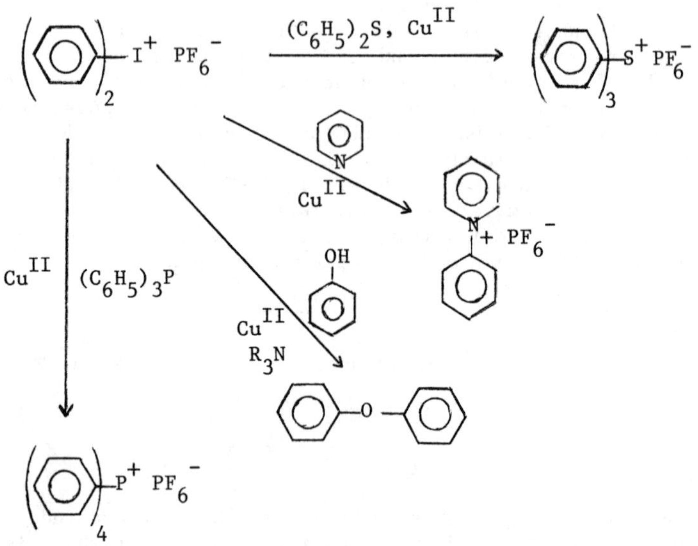

Realizing that cationically polymerizable monomers are, by defini-
tion, nucleophiles, it appeared that it might be possible to initiate cationic polymerization using the same arylation reaction. Indeed, when cationically polymerizable monomers were heated at temperatures in excess of 80°C in the presence of diaryliodonium salts
containing a trace of cupric benzoate as a catalyst, spontaneous
polymerization was observed(9). The polymerization is completely
general with respect to the types of monomers which can be used.
Among those representative monomers which have been thermally
polymerized using this new catalyst system include: cyclohexene-
1,2-oxide, s-trioxane, 2-chloroethyl vinyl ether, ε-caprolactone,
α-methylstyrene, and tetrahydrofuran. Figure 1 gives the relationship between the reaction time and the conversion of monomer to
polymer in the polymerization of ε-caprolactone. In the polymerization of this particular monomer, an inhibition period can be clearly
seen. In Figure 2 is shown the effect of the concentration of the
diaryliodonium salt on the rate of conversion of phenyl glycidyl
ether to polymer. As the diaryliodonium salt is increased, the
rate of polymerization is also correspondingly increased. All diaryliodonium salts examined behaved similarly, provided they possessed the weakly nucleophilic anions mentioned above.

     In contrast to the marked influence of the diaryliodonium salt
concentration on the polymerization rates, the effect of the concentration of the copper compound was found to be catalytic. In
general, 10 mole % with respect to the diaryliodonium salt was found
to be sufficient. Although many different transition and non-
transition metals in various oxidation states were examined, only

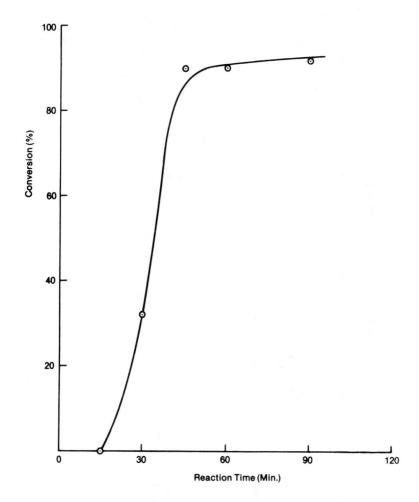

Figure 1.   Study of the bulk polymerization of ε-caprolactone at 70°
with 1 mol % $(C_6H_5)_2I^+$ AsF_6^- and 0.1 mole % Cu(II)
benzoate.

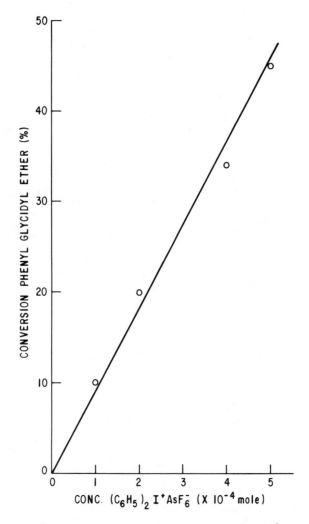

Figure 2.   Effect of the concentration of $(C_6H_5)_2I^+$ AsF_6^- on the
polymerization of phenyl glycidyl ether at 85°C for
30 min catalyzed by 1.6 x 10^{-4} mol Cu(II) benzoate.

copper compounds were catalysts for the polymerization reaction. Virtually any copper compound can be used as a catalyst; however, those compounds such as cupric stearate and cupric benzoate which have appreciable solubility in organic media were most useful.

A number of experiments designed to elucidate the nature of the catalysis by copper were carried out. It was observed that when Cu(II) compounds were combined with diaryliodonium salts in completely unreactive solvents such as chlorobenzene, there was no reaction even at elevated temperatures. In contrast, diaryliodonium salts reacted rapidly and quantitatively even at 25°C in various solvents in the presence of catalytic amounts of a Cu(I) compound. Analysis of the products of this latter reaction shown in Equation 9

$$(9)$$

are consistent with the prior suggestion that arylation of a nucleophile, in this case the solvent methanol, takes place during the reaction. Given the observation that only Cu(I) species are active as catalysts in the above arylation reaction, it appeared that when Cu(II) compounds are employed in these initiator systems, a reduction must occur to generate the catalytically active Cu(I) oxidation state.

In light of the above observations, the mechanism shown in Equations 10-13 has been proposed for the thermal initiation by diaryliodonium salts in the presence of Cu(II) catalysts($\underline{9}$).

$$Red-H + Cu(II)L_2 \longrightarrow Red + Cu(I)L + HL \quad (10)$$

$$Ar_2I^+ X^- + Cu(I)L \longrightarrow [ArCu(III)LX] + ArI \quad (11)$$

$$[ArCu(III)LX] + M \longrightarrow Ar-M^+ X^- + Cu(I)L \quad (12)$$

$$Ar-M^+ X^- + nM \longrightarrow Ar-(M)_n-M^+ X^- \quad (13)$$

The reaction of the Cu(I) species with the diaryliodonium salt results in the formation of a proposed organometallic intermediate, [ArCu(III)LX], whose structure has not been fully elucidated due to its lability. This intermediate undergoes as its primary reaction, an electrophilic attack on the monomer, M, to initiate cationic polymerization. The above mechanism predicts and it has been confirmed that only trace amounts of reducing agents are required to convert a catalytic quantity of Cu(II) compound to its lower valence state. Once the Cu(I) is formed, it is continually recycled between Equations 11 and 12 until all the diaryliodonium salt has been consumed. Another consequence of this mechanism is that polymers prepared using these catalysts should possess aromatic end groups which

originate from the diaryliodonium salt. Indeed, bands due to the
presence of aromatic end groups can be observed in the UV spectra of
polycyclohexene oxide and poly-ε-caprolactone prepared using these
initiators. Additional work with model compounds shown in Equations
14 and 15 has verified that the chief mode of initiation involves
arylation of the monomer.

$$(14)$$

$$(15)$$

The nature of the reducing agent, Red-H, has been the subject
of a considerable amount of research. In most cases, and especially
when oxygen containing heterocyclic monomers are used, the major reducing agents are alcohols which are present in these monomers as
impurities or as a result of hydrolysis. Further, Cu(II) compounds
are known to oxidize aliphatic alcohols at elevated temperatures(10).
Lastly, the addition of small amounts of such alcohols results in a
reduction in the inhibition period at the start of the polymerization and increases the overall rate.

Initiators Active at Room Temperature. The ability of Cu(I) compounds to catalyze the quantitative reduction of diaryliodonium
salts has led to the design of a number of novel initiator systems
which can be used at low temperatures. The most simple of these
systems consists of adding a Cu(I) compound directly to an appropriate monomer containing a diaryliodonium salt. Spontaneous polymerization is observed on mixing. Alternatively, the Cu(I) species
can be generated by an in-situ reduction of the corresponding Cu(II)
compound. This can be accomplished by the addition of easily oxidized alcohols such as benzoin or ascorbic acid, which reduce Cu(II)
compounds at room temperature. Again, when these reducing agents
are added to reactive monomers containing diaryliodonium salts and
Cu(II) catalysts, spontaneous cationic polymerization occurs at
25°C on mixing(11,12).

Another very useful class of reducing agents which can be used
are Sn(II) carboxylates(13). In the presence of a Cu(II) catalyst,
Sn(II)-2-ethylhexanoate quantitatively catalyzes the decomposition
of diaryliodonium salts. Model reactions have shown that the initial step in this reaction is the facile reduction of Cu(II) to Cu(I)
by the Sn(II) compound as depicted in Equation 16.

$$2 \text{ Cu(II)L}_2 + \text{Sn(II)}_2\text{L'}_2 \longrightarrow 2 \text{ Cu(I)L} + \text{Sn(IV)L'}_2\text{L}_2 \quad (16)$$

Free Radical Initiators as Reducing Agents for Diaryliodonium Salts.
A final method by which diaryliodonium salts can be used as thermal
initiators of cationic polymerization has recently been reported by
Ledwith and his coworkers(14,15) and is shown in Equations 17-19.

$$R-R \xrightarrow{\Delta} 2R\cdot \qquad (17)$$

$$R\cdot + Ar_2I^+ \ X^- \longrightarrow R^+ \ X^- + ArI + Ar\cdot \qquad (18)$$

$$Ar\cdot + THF \longrightarrow ArH + THF\cdot \ etc. \qquad (19)$$

Free radicals produced by the thermolysis of typical radical initiators as AIBN, benzopinacole and phenylazotriphenylmethane reduce the
diaryliodonium salt generating aryl radicals and solvent derived
radicals which in a chain reaction induce the decomposition of more
diaryliodonium salt. Through the selection of particular radical
initiators with specific decomposition rates, it is possible to ad-
just the initiation temperature of the cationic polymerization with
considerable latitude.

Conclusions

Diaryliodonium salts are a novel and highly versatile class of
initiators for cationic polymerization. These compounds are efficient photoinitiators of cationic polymerization whose structure may
be readily modified to achieve a wide degree of photosensitivity
and reactivity. While these compounds are unique in that they do
not thermally initiate polymerization even at elevated temperatures,
they can be converted to excellent thermal initiators simply through
the addition of catalytic quantities of a copper(II) compound. The
further discovery that the addition of reducing agents markedly
accelerates the initiation and lowers the initiation temperature
allowed the design of systems which initiate polymerization spontaneously at 25°C on mixing or at any desired temperature. Typical
reducing agents which have been explored are: ascorbic acid,
benzoin, Sn(II) carboxylates in combination with Cu(II) compounds
and common free radical progenitors.

Literature Cited

1. Crivello, J. V.; Lam, J. H. W. *Macromolecules* 1977, 10, 1307.
2. Pappas, S. P.; Gatechair, L. R. *Proc. Soc. Photogr. Sci & Eng.* 1982, 46.
3. Timpe, H.-J.; et al *Z. Chem.* 1983, 3, 102.
4. Crivello, J. V.; Lam, J. H. W. *J. Polym. Sci., Polym. Chem. Ed.* 1978, 16, 2441.
5. Pappas, S. P.; Jilek, J. H. *Photogr. Sci. Eng.* 1979, 23, 140.
6. Crivello, J. V.; Lee, J. L. unpublished results.
7. Crivello, J. V.; Lam, J. H. W. *J. Org. Chem.* 1978, 43, 3055.
8. Crivello, J. V.; Lam, J. H. W. *Synth. Comm.* 1979, 9, 151.
9. Crivello, J. V.; Lockhart, T. P.; Lee, J. L. *J. Polym. Sci., Polym. Chem. Ed.* 1983, 21, 97.
10. Clarke, H. T.; Dreger, E. E. *Org. Syn.*, Coll. Vol. 1 1941, 87.
11. Crivello, J. V.; Lee, J. L. *J. Polym. Sci.,Polym. Chem. Ed.* 1981, 19, 539.

12. Crivello, J. V.; Lee, J. L.   J. Polym. Sci., Polym. Chem. Ed.
    1983, 21 1097.
13. Crivello, J. V.; Lee, J. L.   Makromol. Chem. 1983, 184, 463.
14. Abdul-Rasoul, F. A. M.; Ledwith, A.; Yagci, Y.   Polymer 1978, 19,
    1219.
15. Abdul-Rasoul, F. A. M.; Ledwith, A.; Yagci, Y.   Polymer Bull.
    1978, 1, 1.

RECEIVED September 14, 1984

Polymerization of Substituted Oxiranes, Epoxy Aldehydes, and Derived Oxacyclic Monomers

Z. J. JEDLIŃSKI, M. BERO, J. KASPERCZYK, and M. KOWALCZUK

Institute of Polymer Chemistry, Polish Academy of Sciences, Curie-Sklodowskiej 34, 41-800 Zabrze, Poland

This paper is a review of studies on the ring-opening polymerization of cyclic ethers, e.g., styrene oxide, phenyl glycidyl ethers, epoxy aldehydes and derived oxacyclic monomers. Model reactions involving ring-opening processes occurring in those compounds have also been discussed.

Many papers have been published concerning the structure of the active centers in anionic and cationic ring-opening polymerization reactions of oxacyclic monomers. Recently, attention has been paid in our laboratory to the influence of the structure of complex carbonium salt initiators, especially of the dioxolanylium salts used for initiating the cationic polymerization reactions of trioxane, tetrahydrofuran and dioxolane, on the course of the polymerization (1).

In the present report some examples of the influence of monomer structure, especially of steric hindrance and electronic effects, on the mechanism of polymerization and on the nature of the active centers formed in the anionic polymerization of certain oxacyclic monomers are discussed.

Polymerization of Styrene Oxide

In the polymerization of substituted oxiranes, the direction of opening of the epoxy ring is of great importance because it

0097-6156/85/0286-0205$06.00/0

determines both the kind of active centers present in the
polymerization reaction and the molecular structure and properties
of the polyethers formed. In this regard, three different epoxy-
ring-opening routes are possible: (1) β-ring opening; that is,
opening of the O–CH$_2$ bond; (2) α-ring opening, that is, opening of
the O–CHR bond; and (3) combined α- and β-ring opening, as shown
below:

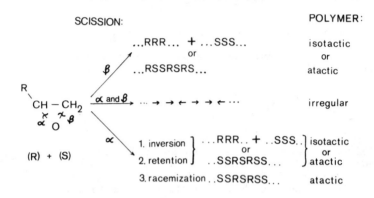

As a result of the β-ring opening, the configuration of the
asymmetric carbon atom remains unchanged. However, the α-ring
opening may take place with either an inversion or retention of the
configuration, depending on reaction conditions.
     Certain general rules determining the conditions for α- or β-
ring-opening processes have been established from the investigations
of C. C. Price (2) and E. Vandenberg (3), and the general view is
that in the anionic polymerization of epoxides, the ring opening
occurs at the β position, while for cationic initiators, both α- and
β-ring opening take place simultaneously with the formation of both
cyclic oligomers and linear oligomers containing irregular head-to-
head and regular head-to-tail sequences. This latter typical course
for these polymerization reactions has been found to occur in the
case of phenyl glycidyl ethers polymerized by Lewis acids, and also
quite surprisingly, polymerized by aluminum alkoxides (4).
     There are, however, numerous exceptions to those rules. For
instance, Tsuruta showed that t-butyloxirane polymerized in the
presence of BF$_3$ gave a polymer with regular head-to-tail sequences,
by an almost exclusively β-ring-opening process (5). Such a course
for the polymerization reaction results from the considerable steric
hindrance provided by the bulky t-butyl substituent.
     In recent studies of styrene oxide polymerization reactions we
found the phenyl substituent to have a significant influence on the
course of the polymerization process, too. In our particular case,
however, the influence is due not only to steric factors, but also
to the inductive effects of the phenyl ring, which influences
directly the course of the oxirane ring-opening reaction.

Anionic Polymerization by Sodium Methoxide Catalyst

The direction of the ring opening in the anionic polymerization of styrene oxide was determined both by analyzing the products of model reactions involving the addition of alcohols to the monomer and by structural studies of the polymers and oligomers obtained. The model addition reactions of alcohols to styrene oxide, catalyzed by the sodium alkoxide, showed that the oxirane ring opens irregularly, both in the α- and β-positions according to the alcohol involved. The results of this study are collected in Table I. In the polymerization reaction, polymer with a number average molecular weight of approximately 3,000 was obtained with CH_3ONa initiator concentration of about 2 mole %. This polymer was found to be atactic with a regular head-to-tail chain sequence as determined by its ^{13}C NMR spectra in Figure 1.

With a high concentration of this initiator (25 mole %), the formation of low molecular mass oligomers resulted. These oligomers contained the linear dimers C and D shown below, in about 30% by weight, and higher linear oligomers, E, in about 70% by weight:

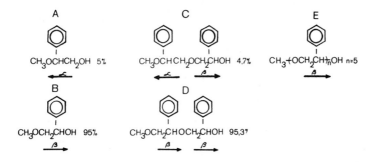

A

CH_3OCHCH_2OH 5%

B

CH_3OCH_2CHOH 95%

C

$CH_3OCHCH_2OCH_2CHOH$ 4.7%

D

$CH_3OCH_2CHOCH_2CHOH$ 95.3%

E

$CH_3(OCH_2CH)_nOH$ n≈5

The dimer C is formed by an initial α-ring-opening reaction followed by a β-ring-opening one, while the dimer D is formed by a double β-ring-opening process. The reaction mixture was also found to contain about 1% by weight of the monomeric alcohols, A and B, formed by both α- and β-ring-opening processes.

From these findings, and from results of the model reactions in Table I, which showed that the bulkier the substituent in the alcohol, the greater the participation of the β-ring opening, it is possible to propose the following mechanism of initiation and propagation for the polymerization of styrene oxide by the sodium methoxide:

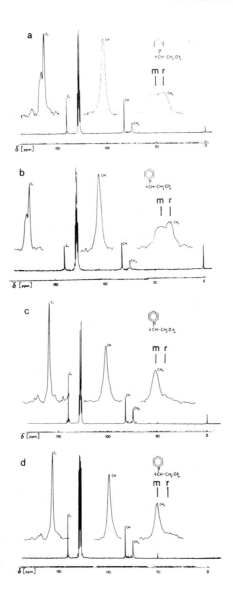

Figure 1. ^{13}C-NMR 20 MHz spectra of poly(styrene oxide) obtained
from the following polymerization reactions:
(a) R,S – styrene oxide initiated with CH$_3$ONa
          as initiator
(b) R,S – styrene oxide catalyzed with Al(OiPr)$_3$
          as initiator
(c) R(+) – styrene oxide initiated with CH$_3$ONa
          as initiator
(d) R(+) – styrene oxide catalyzed with Al(OiPr)$_3$
          as initiator

According to these experimental results, the proposed reaction mechanism for the formation of poly(styrene oxide) with a regular chain structure by anionic polymerization involves the oxirane ring opening exclusively at the β position. However, two kinds of active centers, A and B in the reactions above, occur in the initiation step. The active center A, formed by α-ring opening, adds to a monomer molecule in the next step, but in the second step the oxirane ring is opened at the β position.

Polymerization by Aluminum Alkoxides Catalyst

The polymerization of monosubstituted oxiranes catalyzed by aluminum alkoxides is of particular interest from the point of view of the stereochemistry of the ring-opening reaction. The oxiranes which have been studied to date, including propylene oxide and phenyl glycidyl ethers, were found to polymerize with aluminum alkoxides to yield polymers with an irregular chain structure containing both head-to-head and tail-to-tail linkages (6). That is, these polymers were formed as a result of the oxirane ring opening at both the α and β positions, and they had the same chain microstructure as those obtained by the polymerization reactions initiated by standard cationic catalysts, such as Lewis acids.

The polymerization of styrene oxide by such cationic initiators as BF_3H_2O, $SnCl_4$, $FeCl_3$, etc. does not lead to the formation of polymers with high molecular mass, but only low molecular mass oligomers, both linear and cyclic, are formed. On the other hand, polymerization reactions carried out with aluminum isopropoxide result in the formation of both an oligomeric fraction and a polymer of higher molecular mass ($\bar{M}_n \simeq 2500$). The results of spectroscopic studies indicate that the polymer is both regular and atactic, according to the spectrum in Figure 1b. The model reactions of addition of alcohols to the styrene oxide in the presence of suitable aluminum alkoxides showed that the oxirane ring opens almost exclusively at the α position as seen by the data in Table II.

It was considered of interest, therefore, to investigate the mechanism of polymerization and the kind of active centers which give rise to the formation of the regular, atactic poly(styrene oxide). Optically active monomer was prepared and polymerized for this purpose. The polymerization of R(+)-styrene oxide by $Al(OiPr)_3$ lead to the formation of isotactic poly(styrene oxide), as indicated in Figure 1d, with a positive optical rotation, while the polymerization of R(+)-styrene oxide by CH_3ONa gave an isotactic polymer, Figure 1c, with a negative rotation. As indicated earlier, in the anionic polymerization by sodium methoxide, poly(styrene oxide) is formed exclusively by a β-ring-opening reaction in which the center of asymmetry and the configuration of the asymmetric carbon atom remain unchanged. Therefore, the polymerization of R(+)-styrene oxide by CH_3ONa resulted in the formation of an isotactic R(-)-polymer, while the dextrarotatory poly(styrene oxide) obtained with $Al(OiPr)_3$ had an S configuration of the asymmetric carbon atoms. Consequently the α position of the oxirane ring opened and an inversion of configuration of the center of asymmetry took place for the latter catalyst, while β-ring opening with retention occurred for the former initiator, as shown below:

Table I. Products obtained by the addition of alcohols to
          styrene oxide under the influence of sodium alkoxides

| Initiator | Alcohol | Products | |
		α–Opening	β–Opening
CH_3ONa	CH_3OH	2-methoxy-2-2 phenylethanol 35 mol %	2-methoxy-1-1 phenylethanol 65 mol %
i–PrONa	i–PrOH	2-isopropoxy-2-phenylethanol 12 mol %	2-isopropoxy 1-phenylethanol 88 mol %
t–BuONa	t–BuOH	2-t-butoxy-2-phenylethanol 6 mol %	2-t-butoxy-1-phenylethanol 94 mol %

Table II. Products obtained by the addition of alcohols to
           styrene oxide under the influence of aluminum alkoxides

| Initiator | Alcohol | Products | |
		α–Opening	β–Opening
$Ak*OEt)_3$	EtOH	2-ethoxy-2-2 phenylethanol 100%	–
$Al(OiPr)_3$	i–PrOH	2-isopropoxy-2-phenylethanol 95 mol %	2-isopropoxy-1-phenylethanol 5 mol %
$Al(OtBu)_3$	t–BuOH	2-t-butoxy-2-phenylethanol 91 mol %	2-t-butoxy-1-phenylethanol 9 mol %

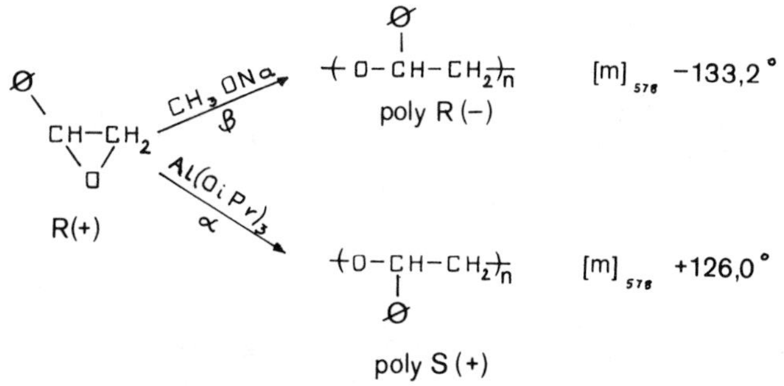

$$\begin{array}{c}
\emptyset \\
| \\
+O-CH-CH_2 \xrightarrow{}_n \qquad [m]_{578} \quad -133.2^\circ \\
\text{poly R} (-)
\end{array}$$

$$\begin{array}{c}
+O-CH-CH_2 \xrightarrow{}_n \qquad [m]_{578} \quad +126.0^\circ \\
| \\
\emptyset \\
\text{poly S} (+)
\end{array}$$

The results of these studies of styrene oxide polymerization carried out in the presence of aluminum isopropoxide catalyst indicate that various kinds of active centers are present during the polymerization process. Therefore, in addition to the active centers leading to the formation of products identical to those obtained in the cationic polymerization of styrene oxide (i.e., cyclic and linear oligomers), there also exist active centers responsible for the formation of a regular polymer having a higher molecular mass. In the latter case an α-opening of the oxirane ring prevails, and an inversion of configuration of the chiral carbon atom takes place. This process, however, is not stereoselective because an atactic polymer is formed starting from a racemic monomer.

Polymerization by ZnEt$_2$/H$_2$O Catalyst

The ZnEt$_2$/H$_2$O catalyst has been used for polymerizing many types of monosubstituted oxiranes including phenyl glycidyl ethers, propylene oxide, butylene oxide, and others (7, 8), to yield partly crystalline, high molecular weight polymers. Detailed studies of the oxirane ring-opening reaction have shown that the β-ring-opening process is predominant and leads to the formation of a regular, highly isotactic polymer. A very limited number of head-to-head and tail-to-tail linkages have also been observed in such polymers and were found to depend on the ZnEt$_2$/H$_2$O ratio and the method of initiator preparation.

The styrene oxide polymerization by the ZnEt$_2$/H$_2$O (1/0.8) catalyst was found to result in the formation of two polymer fractions, a partly crystalline fraction, I, and an amorphous fraction, II. The syntheses of poly(styrene oxide) starting from the R monomer showed both polymers were formed by almost exclusively β-ring-opening reactions, as shown below:

$$\underset{R(+)}{\overset{\varnothing}{\underset{\underset{O}{\diagup}}{\underset{\diagdown}{CH-CH_2}}}} \xrightarrow[\beta]{ZnEt_2 H_2O} \underset{poly\ R\ (-)}{\left(O-\overset{\varnothing}{\underset{|}{CH}}-CH_2\right)_n} \quad [m]_{578} -128,4°$$

From the ^1H–NMR analysis of poly(β,β–d_2–styrene oxide) obtained using $ZnEt_2/H_2O$ as initiator (Figure 2), the formation of the partly crystalline fraction can be described by first order Markov statistics, while that for the amorphous fraction follows Bernoullian statistics. Different chain propagation mechanisms are, therefore, responsible for the formation of the two different polymer fractions obtained from this particular catalyst. Consequently, the existence of two different active centers, responsible for the two polymerization mechanisms and for formation of fractions I and II, are clearly indicated.

When these results are compared with those from the polymerization of the other monosubstituted oxiranes (9), it should be emphasized that the partly crystalline fraction of poly(styrene oxide), I, obtained in the presence of the $ZnEt_2/H_2O$ catalyst is so far the only polyoxirane reported which does not follow Bernoullian statistics. Only one other case, namely that of the polymerization of phenylthiirane by a coordination catalyst system, is known until now which follows first–order Markov statistics (10).

These results for styrene oxide polymerization indicate clearly that when a monosubstituted oxirane is polymerized, one can expect the occurrence of various active centers and of different polymerization mechanisms in the reaction.

Controlled Polymerization of Oxiranes

As discussed in the preceding section, the type of initiator and oxirane monomer can influence directly the structure and properties of polymers obtained by anionic polymerization. In the course of our investigations on the polymerization of chloro–substituted phenyl glycidyl ethers with the following general formula:

$$CH_2\underset{O}{\overset{}{\diagdown\diagup}}CH-CH_2-O-\!\!\left\langle\bigcirc\right\rangle\!\!\overset{Cl_n}{\diagup} \qquad n = 1 \div 5$$

we found that they will polymerize both in the presence of both Lewis acids and aluminum alkoxides to yield linear and/or cyclic oligomers with an irregular chain structure (45% of head-to-tail linkages). These studies also included investigations of the chain microstructure of polymers obtained by the mixed initiator proposed by Price ($\underline{11}$), Al(OiPr)$_3$ + ZnCl$_2$. With an equimolar ratio of those initiators, perfectly regular, isotactic, high molecular weight polymers with a high degree of crystallinity may be obtained. The latter findings coupled with the results of Price's earlier work ($\underline{11}$), allow for the control of the structure and properties of these polymers by varying the composition of the mixed initiator mentioned above.

The chain microstructure of individual fractions of the polymers so obtained was investigated by ^{13}C-NMR techniques described by us earlier ($\underline{4}$). From the results of ^{13}C-NMR measurements, and those of end group analyses of the alkoxide and chlorine-containing end groups, we propose that the probable mechanism for the formation of the high molecular mass, stereoregular polymer was as follows, taking into account the earlier observations of Teyssie ($\underline{12}$) and assuming that the formation of a soluble, equimolecular catalytic ZnCl$_2$-Al(OiPr)$_3$ complex is the initiator for the polymerization reaction studied:

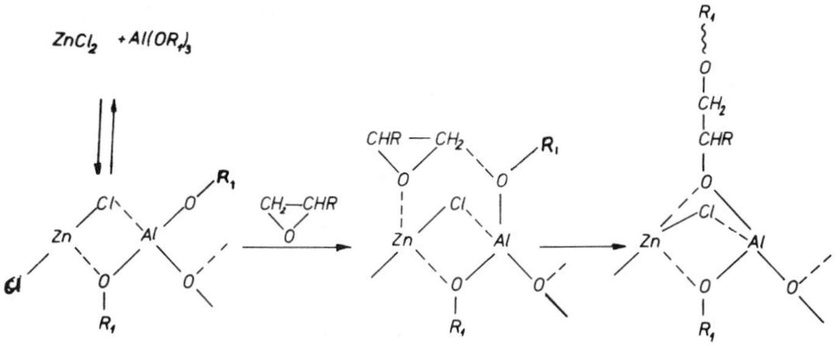

The structure and properties of the polymer obtained may be controlled by varying the quantitative ratios of the catalyst components. Similar phenomena were observed also for the system Al(OiPr)$_3$ + ZnEt$_2$ ($\underline{13}$).

Polymerization of Epoxy Aldehydes and Derived Oxacyclic Monomers
The epoxy aldehyde shown below is an asymmetrically-substituted oxirane in which the aldehyde group is the substituent. It might be supposed that two possible routes for the polymerization of this monomer by either aluminum alkoxides or trialkylaluminum compounds could take place on the reaction temperature, as follows:

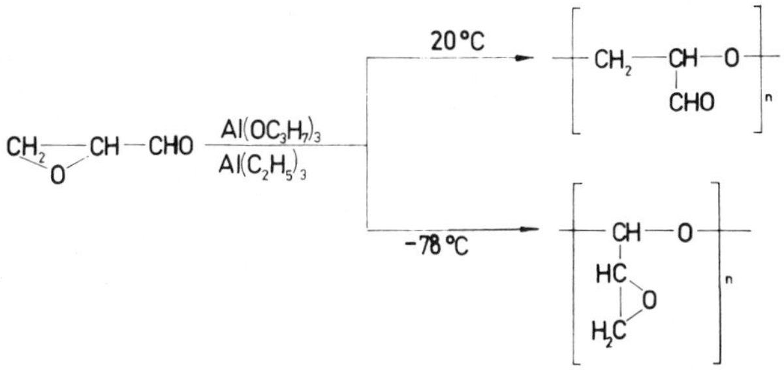

When this polymerization reaction was studied for monomers such as 2,3-epoxypropanal (above) or 2,3-epoxybutanal, however, an unusual initiation step was found to take place by the Tishchenko-Claisen reaction, which is a well-known disproportionation reaction of aldehydes to esters, and an intermediate product, the diepoxyester was identified. This intermediate product on polymerization formed, up to about 20% conversion, linear polyethers with both oxirane and ester side groups by an epoxy ring-opening reaction, as follows:

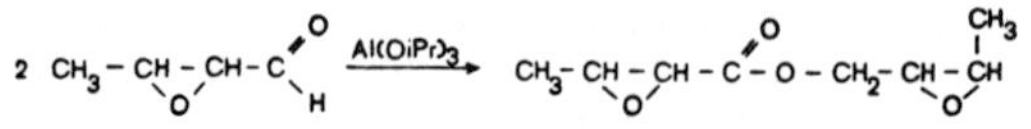

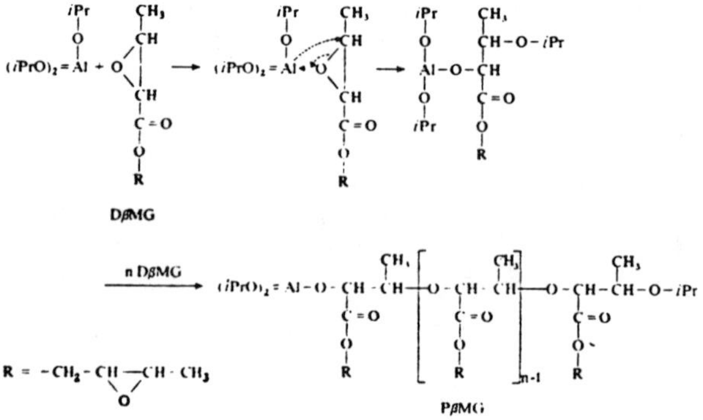

Partial crosslinking was also observed to occur, presumably by the opening of the epoxy rings in the pendant groups of the linear polyether formed (14).

Moreover, with epoxy aldehydes having an electron donating substituent in the α-position (for example, 2-methyl-2,3-epoxypropanal) expansion of the epoxide ring takes place and derivatives of β-propiolactone were found to be the main products in that particular case, as follows:

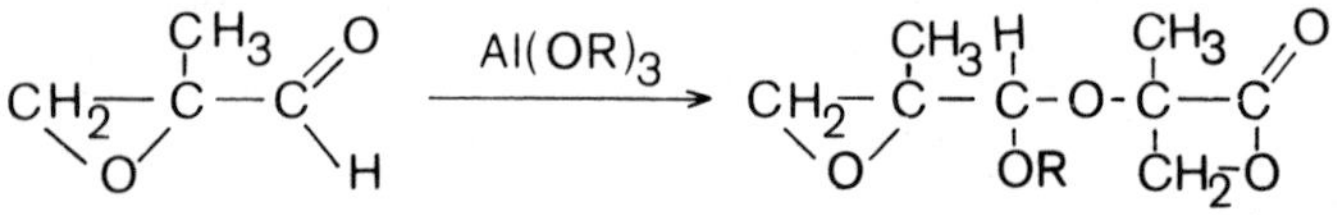

It is worth emphasizing that the course of these reactions could be controlled by choosing both a suitable epoxy aldehyde monomer and the type and concentration of the alkoxide employed (15).

Further polymerization of the epoxy derivatives of the β-lactone intermediate may proceed selectively with the opening of either the epoxy ring or the lactone ring, or with simultaneous opening of both. In this case too, course of the polymerization is also strongly dependent on the type of initiator employed.

It was found that polymerization of these compounds by tertiary amines proceeds selectively, and polyethers with an intact β-lactone ring were formed, as follows:

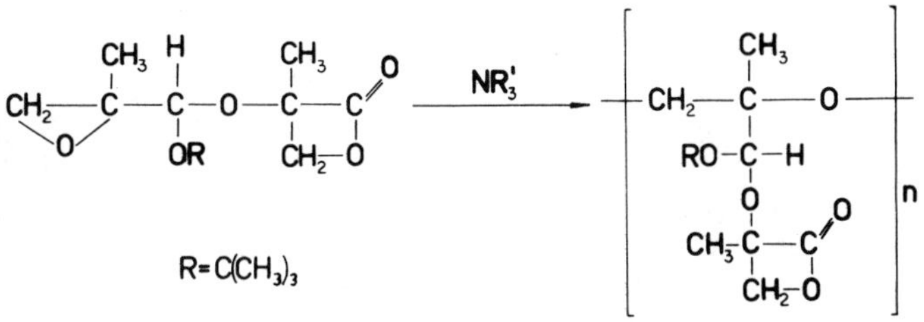

$$R = C(CH_3)_3$$

The selective polymerization of these compounds by the oxirane ring-opening was also observed to take place in the presence of potassium hydroxide and alkali metal solutions. The alkali metal solutions constitute a specific homogeneous initiation system obtained by dissolving a sodium mirror in a tetrahydrofuran solution of crown ethers. The metal anions so formed were found to be the predominant active species of these blue solutions (16). When these alkali metal solutions were used for initiation, the polymerization of unsubstituted β-propiolactone, and of its α or β alkyl substituted derivatives, occurred by an unusual ring opening with cleavage of a C–C bond in the initiation step (17). The ^{1}H-NMR spectra of the β-propiolactone polymers obtained in the polymerizations initiated by potassium solutions and terminated in alcohols or water showed beside the signals corresponding to the polyester chain, an additional singlet at $\delta = 2.06$, which can be ascribed to an acetoxy group (Fig. 3). That finding coupled with the results of quenching of the polymerization reaction either with benzyl chloride:

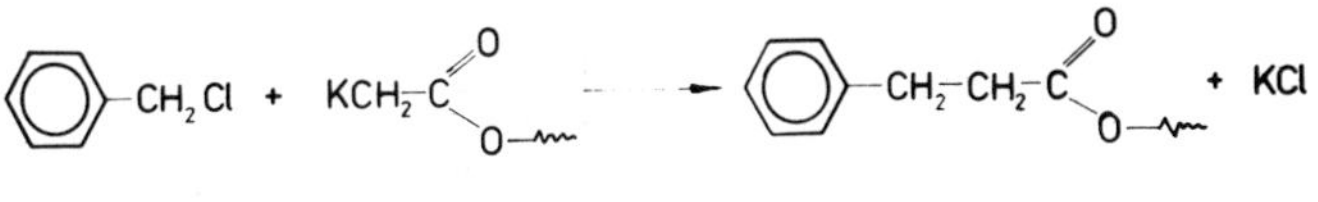

or with deuterated water, made it possible to propose the following reaction mechanism:

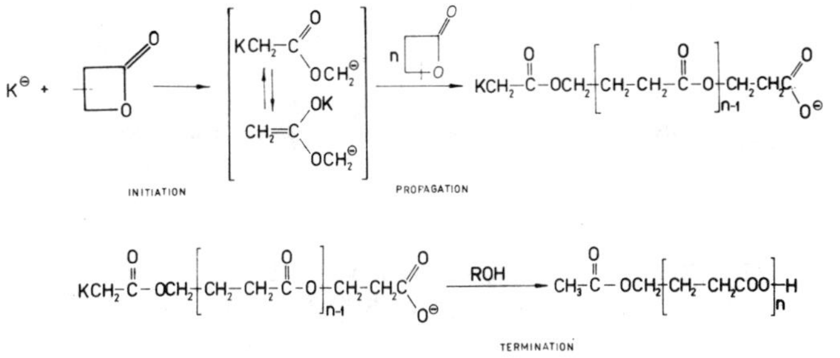

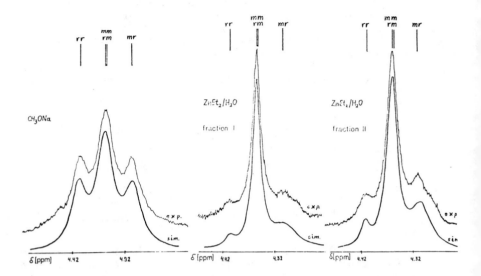

Figure 2. ^{1}H-NMR 100 MHz spectra of poly(β,β-d$_2$-styrene oxide)-
methine proton peaks.

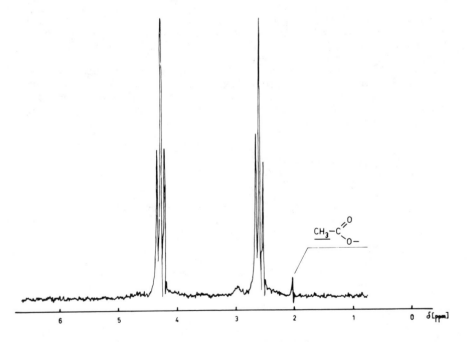

Figure 3. ^{1}H-NMR 100 MHz spectrum of poly(2-oxetanone) obtained
with potassium solution as initiator, $M_n \equiv 1900$.

According to this mechanism, the potassium anion attacks the β-lactone ring to cleave the CH_2-CH_2 bond and form a carbanion. This anion, which can exist in a tautomeric equilibrium with its enolate form, reacts with the next β-lactone monomer by the opening of the alkyl-oxygen bond. The propagation reaction, therefore, proceeds by reaction of the carboxylate anion active centers. Experimental data, to be reported later, provides evidence for this unusual C-C bond-breaking reaction in the monomer and for two types of active species, which are responsible for the course of the lactone initiation and polymerization reactions that occur.

These results provide yet another example of an anionic ring-opening polymerization, in addition to that of styrene oxide polymerization, which can occur with formation of different active centers under the same reaction conditions. These results are in contradiction to those known until now, and to the general opinion that the anionic polymerization of β-lactones proceeds by either the opening of alkyl-oxygen bond or an acyl-oxygen bond.

Literature Cited

1. Jedliński, Z.; Gibas, M. <u>Macromolecules</u> 1980, 13, 1700.
2. Price, C. C.; Osgan, M. <u>J. Am. Chem. Soc</u>. 1966, 78, 4787.
3. Vandenberg, E. J. <u>J. Polym. Sci</u>. 1969, A1, 7, 525.
4. Dworak, A.; Jedliński, Z. <u>Polymer</u> 1980, 1, 93.
5. Sato, S.; Hirano, T.; Suga, M.; Tsuruta, T. Polym. J. 1977, 9, 209 <u>Proceedings of the 26th IUPAC Congress</u>, Tokyo 1977 Preprint 9 Et-03.
6. Jedliński, Z.; Dworak, A.; Bero, M. <u>Makromol. Chem</u>. 1979, 180, 949.
7. Price, C. C.; Spector, R.; Tumolo, A. L. <u>J. Polym. Sci</u>. 1967, A1, 5, 175.
8. Oguni, N.; Shinohara, S.; Lee, K. <u>Polym. J</u>. 1979, 10, 755.
9. Tsuruta, T. <u>J. Polym. Sci</u>. 1972, D6, 179.
10. Cais, R. E.; Bovey, F. A. <u>Macromolecules</u> 1977, 10, 752.
11. Miller, R. A.; Price, C. C. <u>J. Polym. Sci</u>. 1959, 34, 161.
12. Kohler, N.; Osgan, M.; Teyssie, Ph. <u>Polym. Lett</u>. 1968, 6, 559.
13. Bero, M. <u>J. Polym. Sci</u>. 1982, 19, 191.
14. Jedliński, Z.; Majnusz, J. <u>Makromol. Chem</u>. 1972, 155, 111.
15. Jedliński, Z.; Kowalczuk, M. <u>J. Org. Chem</u>. 1979, 44, 222.
16. Szwarc, M.; Jedliński, Z.; Stolarzewicz, A.; Grobelny, Z. in press.
17. Jedliński, Z.; Kowalczuk, M.; Grobelny, Z.; Stolarzewicz, A. <u>Makromol. Chem., Rapid Commun</u>. 1983, 4, 355.

RECEIVED October 4, 1984

Polymerization and Copolymerization of N-Alkylaziridines

E. J. GOETHALS, M. VAN DE VELDE, G. ECKHAUT, and G. BOUQUET

Institute of Organic Chemistry, Rijksuniversiteit-Gent, Krijgslaan 281 (S-4bis), 9000 Gent, Belgium

A number of N-alkylaziridines have been found to give cationic ring-opening polymerizations with a high living character, i.e. the ratios of the rate constants for propagation to those of termination (k_p/k_t) are high.($\underline{1}$) Provided a fast and quantitative initiation, these polymerizations lead to the corresponding well defined poly-amines with predictable molecular weight and low dispersity. This high living character was observed with such monomers which contain a bulky N-substituent (such as tert.butyl) or which carry an additional methyl group on the carbon atoms of the aziridine ring. The high living character was ascribed to the "steric deactivation" ($\underline{2}$) of the amino functions in the polymer chain by the N- and C-substituents which markedly decrease the nucleophilic reactivity of these amino functions towards the electrophilic aziridinium ions which are the active species for the propagation.

In the present communication some new results in the study of these highly living polymerizations are reported. In the first part, the influence of the presence of carbon-substituents on the "reactivity" of the aziridine monomers will be discussed. In the second part, the possibility of sequential polymerization to form block copolymers is described. Finally some preliminary results on the copolymerization of these aziridines with other heterocyclic monomers will be presented.

Results and Discussion

The influence of carbon substitution on the polymerization behavior of N-alkylaziridines. The polymerization of N-alkylaziridines is generally characterized by a fast propagation and a fast termination reaction which consists of a nucleophilic attack of a polymer amino function on the active species, the aziridinium ion. Therefore these polymerizations generally stop at limited conversions producing low molecular weight polymers. N-tert. butyl aziridine (TBA) is an exception to the rule which is ascribed to the "steric deactivation" of the polymer amino functions of the bulky tert.butyl groups.

If the aziridine ring carries an additional substituent on one

0097–6156/85/0286–0219$06.00/0
© 1985 American Chemical Society

of the carbon atoms, the polymerization behavior changes dramatically. Under analogous reaction conditions the rate of polymerization decreases by several orders of magnitude but the rate of termination decreases even more markedly so that the polymerization has a higher living character. In the case of C-substituted N-tert.butyl aziridines, the rate of polymerization is even reduced to zero, in other words these monomers cannot be polymerized.

When two (methyl) substituents are placed on one of the carbon atoms, the aziridine does not polymerize either. This behavior is surprising, taking into account the high reactivity of three membered rings.

A classification of aziridine monomers according to their polymerization behavior is given in Table I.

Table I. Classification of Aziridine Monomers.

Fast Polymerization		Slow Polymerization		No Polymerization	
	(IPA)		(MBA)		(MTBA)
	(TBA)		(PEMA)		(PTBA)*
	(BA)		(CEMA)		(DMBA)
	(PEA)				
	(CEA)				

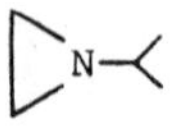

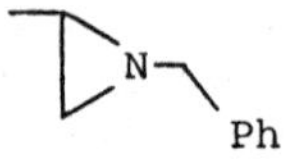

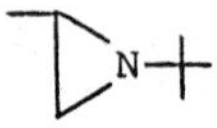

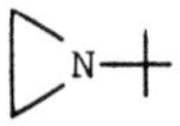

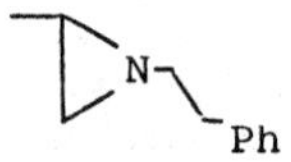

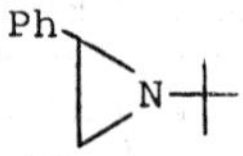

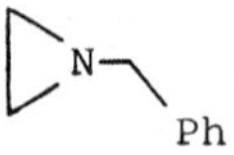

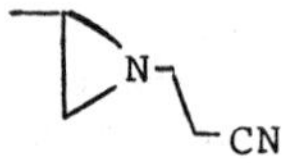

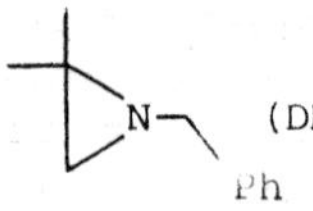

It thus appears that an aziridine is non-polymerizable if it contains a gem.disubstituted carbon atom or one substituted carbon atom and a bulky N-substituent (tert.butyl). The reasons behind this behavior must be sought in the steric structure of the two

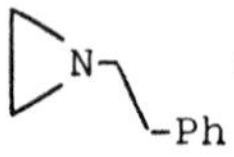

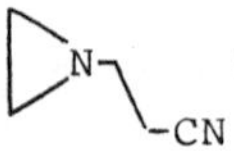

partners in the propagation reaction: the aziridine and the
aziridinium ion.

The aziridinium ion is always sterically hindered at both sides
of the 3-membered ring regardless the presence of additional carbon-
substituents, although it may be assumed that these additional
substituents will render a nucleophilic attack on the ring more
difficult.

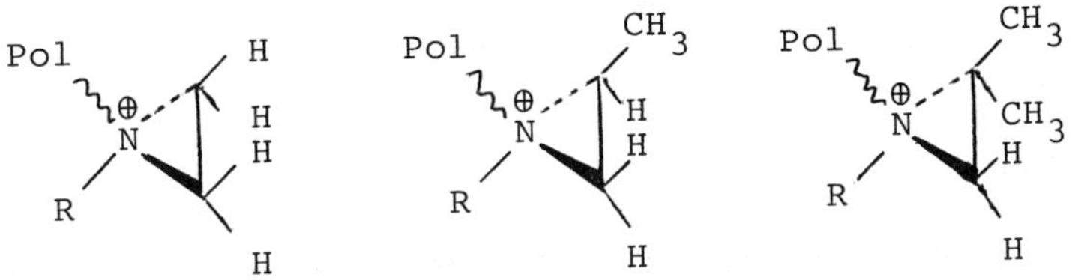

For the monomers the situation differs according to the C-
substitution. In the C-unsubstituted monomers the site of the
electron pair, which becomes the new C-N bond during the progatation
step, is unshielded and it may therefore be expected that these
monomers show a high nucleophilic reactivity. C,N-Disubstituted
aziridines on the other hand, can adopt two conformations due to the
easy inversion of the terminal amine function:

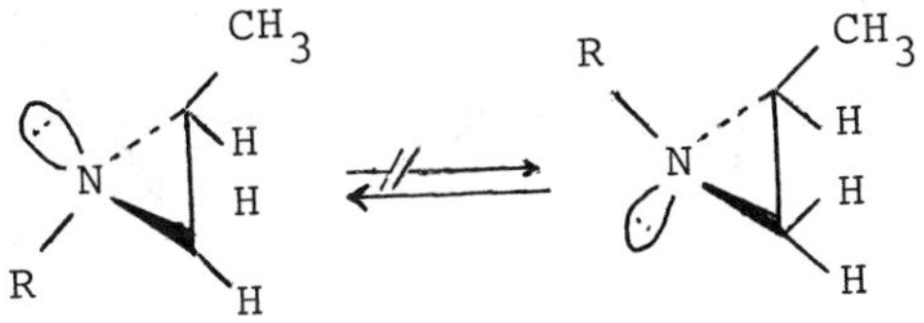

According to different conformational studies(4,5), these compounds
are present "only" as the conformer in which the methyl group is cis
to the electronpair and trans to the N-substituent. In such a
conformer both sides of the ring are shielded by a substituent and
it may be assumed that this causes a lowering of the nucleophilic
reactivity. Now the question arises whether the (observed) low
reactivity is due to the presence of only one (less reactive)
transconformer or is the consequence of an equilibrium between a
predominant non-reactive trans-conformer and the reactive cis-
conformer.

Alkylation of PEMA with methyl triflate leads to more than 95%
of one of the theoretically two possible diastereoisomers as shown
by the NMR spectrum (Figure 1) of the aziridinium salts which shows

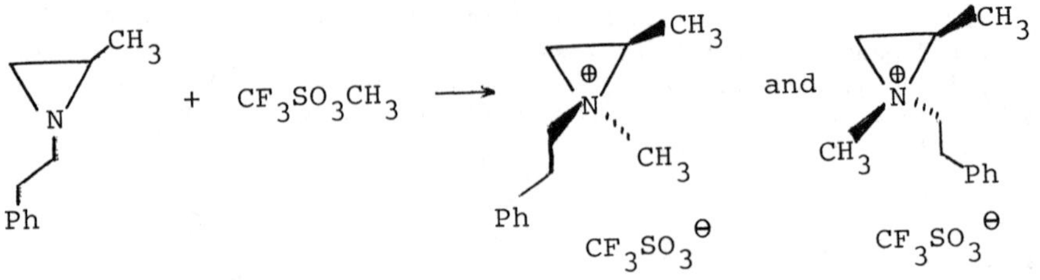

large and a very small doublet for the C-methyl substituent.
Although we don't know yet which isomer is the predominant one, the
fact that almost only one is formed, proves that alkylation of the
aziridine by methyl triflate takes place preferentially with one
conformational state. The ratio of the two reaction products may
reflect the ratio of the conformers of the starting material as was
described earlier for another aziridine (6). The predominant salt
can be recrystallized and is now being analyzed by X-ray diffraction
in order to know its configuration.

DMBA is an example of a non-polymerizing aziridine, although it
is readily alkylated by methyl triflate:

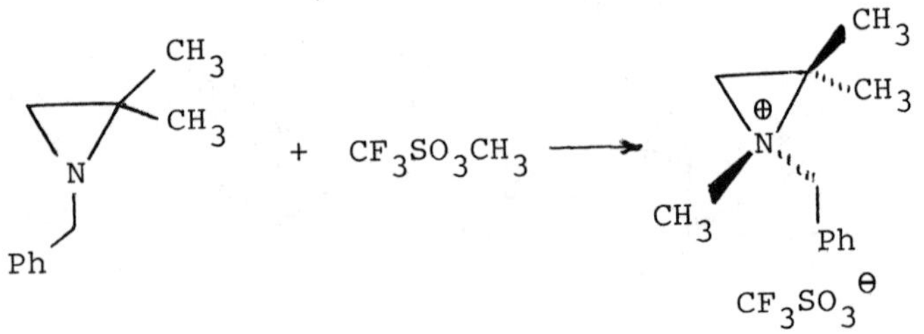

This proves that an aziridine which is shielded at both sides of the
ring is still reactive towards unhindered electrophiles such as
methyl triflate. However, reaction between DMBA and its highly
hindered aziridinium salt (i.e. propagation) is not possible.

MTBA is another non-polymerizing aziridine. Alkylation of MTBA
with methyl triflate leads to only one aziridinium ion since its NMR
spectrum shows only one doublet for the C-methyl group (Figure 2).
This is a strong indication that MTBA is conformationally
homogeneous, i.e. that only the trans-form exists, and it may
therefore be assumed that the aziridinium ion has the structure in
which the N- and C-methyl groups are on the same side of the ring.

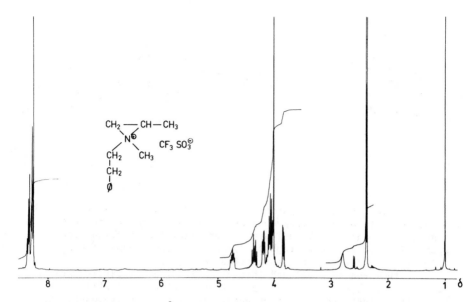

Figure 1 . 360 MHz ¹H-NMR spectrum of 1,2-dimethyl-1-(2-phenylethyl)aziridinium triflate.

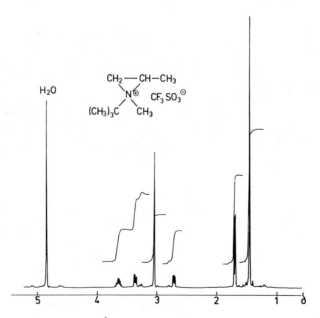

Figure 2 . 360 MHz ¹H-NMR spectrum of 1,2-dimethyl-1-tert. butyl aziridinium triflate.

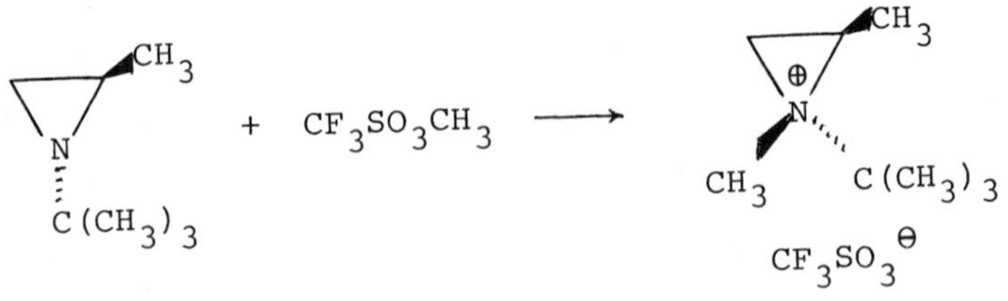

The exclusive trans configuration of the MTBA monomer makes this monomer still reactive towards sterically unhindered electrophilic reagents such as methyl triflate but prevents its reaction with sterically hindered electrophiles such as a highly substituted aziridinium ion (i.e. the propagation reaction).

It thus can be concluded that all aziridines, regardless their degree of substitution, are able to react with a sterically unhindered electrophile such as methyl triflate. Their reactivity towards sterically hindered electrophiles, however, is dramatically influenced by the presence of substituents. It seems that an unshielded face of the aziridine ring is necessary for a reaction to occur. Consequently gem. disubstituted aziridines and those 1,2-disubstituted aziridines which exist only in the trans form, cannot homopolymerize.

Block-Copolymers by Sequential Monomer Additions. The rate constant for propagation of TBA at 0°C in THF is 0.18 $l \cdot mol^{-1} sec^{-1}$. The polymer is "temporarily living". The rate constant for propagation of PEMA at 25°C in CH_2Cl_2 is 0.013 $l \cdot mol^{-1} sec^{-1}$ and this polymer is also "temporarily living". If PEMA is added to a solution of living poly-TBA, a block copolymer poly-TBA-poly-PEMA with narrow dispersity is formed in quantitative yield as is shown by GPC analysis of the end product (Figure 3). This proves that the initiation of the PEMA polymerization by the tert.butylaziridinium end group is quantitative and not slow compared with the PEMA propagation, in other words: $k_{12} \approx k_{22}$.

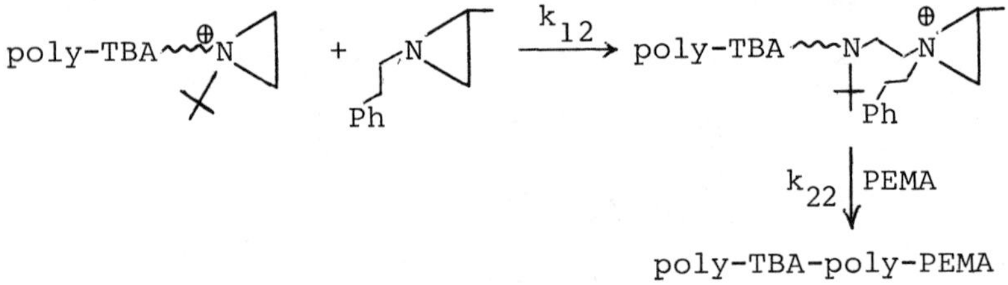

This is not unexpected since we know that the TBA aziridinium ion has a high reactivity towards all kinds of nucleophilics.

Surprisingly, the addition of TBA to a solution of living poly-PEMA also leads to the quantitative formation of a block copolymer as is shown by GPC analysis (Figure 4). This proves that the rate constant of the initiation for the TBA polymerization by the PEMA aziridinium ion is of the same order of magnitude as the homopropagation constant for TBA or $k_{21} \approx k_{11}$:

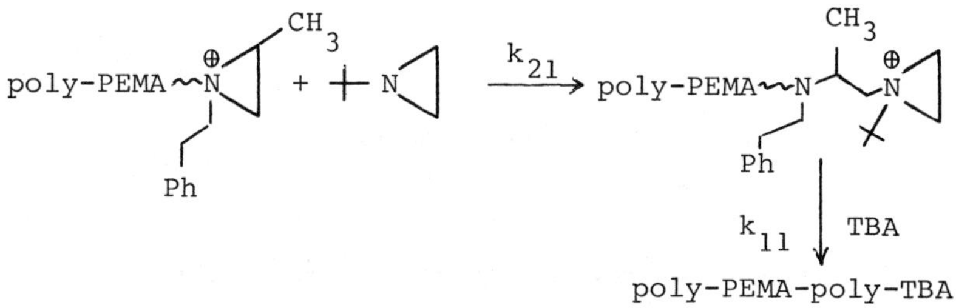

Consequently, the PEMA aziridinium ion and the TBA aziridinium ion must have similar reactivities towards the TBA monomer. This is a strong evidence that the main reason for the much slower polymerization of PEMA compared with TBA must be the lower reactivity of the PEMA monomer rather than a lower reactivity of the PEMA aziridinium ion. The validity of this statement is now further investigated by copolymerization experiments.

Copolymerization of N-alkylaziridines with β-propiolactone. β-propiolactone (PL) reacts with tertiary amines to form the corresponding zwitterion (7). If an N-substituted aziridine is used, a zwitterion containing an aziridinium ion and a carboxylate ion is formed. This zwitterion can initiate a cationic polymerization of the aziridine or an anionic polymerization of the lactone or undergo a coupling reaction with another (monomeric or polymeric) zwitterion.

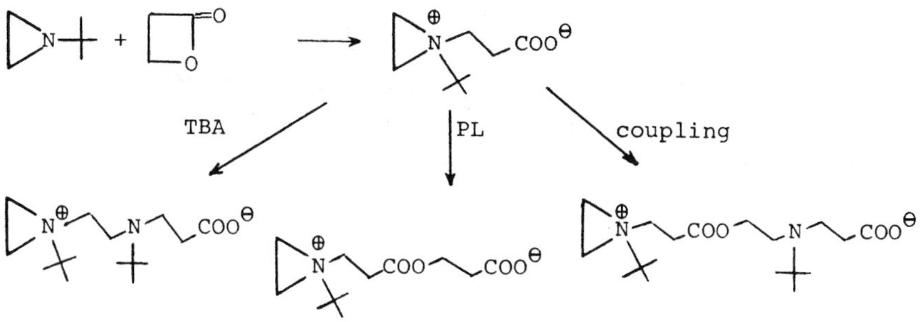

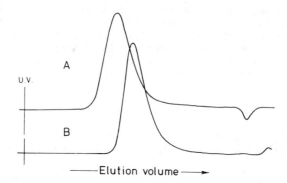

Figure 3 . GPC analysis of poly-TBA-poly PEMA block-
copolymer (A) and of the poly-TBA used as the
macromolecular initiator (B).

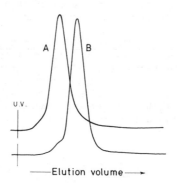

Figure 4 . GPC analysis of poly-PEMA-poly TBA block-
copolymer (A) and of the poly-PEMA used as the
macromolecular initiator (B).

The last reaction is another example of a spontaneous "alternating" copolymerization between an electrophilic and a nucleophilic monomer as described by Saegusa (8). The structure of the resulting polymer is determined by the rate constants of the different reactions and by the concentration of the two monomers. This reaction has been investigated with the monomer pair PL-TBA, in different solvents at 50°C.

The polymers formed at the beginning of the reaction contain a considerable excess of amino units which proves that the cationic aziridine propagation is more important than the anionic lactone propagation under the reaction conditions used. A special feature of this polymerization, however, is that large amounts (up to 50%) of cyclic oligomers are formed, which is obviously due to an intramolecular coupling reaction between a carboxylate ion and an aziridinium ion of oligomeric products. This is clearly demonstrated by the GPC analysis of the reaction mixture as shown in Figure 5.

The oligomers could also be analyzed by gas chromatography and coupling with the mass-spectrometer allows one to determine their structures (Figure 6). Table II gives a survey of the structure and the relative abundance of cyclic oligomers formed in the copolymerization of TBA and PL. The masses all correspond to the general formula A_nL_m with n = 2, 3 or 4 and m = 1 or 2 (A ≡ amine unit, L ≡ lactone unit). Since it is unlikely that zwitterionic species would be volatile, it must be accepted that these oligomers are cyclic.

Table II. Structure and relative abundance of cyclic oligomers formed in the spontaneous copolymerization of equimolar amounts of TBA and PL.

M[1]	Oligomer structure [2]	rel. intensity [3]
342	⌐ALAL⌐	74
342	⌐AALL⌐	10
369	⌐AAAL⌐	100
441	⌐AAALL⌐	7
468	⌐AAAAL⌐	98
441	⌐AALAL⌐	36

[1] M = molar mass of oligomer determined by CI-MS.

[2] A ≡ amine unit, L ≡ lactone unit.

[3] from GC-MS using electron impact ionization.

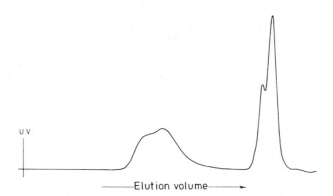

Figure 5 . GPC analysis of the reaction products of equimolar amounts (0.1 mol.1^{-1}) of TBA and PL after 8 hrs at 50° in acetonitrile.

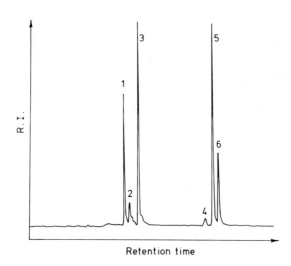

Figure 6 . Gas chromatographic analysis of the oligomers formed from TBA and PL.

In Figure 6, peaks nr. 1 and 2 correspond to oligomers containing two TBA and two PL units. Although the mass spectra could not be used to distinguish between the ALAL and AALL structural isomers, it is reasonable to assume that the much larger peak nr. 1 corresponds to the alternating oligomer since the occurrence of two adjacent PL units by homopolymerization is not very probable due to the low polymerization rate of PL at 25°C. The most abundant oligomers are the -AAAL- and -AAAAL- compounds (peaks nr. 3 and 5) which are 13- and 16-membered rings respectively. These are formed by a ring closure of the zwitterions $AAAL^\ominus$ and $AAAAL^\ominus$. This confirms the assumption that in this copolymerization the TBA monomer is consumed in a considerable proportion by cationic homopropagation.

This investigation is now continued with other N-alkylaziridines and with pivalolactone.

Literature Cited

1. Goethals, E. J.; Munir, A.; Bossaer, P. Pure & Appl. Chem. 53, 1753 (1981).
2. Goethals, E. J. Proceedings of the 28th IUPAC Macromol. Symp., Amherst, 1982, p. 204.
3. Le Moigne, F.; Sanchez, J. Y.; Abadie, M. J. Preprints of the 6th Intern. Symp. on Cationic Polymerization, Ghent, 1983, p. 87.
4. Maat, L.; Wulkan, R. W. Rec. Trav. Chim. 100, 204 (1981).
5. Razumova, E. R.; Kostyanovskii. Izv. Akad. Nauk SSSR, Ser. Khim. 9, 2003 (1974).
6. Bottini, A. T.; Dowden, B. F.; Van Etten, R. L. J. Am. Chem. Soc. 87, 3250 (1965).
7. Jaacks, V.; Mathes, N. Makromol. Chem. 131, 295 (1970).
8. Saegusa T.; Kobayashi, S.; Kimura, Y. Pure & Appl. Chem. 48, 307 (1976).

RECEIVED March 27, 1985

Block Copolymer of Poly(ethylene glycol) and Poly(N-isovalerylethylenimine)

Kinetics of Initiation

M. H. LITT and X. SWAMIKANNU

Department of Macromolecular Science, Case Western Reserve University,
Cleveland, OH 44106

We have investigated the molecular weight distribution
of a block copolymer obtained when poly(ethylene glycol)
ditosylate (MW 3500) is used to initiate the cationic
polymerization of 2-isobutyl oxazoline (living polymer).
In this system the rate of initiation is much smaller
than the rate of propagation. This leads to the formation of a mixture of di- and triblock polymers for
monomer initiator ratios $\leq$ 400.

A theoretical model was developed to correlate molecular
weight distribution of this system. This was compared
to the experimental gel permeation chromatography trace
with the theoretical model modified to include the effect
of the PEG central block, the spreading of the trace as
it went through the columns and the slope of the log MW
versus retention volume line. A good fit was found with
ki/kp = 0.0070. When methyl tosylate was used to polymerize 2-isobutyl oxazoline, a similar treatment of the
data showed ki/kp = 0.22. The effect of the PEG is
explained as due to solvation of the initial adduct by
the neighboring ether group.

2-Oxazolines are known to undergo ring opening polymerization when
initiated by acid catalysts (1-4). The polymerization proceeds
through living oxazolinium ion; in the absence of chain transfer and
termination and when the initiation is instantaneous, the polymer
produced would be monodisperse (5). Alkyl esters of p-toluenesulfonic acid have been used to initiate the polymerization of 2-oxazo-
lines (6-8). Bifunctional toxylate initiators have been used (9-10)
to prepare triblock copolymers of 2-oxazolines. We wanted to prepare
monodisperse triblock polymers of controlled block lengths of poly(N-
isovaleryl ethyleneimine) (PiVEI) from 2-isobutyl oxazoline and poly
(ethylene glycol) (PEG), using PEG ditosylate as initiator. It was
found that at small monomer/initiator ratio when all the monomer had
been consumed, unconsumed initiator was still present. The present
paper describes the synthesis and characterization of a triblock
copolymer. A theory was developed to predict the molecular weight

0097-6156/85/0286-0231$06.00/0
© 1985 American Chemical Society

distribution of the block copolymer and evaluate the relative rates
of initiation and propagation for the PEG ditosylate/2-isobutyl
oxazoline system. A similar study was made for the methyl tosylate/
2-isobutyl oxazoline system.

Experimental

2-Isobutyl oxazoline: This was prepared by the method of Kaiser (11)
b.p. 160°C/760 mm Hg (lit (6), b.p. 76°C/40 mm Hg). The structure
was confirmed by ^1H NMR and IR spectroscopy.

PEG ditosylate: The synthesis of pure PEG ditosylate (MWt = 3500) is
described briefly. The full procedure is given elsewhere (12). PEG
3500 (Dow Chemical) was purified by treatment with $NaBH_4$ in absolute
ethanol followed by several recrystallizations from absolute ethanol
at 4°C. The tosylate was prepared using the Schotten-Bauman procedure: p-toluenesulfonyl chloride/pyridine was reacted with poly(ethylene glycol) at 0°C. Polymer was isolated by recrystallization from
ethanol at 4°C. Four recrystallizations gave pure polymer with 2.0
tosylate groups per chain as determined by ^1H NMR, IR and UV spectroscopy. The PEG ditosylate had the same degree of polymerization as
the starting diol, as found by gel permeation chromatography.

Preparation of the copolymer and the homopolymer: Polymerizations
were run under vacuum in ampoules fitted with high vacuum Teflon
stopcocks. Table I lists the amounts of initiators, monomer and
solvent, (o-dichlorobenzene). Solvent and monomer were distilled
under reduced pressure into the ampoule containing initiator. After
degassing, the ampoules were heated at temperatures and times given
in Table I. The ampoules were removed and samples were taken for GPC.
Polymers were isolated by precipitation into mixed hexanes (boiling
range 38°-55°C) dried at 60°C. Yields were quantitative except for
small handling losses. 2.45 of the product from run 1 was stirred
twice with 10 ml. of acetone and the undissolved polymer was separated from the supernatant by decantation. The polymer was dried
under vacuum (< 1mm Hg) at 60°C. Yield 2.15 g.

^1H NMR ($CDCl_3$) of the copolymer: δ 3.65 (s,$-OCH_2CH_2$), δ 3.45
(broad-NCH_2CH_2-), δ 2.2 (broad, $(-\underline{CH}_2\underline{CH}(CH_3)_2)$, δ 0.92 (d,$(C\underline{H}_3)_2CHCH_2$-).

Table I. Polymerization Conditions

Run	Initiator, g	Monomer, g	M/I_o	Solvent/ml	T,°C	Time
1	PEGTs, 0.952	2.4	38	10	100	4 h.
2	MeOTs, 0.927	3.75	5.92	14	130	10 min.

Gel permeation chromatography: Molecular weights of the polymers and
their distributions were measured using a gel permeation chromatograph, with THF as the elution solvent. The GPC was equipped with a
differential refractomonitor, Model LDC 1107 and a UV (254 nm) absor-
ance meter, Model 153 (Beckman), in series. The GPC traces given in

Figure 1 were obtained using μ-Spherogel (Altex) columns of nominal porosities 10^4, 10^3, and 500Å and a 100Å μ-Styragel (Waters Assoc.). The GPC trace in Figure 2 was obtained using μ-Styragel columns of porosities 10^5, 10^4, and 10^3 Å. μ-Spherogel columns of porosities 500 and 100 Å and a 100 Å Ultrastyragel column were used to obtain the GPC trace in Figure 3. The GPC columns were calibrated with monodisperse standards of polystyrene and polyethylene glycol.

$\overline{M}n$ and $\overline{M}w$ were calculated using Mn = $\Sigma Hi/\Sigma Hi/Mi$ and Mw = $\Sigma HiMi/\Sigma Hi$ where Hi is the height of the GPC trace at a given elution volume and Mi the corresponding molecular weight. The latter was obtained from the calibration curve. Heights were taken at equal intervals (0.2 ml). The ratio of peak intensities of UV and ΔRI peaks (UV/ΔRI) was used as a measure of UV absorbing (e.g., tosylate) functionality of the materials.

The area-to-mass relations for the homopolymer and methyl tosylate were obtained from the areas under the GPC traces (ΔRI) of solutions having known concentrations of isolated homopolymer and MeOTs. Areas were measured by the paper-weighing method. ^{1}H NMR spectra were obtained on a Varian EM360 or an A60, 60 MHz spectrometer in C_6D_6, $CDCl_3$ or o-$C_6H_4Cl_2$ solutions. TMS was used as an internal standard. For infrared spectra, thin films of the samples were cast from $CHCl_3$ solution on NaCl plates and the solvent evaporated under vacuum. Spectra were recorded on a Digilab FTS 14 FTIR spectrophotometer.

Results

Synthesis of block copolymer: GPC analysis of the polymerization solution indicated that all the monomer was consumed. The presence of unconsumed initiator was observed. PEG tosylate eluted at 26.2 ml and had UV/ΔRI of 8000. The chromatogram of a solution of run 1 is given in Figure 1a. The peak at 24.2 ml corresponds to that of the block copolymer. The peak at 26.2 ml has UV/ΔRI ∿ 8000 and should be the unreacted PEG ditosylate. Figure 1b shows the elution profile of the of the acetone-washed copolymer and it is free from PEG tosylate. The presence of unconsumed initiator indicated slow initiation. Also, since the two ends of PEG tosylate are identical in reactivity by random statistics (12) we expect the copolymer to be a mixture of ABA type triblock (initiation at both ends) and AB type diblock (initiation at one end) copolymers. This was observed, as shown by the solid line of Figure 2 when the copolymer was eluted through a different combination of columns. The peak at 24.0 ml corresponds to the triblock copolymer and the peak at 25.4 ml corresponds to the diblock copolymer.

Composition of the block copolymers: The average lengths of the PiVEI blocks was obtained from ^{1}H NMR spectroscopy on the acetone washed copolymer. The PEG CH_2's have a sharp resonance at δ 3.65 ppm and the methylene protons of the PiVEI part occur as a broad peak centered at δ 3.45 ppm which overlaps the PEG resonance. The areas under these peaks were separated as follows. The protons of (CH_3) of the PiVEI occur as an isolated doublet at δ 0.92 ppm. Since there are six methyl protons for four >N-CH_2CH_2- protons in PiVEI, the area of >NCH_2CH_2- is two-thirds the area of the methyl

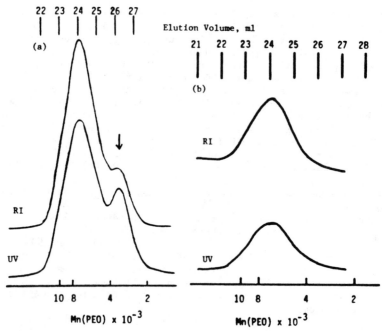

Figure 1. GPC traces of (a) copolymer 1 and (b) acetone-washed copolymer 1. Columns used: 10^4, 10^3, 500 Å μSpherogel and 100 Å μStyragel.

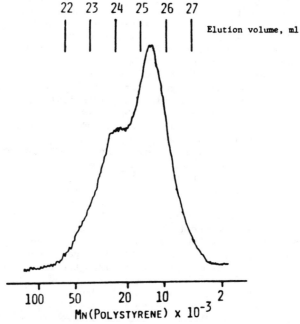

Figure 2. GPC trace of acetone-washed copolymer 1. Columns used: 10^5, 10^4, 10^3 Å μStyragel.

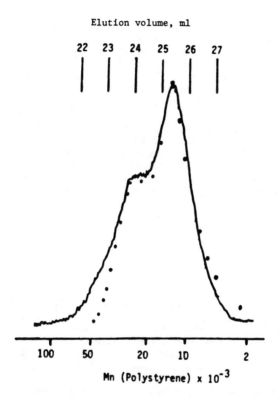

Figure 3. Comparison of theoretical and experimental distributions of the copolymer of PiVEI and PEG.

protons. The area due to OCH_2CH_2 is the total area in the region δ
3.00-3.85 ppm minus the area due to $>NCH_2CH_2-$. From the areas of
OCH_2CH_2 and NCH_2CH_2, the PiVEI/PEG block length ratio was 1:1.50.
Since the original block length of PEG was 80, the total PiVEI block
length is 120 which is larger than the block length expected from
M/I_0 value.

<u>Unreacted initiator</u>: After the acetone wash, the copolymer lost
12.3% by weight. Since polymers containing a large number of N-iso-
valeryl ethyleneimine units are insoluble in acetone, the loss in
weight is mainly due to the removal of unreacted PEG plus some singly
reacted ditosylate initiator. In the following, a theoretical model
is presented. It correlates the molecular weight distribution of the
block copolymer and reproduces the GPC trace.

<u>Molecular Weight Distribution Theory</u>: Bifunctional initiator -- In
the case of a bifunctional initiator such as poly(ethylene glycol)
tosylate, each initiator end reacts with monomer independently. When
the rate of initiation is smaller than the rate of the propagation,
initiation proceeds side by side with propagation. If z monomer
units have added to an active end which initiated at time zero, then:

$$I = I_0 \exp(-Kz) \tag{1}$$

$$C = I_0 (1-\exp(-Kz)) \tag{2}$$

$$z = \int_0^t kp(M)dt \tag{3}$$

where $K = ki/kp$: I is the concentration of initiator ends at time t
and C is the concentration of oxazolinium ends. $z_{max} = Z$, where z_{max}
is the value of z when all monomer is consumed.

We have postulated that all oxazolinium ends react at the same
rate. Therefore, if an initiator end reacts with monomer at a time
when $z = Z-i$, it will have a degree of polymerization of i when all
the monomer is consumed. The change in C when z goes from $Z-i$ to
$Z-i+1$ is the number of molecules with a final degree of polymerization of i, $N(i)$. The differential equation can be written as Equation 4.

$$dC = dN(i) = KI_0 \exp(-K(Z-i))di \tag{4}$$

For a bifunctional initiator with a total of x monomer units
added, we can write:

$$F(x) = 2N(0) N(x) + \int_0^x N(x-i) N(i)di \tag{5}$$

where $F(x)$ is the distribution function of polymers produced. In
Equation 5 the first term represents the polymer molecules produced
by addition of monomer at one end of the initiator only, and the
second term the polymer produced by addition of monomer at both ends
of the initiator.

For a bifunctional initiator, x can take values up to $2Z$ for
$1 \leq i \leq Z$, (singly and doubly initiated).

$$F(x) = 2 \exp(-KZ) \, K \, \exp(-K(Z-x)) + K^2 \int_0^x \exp(-K(Z-x+i)) \exp(-K(Z-i)) di$$
(6)

$$= 2K \exp(-K(2Z-x)) + K^2 x \exp(-K(2Z-x))$$
(7)

$$= K \exp(-K(2Z-x))(2+Kx)$$
(8)

For $Z \le x \le 2Z$, (doubly initiated)

$$F(x) = K^2(2Z-x) \exp(-K(2Z-x))$$
(9)

From Equation 4

$$\frac{dC}{dZ} = KI_0 \exp(-K(Z-x))$$
(10)

We can also write

$$\frac{-dM}{dz} = C$$
(11)

where M is the concentration of monomer at any time. Therefore,

$$M_0-M = zI_0-I_0[1-\exp(-Kz)]/K$$
(12)

or

$$(M_0-M)/I_0 = z -[1-\exp(-Kz)]/K$$
(13)

where I_0 is the initial concentration of initiator ends.

Also

$$1-\exp(-KZ) = f$$
(14)

where f is the fraction of initiator consumed when all monomer has been used up.

Then:

$$\exp(-KZ) = 1 - f \;\therefore\; Z = -\ell n(1-f)/K$$
(15)

Hence when all the monomer has been consumed,

$$(M_0-M)/I_0 = M_0/I_0 = Z-[1-\exp(-KZ)]/K$$
(16)

Equation 16 shows that the maximum number of monomer units added is controlled both by K and the initial monomer/initiator ratio.

Application of Theory to GPC Trace: Equations 6 and 9 describe the molecular weight distribution of molecules produced during polymerization. Since in gel permeation chromatography the refractive index difference is proportional to the mass of material present, the equations have to be modified to express the polymer weight and also the differential refractive increment. This requires that besides changing to polymer mass, the change in refractive index of the polymer as a function of x has to be included. Since every copolymer molecule has a polyethylene glycol molecule in it, the total increment in refractive index is the sum of the PEG refractive index increment plus the $\Delta n(x)$ of the added isovaleryl ethyleneimine units.

Table II lists the refractive indices and Δn with respect to the eluant (THF) of the different blocks.

Table II.   Refractive Indices of the Polymers

Polymer	n	(Ref)	Δn (Polymer-THF)
PiVEI	1.512	(13)	0.105
PEG	1.475	(14)	0.068

For the copolymer,

$$\Delta n(x) = \frac{0.068 \cdot 3500 + 0.105 \cdot 127x}{3500 + 127x} \tag{17}$$

Therefore,

$$H_o(x) = F(x) \cdot \Delta n(x) \cdot (3500 + 127x) \tag{18}$$

or,

$$H_o(x) = F(x) \cdot 13.3(x + 17.82) \tag{18a}$$

$H_o(x)$ represents the height of the ΔRI peak for a copolymer containing x monomer units of iVEI at the elution volume $V(x)$,

$$V(x) = 43.8 - 8.5 \log (x + 46.2) \tag{19}$$

$H_o(x)$ would be the actual height at $V(x)$ if there were no peak broadening. Since peaks do spread, the height at $V(x)$, $H(Vx)$, must be considered as the sum of the contributions of all the fractions with peak maxima in the neighborhood of $V(x)$. If a gaussian shape is assumed for the elution curve, the contribution of the polymers with y monomer units (whose maxima appear at $V(y)$) to the absorption at $V(x)$ is shown below:

$$H(V(x)) = \sum_{y=0}^{2z} H_o(y) \exp(-(V(y) - V(x))^2/\sigma^2) \tag{20}$$

Values of x and y were taken so that $V(x)$ points were spaced uniformly at 0.2cc. σ was taken as 0.4cc, the experimentally determined value for o-dichlorobenzene. The factor 8.5 in Equation 19 is the inverse of the slope of the log M_n versus elution volume for the GPC columns.

K and Z values were calculated for f values from 0.52 to 0.58 using Equations 15 and 16. A good fit was obtained when f was 0.56, K = 0.0070 and Z = 118, shown in Figure 3 as points. The positions of the two peaks and their relative heights are the same as those in the experimental plot. The two peaks are separated by the same distance as in the experimental. The theoretical distribution shows a greater low molecular weight component than we found. This is probably due to the selective extraction of the low molecular weight fraction by the acetone wash. If only PEG ditosylate were soluble, 5.5% of the total polymers should have been removed. Actually 12.3% was lost.

At the high molecular weight end the theoretical curve lies below the experimental one. The present theory does not consider chain transfer and repolymerization (5) which must occur to some degree. For small amounts of chain transfer, the effect of repolymerization is to generate a high molecular weight tail on the initial distribution, which can be seen here.

Formation of ionic species from covalent alkyl tosylate and monomer will be favored in the initiation reaction by a polar medium which can solvate and stabilize the resulting ion pair. In these polymerizations, the initial medium was 80% o-dichlorobenzene with the rest monomer and initiator, which was of low polarity (ε(O-DCB) = 9.8, ε(2-isobutyl oxazoline) $\approx$4). The relatively nonpolar medium slows down the initiation considerably. The propagation reaction requires the coordination of the monomer to the oxazolinium ion. The propagation is faster when the solvent is non-nucleophilic (15). Since o-DCB is relatively non-nucleophilic, kp must be relatively high. There could also be an influence due to the PEG part on the relative rates of propagation and initiation.

Methyl tosylate initiated polymerization: In order to determine if the PEG part plays any significant role in initiation, a polymerization was carried out with methyl tosylate as initiator using M/I of 5.92. At complete consumption of monomer, unreacted methyl tosylate was still present (see Figure 4). The polymer produced had a broad molecular weight distribution, $\overline{x}_w/\overline{x}_n$ = 1.39. kp/ki was estimated by two methods.

Estimation of kp/ki: For polymerizations where initiation is slower than propagation and where the polymer chain has one active end, we can write from Litt (16):

$$M/I_o = - [f + \ln(1-f)]/K + f \qquad (21)$$

and

$$\overline{x}_n = M/I_o f \qquad (22)$$

where f is the fraction of initiator consumed and $\overline{x}_n$ is the number average degree of polymerization when all monomer has reacted. f was obtained from GPC as well as from ^{1}H NMR spectrum.

The fraction f was calculated by comparing the original amount of initiator with the amount of unconsumed initiator, which was obtained from its area under the GPC trace and its area-to-mass relation. The fraction f and kp/ki were 0.86 and 4.6 respectively.

The polymer produced with methyl tosylate as initiator has one N-CH$_3$ group per chain and this methyl group could be observed in the ^{1}H NMR spectrum. By comparing the integrals of N-CH$_3$ (a) and CH$_3$ (b) of the isobutyl side chains, the degree of polymerization was calculated.

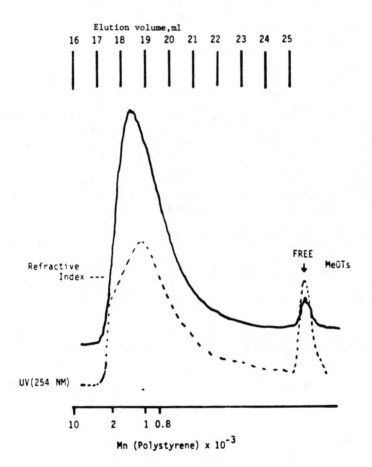

Figure 4. GPC trace of methyl tosylate initiated polymerization. Columns used: 500 Å, 100 Å µStyragel and a 100 Å Ultrastyragel.

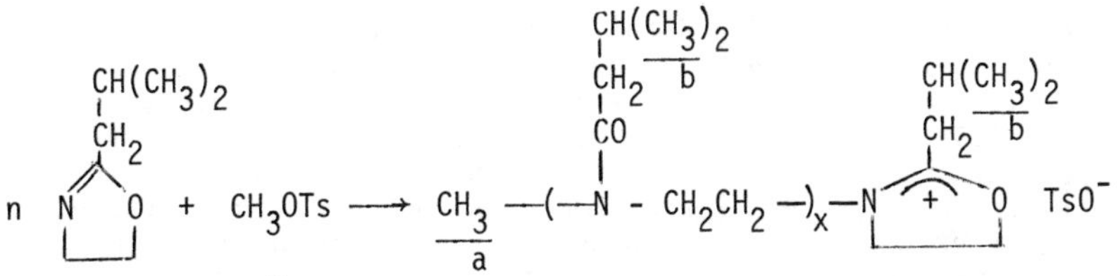

$$n \quad \underset{\underset{\text{N}\quad\text{O}}{\overset{\overset{\text{CH}_2}{\overset{\text{CH(CH}_3)_2}{|}}}{\bigwedge}} \quad + \quad CH_3OTs \quad \longrightarrow \quad \underset{a}{\underline{CH_3}} - (-N - CH_2CH_2 -)_x - \underset{\text{N}\quad\text{O}}{\overset{\overset{\text{CH}_2}{\overset{\text{CH(CH}_3)_2}{|}}}{\bigwedge}} \quad TsO^-$$

$$\bar{x}_n = \frac{I_b}{2I_a}$$

where I is the integral intensity of a group in ^{1}H NMR spectrum of the polymer. x_n, f and kp/ki were 6.64, 0.89 and 4.4 respectively.

kp/ki for 2-isobutyl oxazoline is compared with literature values for 2-methyl oxazoline and 2-phenyl oxazoline in **Table III**. kp/ki for 2-isobutyl oxazoline is of the same order of magnitude as kp_n/kp_1 for the other monomers. kp_n is the polymerization rate constant for an active center with at least two units attached. According to the Saegusa, et al. (17, 18), the slowest step was the addition of the second monomer. By GPC, we observed only MeOTS and polymer and no N-methyl oxazolinium tosylate. Since DC_3CN, the solvent for the polymerizations of 2-methyl and 2-phenyl oxazolines is polar, it promotes the formation of oxazolinium ion from MeOTS and the oxazoline more than the less polar o-DCB. In o-DCB the slowest step seems to be the addition of the first monomer.

Table III. Kinetic Parameters for 2-R Oxazolines Initiated by MeOTS

R	T°C	kp_n/kp_1	kp_n/ki	Reference
CH_3	130*	9.8	1.81	17
CH_3	100*	6.1	1.36	17
C_6H_5	130*	2.87	2.59	18
iC_4H_9	130		4.6 (GPC)	Present Study
			4.4 (NMR)	Present Study

* Extrapolated from data at lower temperatures.

The large value of kp/ki for PEGTs/-2iBUOxz must be due to the PEG part. The ether oxygen of PEG is basic and has been known to complex with alkali metal cations (Equation 19). Probably, the ether oxygen of the terminal ethylene oxide unit coordinates with the oxazolinium positive center (at 2-carbon) thereby reducing the reactivity of the oxazolinium ion towards the monomer.

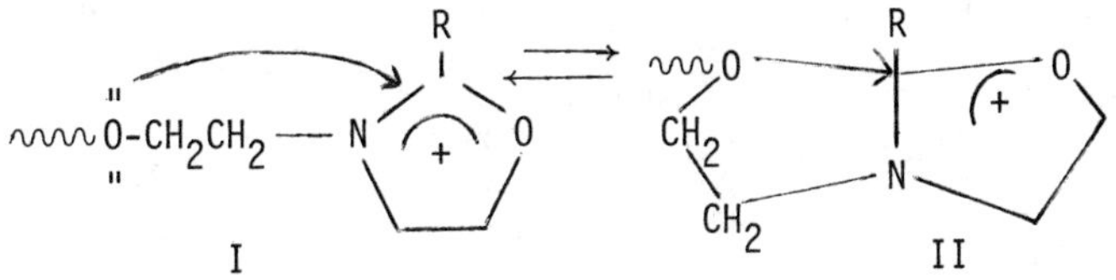

The open chain form I and the bicyclic complex form II are in equilibrium. Ring opening by monomer, however, takes place by reaction with I. The large value of kp/ki,143 (kp/ki for MeoTs/2-isobutyl oxazoline, 4.5) indicates that the equilibrium is in favor of II.

The peak at 26.2 ml in Figure 1 having UV/ΔRI $\approx$ 8000 could be due to the complex produced by adding an oxazoline unit at one or both ends of PEG ditosylate. The absorbence of the tosylate ion is lower than that of the ester. [The molar extinction coefficient of ethyl tosylate is 440 and that of silver tosylate is 223 at 254 nm (20).] However, oxazolinium ions also absorb strongly at 254 nm. Even though the tosylate ion absorbs less than the ester, the absorbence could be made up by the oxazolinium ion. As can be seen in Figure 1, the experimental error involved is quite large and differences in UV/ΔRI of about ± 25% could go undetected.

Summary and Conclusion

For the PEG ditosylate/2-isobutyl oxazoline system, the rate of initiation was much smaller than the rate of propagation; this produced a mixture of di and triblock copolymers and unconsumed initiator. A theoretical model was developed to correlate the molecular weight distribution of the copolymers. The model was successful in correlating the shape of the GPC trace of copolymer produced with $M/I_0 = 38$. ki/kp was found to be 0.0070. In the polymerization of 2-isobutyl oxazoline, when initiated by MeOTs, ki/kp was found to be 0.217. The terminal unit of ethylene oxide in PEGTs is responsible for the substantial decrease in ki compared to kp by solvating the positive center C-2 of the oxazolinium ion.

Acknowledgments

The authors wish to acknowledge the support of the Department of Energy, Lawrence Berkeley Laboratory under Contract 4520610.

Literature Cited

1. Tomalia, A; Sheetz, D. P. J. Polym. Sci. 1966, A-1, 4, 2253.
2. Seeliger, W. Angew. Chem. 1966, 78, 613, 913.
3. Kagiya, T.; Maeda, T.; Fukui, K. J. Polym. Sci. 1966, B, 4, 441.
4. Litt, M. Belgian Pat. 666,828; 666,831, 1965.
5. Szwarc, M. "Carbanion Living Polymers and Electron Processes"; John Wiley, New York, 1968.
6. Bassiri, T. G.; Levy, A.; Litt, M. H. Polym. Lett. 1967, 5, 871.
7. Saegusa, T.; Ikeda, H.; Fujii, H. Polym. J. 1972, 3, 35.
8. ibid, ibidem, 1973, 4, 1, 87.
9. Percec, V. Polym. Bull 1981, 5, 643.
10. Saegusa, T.; Ikeda, H. Macromol. 1973, 6, 805.
11. Kaiser, M. U.S. Patent 4 203 900, 1980.
12. Swamikannu, X.; Litt, M. H. J. Polym. Sci., Polym. Chem., in press.
13. Weast, R. C., Ed., Handbook of Chemistry and Physics, 59th Ed. 1978-79, CRC Press Inc., Boca Raton, Florida.
14. Strazielle, C., Makromol. Chem. 1968, 119, 50.
15. Matsuda, T. Ph.D. Thesis, Kyoto University, Japan, 1972, 107.
16. Litt, M. J. Polym. Sci. 1962, 58, 429.

17. Saegusa, T; Ikeda, H; Fujii, H. Macromol. 1972, 5, 359.
18. Kobayashi, S.; Tokuzawa, T.; Saegusa, T. Macromol. 1982, 15, 707.
19. Armand, M. B.; Chabagno, J. M.; Duclot, M. J. In "Fast Ion Transport in Solids"; Vashista, P.; Mundy, J. N.; Shenoy, G. K., Eds., North Holland, N.Y., 1979, 131.
20. From SAD 2302; SAD 18764, The Sadtler Standard Spectra, Sadtler Res. Laboratories, Philadelphia, 1968.

RECEIVED October 15, 1984

Synthesis and Applications of Polysiloxane Macromers

YUHSUKE KAWAKAMI and YUYA YAMASHITA

Department of Synthetic Chemistry, Faculty of Engineering, Nagoya University, Chikusa, Nagoya 464, Japan

The surface of polymers have been known to be of importance in many applications of polymers. Polyorganosiloxanes are industrially produced in large quantity for various purposes depending molecular weight and copolymer composition. The inherent hydrophobic nature of siloxanes coupled with their ability to accumulate on the surface of polymer blends has allowed the polyorganosiloxanes to be used widely to impart hydrophobicity to the surfaces of other materials, particularly in the case of polymers which by themselves are devoid of such properties.(1-3)

Among the many physicochemical methods employed for the surface modification of polymers, the use of graft copolymers for the purpose is being recognized to be promising.(4-8) This is because the graft copolymers containing surface active segments can be tailor-made for a specific need. However, the availability of well characterized graft copolymers is essential not only for understanding of their action, but also for maximum efficiency.

Polyorganosiloxanes are also noteworthy for their high permeability of gases, but their films are not mechanically strong enough for use as gas permeable membranes. The disadvantage is overcome by making thin films of block copolymers with polycarbonate.(9) However, the selectivity towards different gases was not necessarily high. The low selectivity seemed to come from the inherent nature of the rather long polyorganosiloxane segment in the block copolymer. In order to achieve high selectivity of oxygen permeation, it seems advantageous to use short siloxane sequences in block copolymers or short siloxane branches in graft copolymers. The structure of a multi block copolymer does not seem suitable for the purpose. On the other hand, the structure of graft copolymers with short polysiloxane branches is considered to be the exact structure for the purpose, since one may maintain film forming properties, flexibility of siloxane chain (high permeation rate), and high selectivity. The flexibility of the siloxane chain can be maintained in side chains of graft copolymers and film forming properties and selectivity of permeation can be provided by backbone For this purpose, polystyrene was chosen as the backbone material

As an approach to the designing of such materials, we report

0097-6156/85/0286-0245$06.00/0

here on our studies concerned with the synthesis of well-characterized ω-styryl or methacrylyl polysiloxane macromers and oligomers,
and their copolymerizations with suitable comonomers to give well
characterized graft copolymers. Furthermore, we discuss the surface
modification of bulk polymers by adding small amounts of siloxane
graft copolymers. At the same time, selective oxygen permeation
through the films of polystyrene graft copolymers with short
siloxane chains was also studied.

Experimental

Preparation and Characterization of Polysiloxane Graft Copolymers

The polyorganosiloxane macromers having methacrylyl or styryl
function at one end were synthesized by terminating the anionic
polymerization of hexamethylcyclotrisiloxane with a functionalized
chlorosilane terminating agent.

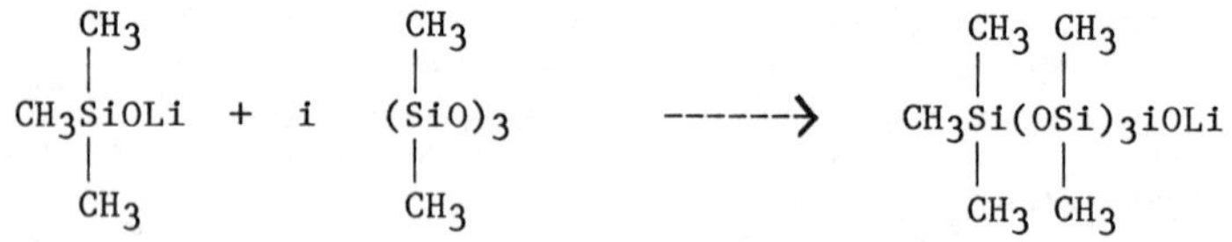

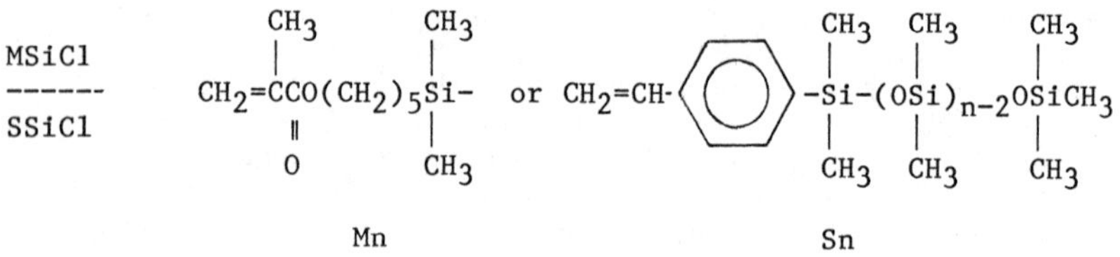

Polysiloxane macromer

The terminating agents, p-oligosiloxane substituted styrenes,
and methacrylate type oligosiloxanes were synthesized according to
Schemes 1 and 2.

Oligosiloxane monomers and macromers are abbreviated as Sn, Mn
(n=1,2,5,8, and n=average number of silicon atoms in the macromer).
Fluorosubstituted analogues of S2 were synthesized similarly and
abbreviated as FS2, FF2, ans SF2 as shown in the scheme.

Graft copolymers were obtained by the ordinary copolymerization
with suitable comonomers, and characterized mainly by GPC and NMR.
The monomer reactivity ratios were estimated by Equation 1 under the
condition [B]>>[A].

$$r_B = \frac{d[B]}{[B]} \quad \frac{d[A]}{[A]} \quad \ldots\ldots(1)$$

[A]: macromer

[B]: comonomer

Comonomers used are styrene(St), methyl methacrylate (MMA),
fluoroalkyl acrylate (FA: $CH_2=CHCO_2CH_2CH_2(CF_2)_{9.5}CF_3$),

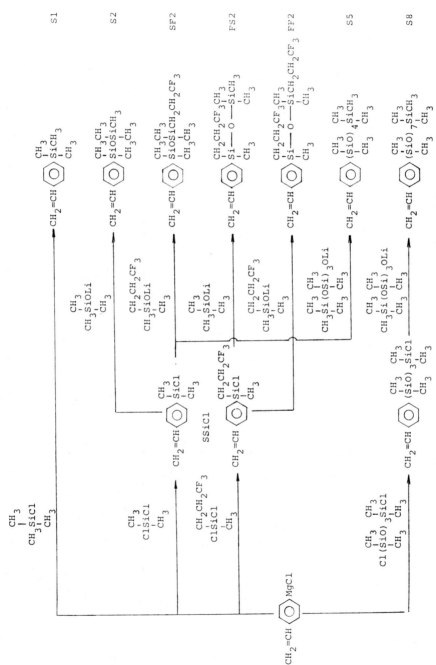

Scheme 1. Synthetic Routes to Styrene Type Oligosiloxane Monomers.

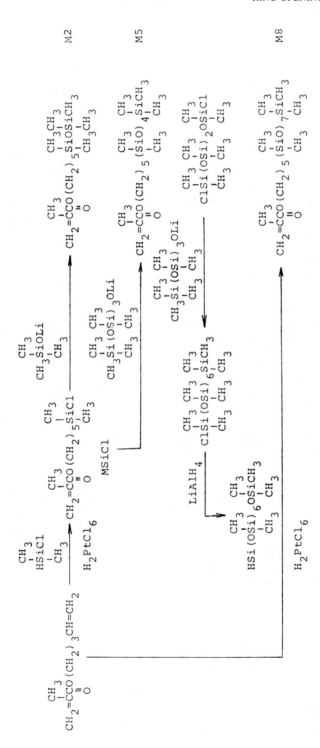

Scheme 2. Synthetic Routes to Methacrylate Type Oligosiloxane Monomers

tris(trimethylsiloxy)silylpropyl methacrylate (MTS), and 3-[3-bis
(trimethylsiloxy)methylsilylpropoxy]-2-hydroxypropyl methacrylate
(BTM).

Contact Angle Measurement. Binary blends of commercial poly MMA
($Mn=6.7 \times 10^4$, $Mw=14.8 \times 10^4$) with one of the siloxane polymers were
prepared. The polymer content varied from 0-10% w/w. Films were
prepared by casting a 6-8% w/v solution in THF on micro glass slides
and allowing the solvent to evaporate during 24-36 h. The films
were detached from the slides by immersing in water, and dried under
vacuum for 48-72 h. Contact angles for both air-side and glass-side
surfaces were measured at 20°C for the water droplet.

FT-ATR-IR amd ESCA. These spectra were also measured on both sides
of the films.

Gas Permeation. Permeability of oxygen and nitrogen was evaluated
for supplied air by a gas chromatographic method (Yanaco GTR-20).

Results and Discussion

Preparation and Characterization of Polysiloxane Macromers and
Oligosiloxane Monomers

All of the reactions in the schemes can be safely handled under N_2
atmosphere without the need of high vacuum technique. The formed
siloxane polymers initiated by lithium trimethylsilanolate (LTMS)
have narrow molecular weight distribution, which is a characteristic
of living polymerization (Table I). The initiation reaction is
faster than propagation reaction, and oligosiloxane monomers of
degree of polymerization i=1 (S5,M5) can be easily obtained in
almost quantitative yield. However, oligosiloxane monomers of
degree of polymerization i=2 (S8,M8) could not be selectively
synthesized. These monomers were synthesized by connecting building
blocks as exemplified in the schemes.
    In the synthesis of terminating agents, ultrasonic irradiation
(10-12) (Bransonic 221) was found effective to obtain reproducible
yield. By the ultrasonic irradiation at the Grignard formation
stage, SSiCl was obtained in 76% (reproducible) yield. No
polymerization was observed. The presence of ultrasonic irradiation
in activating the platinum catalyst (small amounts of
dimethylchlorosilane was added to the chloroplatinic acid in 4-
pentenyl methacrylate in order to activate chloroplatinic acid) gave
MSiCl in 55-60% (reproducible) yield. This is also true in the
synthesis of M8.
    The conditions of the coupling reaction of MsiCl with the
living polysiloxane need special comments. The addition of the
terminating agent at the end of the polymerization of
cyclotrisiloxane, and subsequent stirring either at room temperature
or at low temperature (0°C) always afforded macromers with
functionalities appreciably lower than 1. In many cases the results
were not even reproducible. This problem was also successfully cir-
cumvented by ultrasonically irradiating the reaction mixture after
the addition of the terminating agent. A sonication period of 15-20

Table I. Synthesis of Polysiloxane Macromers

| No. | D_3 (mol/l) | LTMS (mmol/l) | Macromer | | | |
			Yield(%)	Mn,calc ($\times 10^{-3}$)	Mn,UV ($\times 10^{-3}$)	Mn,VPO ($\times 10^{-3}$)	f[b]
SM1	0.911	63.3	79	3.00	2.80	3.00	1.07
SM2	0.613	40.5	80	3.30	3.40	3.50	1.03
SM3	0.651	21.7	85	6.30	6.70	5.90	0.88
SM4[c,d]	0.651	21.7	85	6.30	6.60	5.90	0.89
MM5	1.00	28.9	94	8.00	7.60	5.50	0.72
MM6[c]	0.534	32.8	76	3.90	3.70	3.80	1.03
MM7	1.04	43.3	83	5.70	6.40	5.40	0.84
MM8[c]	0.499	28.0	90	4.20	4.15	4.20	1.01
MM9[c]	1.181	27.8	90	8.80	9.07	8.96	0.99

a) Polymerizations were carried out at 0°C in THF, and terminated
   with three-fold excess SSiCl(SM-series) or MSiCl(MM-series).

b) f=Mn,VPO/Mn,UV

c) Under ultrasonic irradiation.

d) Functionalization by SSiCl was satisfactory as long as the
   concentration of LTMS was kept higher than about 30 mmol/l,
   and any significant effect of ultrasound irradiation was
   observed.

minutes seemed sufficient, and functionalities approached unity.
These results are shown in Table I. Macromers of Mn=9,000-10,000
could be easily obtained by manipulating the monomer/initiator
ratio. The presence of the terminal unsaturation in the macromers
was confirmed by ^{1}H-NMR, IR, and by UV. These macromers had a quite
narrow molecular weight distributions.

Copolymer Synthesis. The conditions used for the synthesis of
copolymers and the characterization data of the copolymers are
presented in Table II with some results of homopolymerizations. The
combination of GPC and ^{1}H-NMR was helpful in showing the effective
incorporation of the macromer units into the copolymers. The
molecular weights of the graft copolymers were not very high.
Consequently, the number of the grafts attached to a given backbone
is generally small.

The monomer reactivity ratios of conmonomers in the copolymerization with styryl type macromer are shown in Table III. Macromers
seem to have similar reactivity with ordinary monomer in
copolymerization. (13-15)

Film Forming and Surface Active Properties of Graft Copolymers. The
use of tailor-made graft copolymers in the surface modifications of
polymers is quite promising because of their efficiency and also
because the graft copolymer structure can be varied for any
particular application.(4-8) All the graft copolymers (#13-17 in
Table II), despite their rather low molecular weights, are capable of
forming transparent, although brittle, films. Binary blend films
were prepared by casting the blends of siloxane polymers with
polyMMA or polySt. The content of the siloxane polymers varied from
0.1-10%. The surface active properties were evaluated/measured by
changes in the contact angle of the water droplet placed on the air-
side surface and glass-side surface of the films. For the
comparison of the efficiency of the surface modification, the
results of random copolymers and homopolymers are also shown in
Figure 1. An examination of the results shows that the siloxane
polymers exhibit pronounced surface accumulation on that side. It
can be seen that quite small amounts of siloxane polymers, e.g. 0.5-
1.05 depending on their siloxane contents (except polydimethylsiloxane), are capable of producing sufficient surface enrichment
and therefore appreciable hydrophobicity on that side of the film.
However, on the glass side, the contact angle variation is not much,
indicating the enrichment of siloxane polymer might be suppressed.
This is in keeping with the well known hydrophobic nature of the
siloxanes.

FTATR-IR and ESCA Measurement. In order to know the distribution of
the graft copjolymer in the blend film, it is necessary to estimate
the concentration at different depths. This can be achieved by
employing different spectroscopic methods detecting at different
depths.

FTATR-IR is one of these spectroscopies. Although the
sensitivity is not so high, the average concentration of graft
copolymers to about 5,000 Å and 15,000 Å depths from the surface can
be estimated by changing the prism for the blend system containing

Table II. Copolymer Syntheses

No.	feed monomer (mol %)	feed comonomer (mol %)	composition[b] monomer (mol %)	composition[b] comonomer (mol %)	Yield (%)	M_w[c] $\times 10^{-4}$	M_n[c] $\times 10^{-4}$	M_w/M_n[c]	Si[d] number	Siloxane[e] (wt %)	no of grafts	Appearance[f]
1	S1,100	-	S1,100	-	96	54.4	27.4	1.74	1.0	41	-	p
2	S2,100	-	S2,100	-	90	23.8	8.4	2.78	2.0	59	-	p
3	S2, 90	FA, 10	S2, 85	FA, 15	71	16.5	6.7	2.47	1.7	45	-	w
4	S5,100	-	S5,100	-	94	34.9	9.6	3.64	5.0	78	-	w
5	S5, 56	FA, 44	S5, 59	FA, 41	83	6.6	2.1	3.12	3.0	41	-	w
6	S5, 40	St, 60	S5, 39	St, 61	64	15.6	7.3	2.14	2.0	58	-	w
7	S5, 20	St, 80	S5, 20	St, 80	73	31.7	6.5	4.90	1.0	41	-	p
8	S8,100	-	S8,100	-	90	12.8	6.4	2.00	8.0	85	-	v
9	S8, 25	St, 75	S8, 23	St, 77	38	5.1	2.7	1.88	1.8	52	-	w
10	S8, 12.5	St, 87.5	S8, 13	St, 87	58	13.4	5.3	2.51	1.0	39	-	p
11	M5,100	-	M5,100	-	89	24.2	8.4	2.87	5.0	70	-	w
12	M5, 64	St, 36	M5, 60	St, 40	58	12.5	8.1	1.55	3.0	42	-	w
13	M60[h], 2.2	MMA, 97.8	M60, 1.6	MMA, 98.4	60	7.8	4.2	1.84	1.0	41	4	p
14	M60, 0.7	MMA, 99.3	M60, 0.6	MMA, 99.4	60	6.1	3.1	1.96	0.4	21	1.5	p
15	M60, 2.2	St, 97.8	M60, 2.7	St, 97.3	57	5.8	4.0	1.45	1.6	53	4	p
16	M120[h], 1.1	MMA, 98.9	M120, 0.8	MMA, 99.2	60	8.4	5.8	1.46	1.0	41	2.7	p
17	M120, 0.4	MMA, 99.6	M120, 0.3	MMA, 99.7	55	5.6	3.2	1.73	0.4	23	0.7	p

No.										
18	MTS,100	—	MTS,100	78	27.5	12.5	2.20	4.0	70	—	w
19	MTS, 12	MMA, 88	MTS, 12 MMA, 88	85	13.8	10.2	1.36	1.0	25	—	p
20	MTS, 30	MMA, 70	MTS, 30 MMA, 70	68	17.3	12.3	1.41	1.8	45	—	p
21	Polydimethylsiloxane			—	7.3	4.0	1.83	4.0	100	—	v
22	FA, 80	BTM, 20	BTM, 16 FA, 84	84	—	—	—	0.5	13	—	w
23	SF2,100	—	SF2,100	77	29.7	16.1	1.84	2.0	68g)	—	p
24	FS2,100	—	FS2,100	56	31.0	15.8	1.96	2.0	68g)	—	p
25	FF2,100	—	FF2,100	68	37.3	17.6	2.12	2.0	74g)	—	w

a) Polymerized at 60°C in THF with AIBN as an initiator(0.1-0.2 mol % to monomer).

b) Determined by NMR.

c) Determined by GPC and correlated to standard polySt.

d) Number of silicon atoms per monomer unit.

e)
$$(SiO)_{n-1}-SiCH_3$$
with CH_3 side groups, weight percent in polymer.

f) p=powdery, w=waxy, v=viscous oil.

g) Weight percent including fluoropropyl groups.

h) M60 and M120 are macromers from No. MM8 and MM9 in Table 1, respectively.

Table III. Monomer Reactivity in the Copolymerization of
            Polysiloxane Macromers

Macromer(A), Mn	Comonomer(B)	r_B	$r_A^{a)}$	$r_B^{a)}$
ω-Stylyl polysiloxane, 3,400	St	1.1	1(St)	1(St)
6,700	MMA	0.60	0.52(St)	0.46(MMA)

a) r_A and r_B are monomer reactivity ratios of low molecular

weight analogue pairs taken from Polymer Handbook.

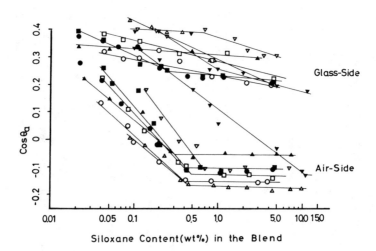

Figure 1.   Contact Angle of Water Droplet at 20°C for Various
            Siloxane Polymers-PolyMMA Blend Films.
            ●: GM211,   ■: GM213,   ▲: GM411,   ▼: GM413,
            ▽: poly-MTS, △: MTS25, ▲: MTS45, ▽: polyDMS are
            copolymers and polymers synthesized in Nos. 13,14,16,17,
            18,19,20, and 21, respectively in Table II.

10% w/w graft copolymer. The results are shown in Table IV.
Contrary to the results of contact angle measurement, by which
higher concentration of graft copolymer on the air-side surface was
clearly seen even at the concentration of 0.1% graft copolymer,
concrete differences between the surfaces of the two sides to 5,000
Å could not be observed until the concentration of 10% graft
copolymer reached. However, the measurement to 15,000 Å depth
clearly indicated that even at the air-side surface of the modified
polyMMA film by adding 10 wt% graft copolymer, the average
concentration to 15,000 Å depth from the surface is much lower than
that to 5,000 Å as shown in Figure 2. This fact suggests that at
the concentration of 10 wt% graft copolymer to bulk polyMMA, the
graft copolymer may exist between the surface and about 10,000 Å
depth, and the relative concentration is in the order:

air-side surface > glass-side surface >> inner part of the film

Information closer to the surfaces can be obtained by ESCA
measurement. The results, with Al $K\alpha1,2$ source which gives the
average concentration to about 20 Å depth, are shown in Figure 3.
Although there is some scatter of the measured points, the results
are consistent with the contact angle data at the points that the
graft copolymer concentration at air-side surface is higher than
that at the glass-side surface, and that the change in the
concentration at the glass-side surface is smaller compared to air-
side surface. Furthermore, the concentration of the graft copolymer
at both surfaces are higher than the average concentration of
the graft copolymer in the bulk, which was calculated from the
weight percent of the added graft copolymer, detected by ESCA.
From the results mentioned above, the following picture of how
the graft copolymer exist in the blend may be proposed.

1. Polysiloxane graft copolymers phase-separate to both
   air- and glass-side surfaces. The concentration of
   graft copolymers seem to be higher near the surface.
2. At the air-side surface under surface active condition,
   the polysiloxane graft copolymers will accumulate at
   the surface resulting in sharp change in contact angle
   and ESCA. At the glass-side surface, although the
   bulk concentration of the graft copolymer seems to
   be similar to that at the air-side surface, the
   concentration at the skin of the surface seems
   different from that at the inner of the surface
   layer. The polar groups (presumably carbonyl groups
   of methyl methacrylate comonent) might be oriented
   near the surface by the influence of glass surface
   environment. This seems to be reflected in the
   slight change in contact angle measurement at the
   glass-side surface. FTATR-IR can not distinguish the
   difference in concentrations at such shallow layer.

Durability of Surface Modification toward n-Hexane Treatment. The
durability of such surface modification is very important from a
practical view point. Thus, there may be some chance of loss of the
surface-modifying polysiloxane component from the surface under

Table IV. ATR-IR Relative Absorbance of Various Polysiloxane-
           PolyMMA Blend Films.

GM411[a)]/PolyMMA (%)	A795/A740 [b)]	
	Air-Side	Glass-Side
0	0.23	0.23
0.1	0.30	0.22
0.5	0.27	0.28
1.0	0.44	0.43
10.0	2.45	2.10
10.0 [c)]	1.48	

a) GM411 is the graft copolymer synthesized in No.
   16 in Table II.

b) The absorptions at 740 and 795 cm^{-1} are coupling of
   CH_2 and skeltal vibration in polyMMA units ,and
   $SiCH_3$ rocking modes, respectively. The spectra
   were recorded on Ge prism.

c) Recorded on KRS5 prism.

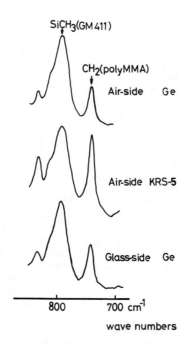

Figure 2. FT-ATRIR Spectra of polymMMA Blend Film Containing
10% Siloxane Graft Copolymer GM411 at Air-, and
Glass-side Surfaces.
GE: Ge prism, KRS-5: KRS-5 prism.

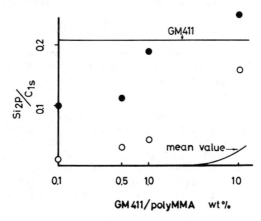

Figure 3. ESCA Relative Intensity of polyMMA Blend Films Containing
Various Amounts of Siloxane Graft Copolymer GM411.

certain conditions. This possibility was realized by the treatment with n-hexane, as shown by the results in Figure 4.

n-Hexane was selected because it is a good solvent for polysiloxane polymers. As can be seen from Figure 4, the poly-dimethylsiloxane was completely washed away due to its solubility in n-hexane, and consequently, the blend film was deprived of any surface hydrophobic modification. Similarly, and most surprisingly, we also observed the complete removal of PMTS and the random copolymer of MTS from the surface of the blend films. This was also found to be due to their solubility in n-hexane. The polysiloxane graft copolymers, on the other hand, were resistant to solvent extraction from the blend film surface as these polymers were insoluble in n-hexane. It is interesting to note that graft copolymers, having siloxane content similar to random copolymers were not extracted from the blend film surface by n-hexane treatment. This phenomenon is significant, and implies what the behavior of polysiloxane polymers is on the surface. The surface active properties of polysiloxane graft copolymers, and their resistance to n-hexane extraction may indicate that they are accumulated on the surface with backbone polyMMA chain remaining in bulk polyMMA as anchor segment. Easy removal of siloxane random copolymer from the polyMMA indicates the random copolymer is a poor anchoring segment. Furthermore, this seems to indicate that in the siloxane random copolymers tested in this study, the resistance to solvent extraction depends on MMA content, unlike that in the graft copolymers.

A comparison of the siloxane graft copolymers with random copolymers makes the former more attractive not only because of the possibility of synthesizing such tailor-made graft copolymers by the use of easily accessible siloxane macromers, but also because of the durability of surface modification characteristics imparted by them.

Film Forming and Selective Oxygen Permeation. Selective oxygen permeation was evaluated mainly for the polymers having short oligosiloxane side chains. Oxygen permeation constants (P_{O_2}) and

separation factor ($\alpha = P_{O_2}/P_{N_2}$) of various polymers are plotted in

Figure 5. Among the polystyrene polymers examined, poly(S) showed the best results. This polymer contains 59% dimethylsiloxane unit in weight, which make it possible to maintain high permeability ($P_{O_2} = 1.0 \times 10^{-8}$ cc.cm/sec.cm^2.cmHg), and the backbone polystyrene

seems to contribute to the film forming property and rather high separation factor ($\alpha = 2.8$). Contrary to this, reported block or cross-linked (16) polysiloxane containing similar amounts of siloxane units showed only low α value (2.0). The phase separated structure of these polymers does not seen adequate to achieve high separation factor. Poly(S1) contains 41 wt% of $-Si(CH_3)_3$, and shows lower P_{O_2} but higher α than poly(S2). The content of siloxane seems

important for oxygen permeability, whereas backbone is important for selectivity. Poly(S5) or poly(S8), although having higher molecular weight than poly(S2), which presumably show higher PO_2 than

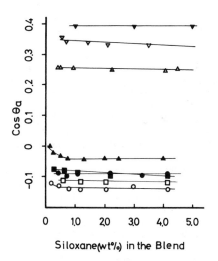

Figure 4.   Contact Angle of Water Droplet at 20°C on Air-side
Surface of Various Siloxane Polymer-polyMMA Blend Films
After Treatment with n-Hexane. (For an explanation of
symbols, and for contact angle values before n-hexane
treatement, see Fig. 1.)

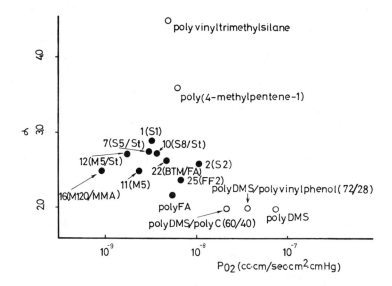

Figure 5.   Oxygen Permeation Behavior of Various Membranes.
Numbers in the Figure correspond to those in Table II.
PolyDMS/Polyvinylphenol (72/28):   cross-linked
polydimethylsiloxane, (16) PolyDMS/PolyC(60/40):
polydimethylsiloxane/polycarbonate block copolymers.(9)

poly(S2), are waxy and not good film forming materials. Longer
siloxane side chains do not seem able to maintain the film forming
property.

In order to impróve the film forming property of poly(S5) and
poly(S8), some copolymers with short siloxane side chains were
synthesized (# 6,7,9,10 in Table II). However, the graft copolymers
which contain similar siloxane percent with poly(S2) are still waxy
and lack in film forming characteristics. The graft copolymers where
siloxane content corresponds to that of poly(S1), showed film
forming properties. However, the separation factor was lower than
poly(S1). These facts may indicate the superiority of polymers of
many short chains over polymers with few long chains for the film
forming property and better oxygen permeation behavior. The graft
copolymers with long polysiloxane side chains showed as low a
separation factor as polydimethylsiloxane. Oxygen permeation may
proceed through the phase separated polysiloxane domains, which
would result in low selectivity.

Polymers of methacrylate type monomers were all waxy or oily
and lacked film forming properties.

As an attempt to improve the selectivity of oxygen permeation,
copolymers with FA, and polymers of fluoroalkyl substituted S2 were
synthesized. The results are also shown in Figure 5. Introduction
of fluoroalkyl groups, at present, does not improve the selectivity
significant.

Surface Modification and Oxygen Permeation. In order to investigate
the effects of surface accumulation of siloxane polymers on
permeation behavior, the permeation through the polystyrene film
containing 1% polysiloxane graft copolymer was studied. The contact
angle and ESCA measurement indicated that the surface of polystyrene
film was completely modified to a siloxane like surface by the
surface accumulated polysiloxane graft copolymer. The addition of
1% polysiloxane graft copolymer to polystyrene resulted in higher
P_{O_2} and lower α which is opposite to our expectation. The addition

of 1% graft copolymers may make the film just as "layered" film.
Much less amount of graft copolymer might have to be used to obtain
high P_{O_2} maintaining high α [α of polyST = 3.6).

Further work is needed to obtain a detailed insight of this
phenomenon.

Literature Cited

1. LeGrand, D. G.; Gaines, G. L. Jr. Polym. Prep. 1970, 11, 442;
   G. L. Gainès, Macromolecules, 14(1), 208 (1981).
2. Azrak, R. G. Colloid. Interfac. Sci. 1974, 47, 779.
3. Pennings, J. F. M.; Bosman, B. Colloid. Polym. Sci. 1980, 258,
   1109.
4. Lee, L. H. ed.: Characterization of Metal and Polymer
   Surfaces, Academic: New York, 1977: Vol. 1,2.
5. Yamashita, Y. J. Appl. Polym. Sci. Appl. Polym. Symp. 1981,
   36, 198.

6. Kawakami, Y.; Murthy, R. A. N.; Yamashita, Y. Polym. Bull.
    1983, 10, 368.
7. Kawakami, Y.; Murthy, R. A. N.; Yamashita, Makromol. Chem.
    1983, 184, in press.
8. Yamashita, Y.; Tsukahara, Y. Ito, H. Polym. Bull. 1982, 7,
    289.
9. Ward, W. J., III; Browall, W. R.; Salame, R. M. J. Membr. Sci.
    1982, 7, 289.
10. Luche, J. L.; Pefrier, C.; Gemal, A. L. Zikra, N. J. Org.
    Chem. 1982, 47, 3806.
11. Mason, T. J.; Lorimer, J. P.; Mistry, B. P. Tetrahedron Lett.
    1982, 23, 5363.
12. Han, B-H.; and Bonduouk, P. J. Org. Chem. 1982, 47, 5030.
13. Kawakami, Y.; Miki, Y.; Tsuda, T.; Murthy, R. A. N. Polym. J.
    1982, 11, 913.
14. Ito, K.; Usami, N.; Yamashita, Y. Macromolecules 1980, 13,
    216.
15. Takaki, M.; Asami, R.; Hanahata, H.; Sukenaga, N. Polym. Prep.
    Jpn. 1981, 30, 860.
16. Asakawa, S.; Saito, Y.; Kawahito, M.; Ito, Y.; Tsuchiya, S.;
    Sugata, K. National. Tech. Report 1983, 29, 93.

RECEIVED April 19, 1985

Homopolymerization of Epoxides in the Presence of Fluorinated Carbon Acids
Catalyst Transformations

J. ROBINS and C. YOUNG[*]

Specialty Chemicals Division, 3M Center, St. Paul, MN 55144

Cationic polymerization of epoxides has been a commercially important means for the preparation of low molecular weight polyethers and curing epoxy adhesives for many years.[1] In order to maximize the molecular weight of the formed polymers, the polymerizations are typically run at sub-ambient temperatures in very polar solvents[2]. However, these conditions are not applicable to the curing of polyfunctional epoxy resins in industrial applications either as adhesives or coatings. Curing of epoxy resins with cationic initiators at ambient temperatures typically results in some monomer rearrangement, formation of cyclic oligomers and catalyst decomposition in addition to the desired linear epoxy homopolymerization[3]. These side reactions can dramatically affect the ultimate properties of the cured resin, making the use of initiators with large counterions more desirable. Previous work, however, has been focused on the polymer formation and the fate of the catalysts has been somewhat neglected.

We report here on the thermo-kinetic analysis of the homo-polymerizations of several epoxide monomers using a unique class of cationic initiators, bis-trifluoromethanesulfonyl methane and its derivatives. In addition we have included standard Lewis and Bronsted acids, which has allowed us to observe some interesting catalyst transformations.

Bis-trifluoromethanesulfonyl methane, disulfone (DS), is a compound having two trifluoromethanesulfonyl groups attached to a methylene group. Because of the strong electron withdrawing properties of trifluoromethanesulfonyl group, the proton on the methylene group becomes very acidic.

Disulfones with different substituents on the methyl group are listed in Table I. Although the acidity is not fully characterized, we believe it to increase in the following order: methyl disulfone (MDS), disulfone (DS), phenyl disulfone (TDS), bromodisulfone (BrDS), chlorodisulfone (ClDS) and tetrasulfone (TS).

The compounds are generally prepared by Grignard reaction, where two moles of an alkyl magnesium chloride are reacted with two moles of trifluoromethanesulfonyl fluoride in ether. These compounds and their synthesis have been reported by Koshar (8,9) and Mitsch (8).

---

[*] Author to whom correspondence should be directed.

0097–6156/85/0286–0263$06.00/0

These catalysts offer some advantages for studying the structure effects of the counter anion in cationic homopolymerization of epoxides.

Experimental

The epoxides used for study included 1,2-butene oxide, cyclohexene oxide and styrene oxide, which were commercially purchased (Aldrich Chemical Co.). Reagent grade solvents were used: 1,2-dichloroethane (1,2 DEC, MCB), 1,1-dichloroethane (1,1 DCE, Aldrich) and nitrobenzene (Aldrich). Acidic catalysts commercially purchased included CF_3SO_3H (3M Co.) perchloric acid (70% aq., Baker Chem. Co.) HPF_6 (60% aq., Alfa) $HSbF_6 \bullet 6H_2O$ (Alfa HBF$_4$ (48% aq., MCB) CF_3COOH (3M Co.) and PTSA$\bullet H_2O$ (Aldrich).

Fluorinated carbon acids were obtained from Koshar prepared according to previously published procedures (8,9).

Calorimetric probes were done as previously described (10), using a 1-1.5M solution of epoxide in solvent in the presence of a relatively small amount (approximately 10^{-4}M) of catalyst. The reaction was run in an insulated plastic-coated paper cup fitted with a magnetic stirring bar, cover and thermocouple. Temperature-rises were recorded versus time using a Hewlett-Packard 7127A strip-chart recorder.

Representative exotherm curves are shown in Figure 1 where four types of catalytic activity are observed:

1. Reaction goes to completion
2. Deactivation that leads to incomplete reaction
3. Induction and then acceleration
4. Poor activity

Results and Discussion

It has been reported previously that bis-trifluoromethane-sulfonyl methane ("disulfone") and its derivatives are good epoxy homo-polymerization catalysts (11,12). Calorimetric studies have shown that 1,1,3,3-tetrakis(trifluoro methanesulfonyl) propane ("tetra-sulfone") is probably the only effective catalyst for homopolymerization of an aliphatic epoxide, e.g. butene oxide, in a non-polar solvent at room temperature (Figure 2). The other disulfone catalyst along with Bronsted acids are poor catalysts.

On the other hand, cyclohexene oxide is polymerized more readily by a number of acidic catalysts, but some lead to an incomplete reaction (e.g. $BF_3C_2H_5O$) and in general, their activity cannot be correlated to their acid strength, since "disulfones" have a pK_a around -1 (Figure 3).

Styrene oxide presents further complications and is polymerized most readily by catalysts which are poor for homopolymerization of cyclohexene oxide such as triflic acid, but phenyl disulfone, which is an active catalyst for cyclohexene oxide, is surprisingly slow for styrene oxide (Figure 4).

The combination of DS and styrene oxide in 1,2-dichloromethane and 1,1,2-trichloroethane, gave another interesting catalytic

Table I. Bis-trifluoromethanesulfone methane (DS) and its derivatives in the order of increasing acidity.

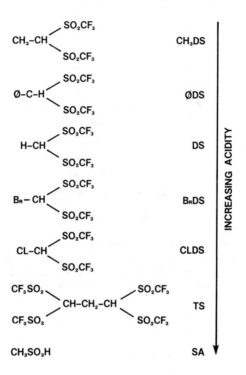

Figure 1. Types of calorimetric behavior exhibited by various catalysts. Reaction conditions: 4-6 mL of an epoxide in 30 mL of solvent in the presence of 1-2 mg of catalyst. Starting temperature 20 °C.

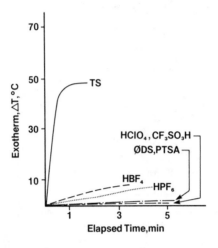

Figure 2. Calorimetric determination of 1,2 butene oxide
reactivity in carbon tetrachloride in the presence of various
acids. Reaction conditions 4.1 mL of butene oxide, 30 mL of CT,
and 50 mg of catalyst. Starting temperature 20°C.

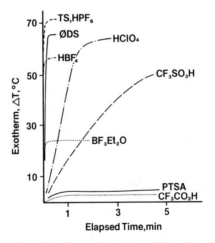

Figure 3. Calorimetric determination of cyclohexene oxide
reactivity in nitrobenzene in the presence of various acids.
Reaction conditions: 5 mL of cyclohexene oxide, 30 mL of
nitrobenzene and 50 mg of catalyst. Starting temperature 20°C.

behavior-induction followed by acceleration (Figure 5). We believe, in this case, the catalyst is being transformed during the reaction, to form a more active catalyst, such as triflic acid.

Table II summarizes the activity of various catalysts with number of different epoxides.

1. BF$_3$ as an example of a Lewis acid gave an incomplete homopolymerization of cyclohexene oxide.

2. Trifluoromethanesulfonic acid as an example of a Bronsted acid was a very active catalyst with styrene oxide, but was a poor catalyst for the butene oxide and cyclohexane oxide at room temperature.

3. Tetrasulfone was the most active catalyst for epoxides studied.

4. DS and MDS, on the other hand, were partially deactivated when cyclohexene oxide was used.

5. Combination of MDS with styrene oxide gave induction followed by acceleration.

We have attempted to account for these observations by proposing the following simplified reaction path (Figure 6). Protonation of the epoxide is followed by the rearrangement of the protonated species to a covalent intermediate or a solvent stabilized ion pair. Homopolymerization then proceeds through the ion-pair intermediate, but the covalent intermediate leads to catalyst transformation and, in most cases, loss of catalytic activity.

The relative stabilities of carbocations and their counterions have been used to explain this behavior (Figure 7).   The least stable carbocation -- protonated butene oxide -- is expected to form covalent intermediates with a large number of counter anions, such as triflate, disulfone and methyl disulfone. However, tetrasulfone being an excellent catalyst, is probably unable to form a covalent intermediate due to internal hydrogen bonding and steric factors (Figure 8).

The most stable carbocation -- protonated styrene oxide -- will form a covalent intermediate only with the most nucleophilic anions such as MDS anion, which exhibits a behavior of induction preceeding activation (Figure 5).

The carbocation of intermediate stability -- protonated cyclohexene oxide -- on the other hand, will form a covalent intermediate with DS catalyst, which leads to deactivation, but not with tetrasulfone, which allows the reaction to go to completion.

One would expect the polarity of the solvent to also have a strong effect on the covalent intermediate formation. Nonpolar solvent should favor the formation of covalent intermediate which would lead to catalyst transformation. A study of the homopolymerization of cyclohexene oxide in solvents of different polarities is shown in Figure 9 using bromodisulfone as a catalyst, and the data shown in Table III illustrate this idea. The extent of the exotherm is dependent on the dielectric constant of these solvents, leading to incomplete reaction in all cases. Even in the most polar solvent, 1,2-dichloroethane, 30% unreacted epoxide was found after the reaction was terminated. In the least polar solvent, carbon

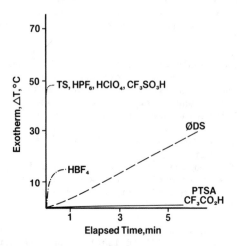

Figure 4. Calorimetric determination of styrene oxide in
1,2 dichloroethane in the presence of various acids. Reaction
conditions: 6 mL of styrene oxide, 50 mL of 1,2 DCE, and 50 mg
of catalyst. Starting temperature 20 °C.

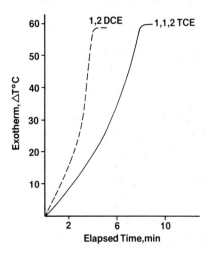

Figure 5. Calorimetric determination of styrene oxide reactivity
in the presence of "methyldisulfone" as a function of solvent.
Note activation as the reaction proceeds. Reaction conditions:
6 mL of styrene oxide, 30g of solvent, and 15 mg of methyl
disulfone. Starting temperature 20 °C.

Table II. Summary of catalyst performance in the presence of various epoxides. CHO–cyclohexene oxide; BO–butene oxide; SO–styrene oxide.

CAT ACTIVITY	BF₃	CF₃SO₃H	TS	DS	MeDS
reaction complete		SO	BO SO CHO		
reaction incomplete cat. deactivation?	CHO			CHO	CHO
induction/acceleration cat. activation?					SO
poor cat.		BO		BO	BO

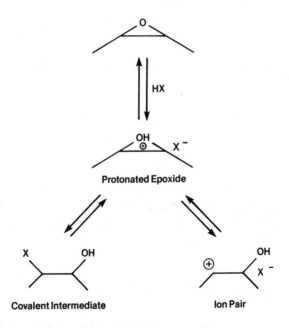

Figure 6. Simplified reaction path for acid catalyzed epoxy systems.

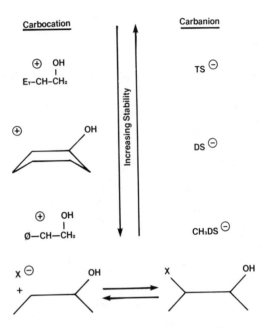

Figure 7. Epoxide-catalyst equilibria as a function of ion stability.

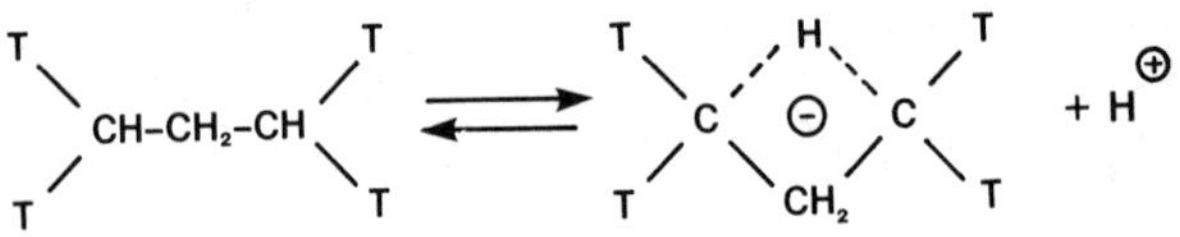

$T = CF_3SO_2^-$

Figure 8. Acid dissociation equilibrium for 1,1,3,3 tetrakis (trifluoromethanesulfonyl) propane (TS).

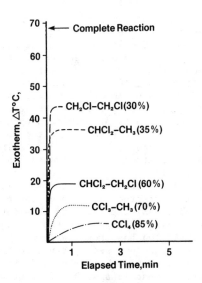

Figure 9. Calorimetric determination of cyclohexene oxide reactivity in the presence of bromodisulfone as a function of solvent. Percent unreacted epoxide after reaction in parenthesis. Reaction conditions: 5 mL of cyclohexene oxide, 30g of solvent, and 15 mg of bromodisulfone.

Table III. Unreacted cyclohexene oxide as a function of catalyst and solvent. Determination done by GC one hour after the addition of catalyst.

Solvent Dielectric C. Catalyst	1,2 DCE 10.5	1,1 DCE 10	1,1,2 TCE	1,1,1 TCE 7.5	CT 2.2
			% Unreacted Epoxide		
TS	0	0	0	0	0
∅DS	0	0	0	0	5
DS	5	20	40	40	70
CH₃DS	15	30	40	50	75
ClDS	15	30	70	50	90
BᵣDS	25	35	60	70	85
DSI	0	20	25	70	80

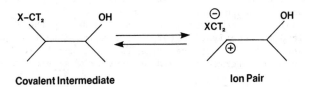

Covalent Intermediate   Ion Pair

tetrachloride, most of the epoxide (85%) was found unreacted after
allowing the reaction to continue for 24 hours.

Table III lists the percentage of the unreacted cyclohexene
oxide for different solvents and different catalysts. Solvents with
higher dielectric constant such as 1,2-dichloromethane (1,2 DCE), or
1,1,-dichloroethane (1,1 DCE), left less unreacted epoxide with all
catalysts studied. Solvents of lower dielectric constant such as
1,1,2 trichloroethane (1,1,2 TCE), 1,1,1-trichloroethane (1,1,1 TCE)
and CCl_4 (CT) left larger amounts of unreacted epoxides. Tetrasulfone (TS) and phenyldisulfone (ODS) exhibited no deactivation.
Disulfone (DS), Methyldisulfone (MeDS), chlorodisulfone (ClDS) and
bromodisulfone (BrDS) were strongly deactivated leading to large
amounts of unreacted epoxides.

These results are in agreement with the speculation that the
deactivation of the catalyst proceeds through the covalent
intermediate, the nonpolar solvent favoring this intermediate
formation, therefore leaving a large amount of unreacted epoxide.

J. B. Hendrickson(13-15) has shown various ways a trifyl group
can be either displaced or decomposed by a nucleophile when attached
to an electron poor site. Hendrickson observed three types of
nucleophilic substitution reactions with different nucleophiles.
This work allows us to propose a mechanism for the deactivation of
disulfone catalysts.

Equation 1 shows the replacement of ditrifyl imide by a bromide
ion. By comparison, Equation 2 shows the replacement of trifluoromethanesulfonyl group on a nitrogen by an alkoxide and Equation 3
demonstrates that a substitution reaction can take place on a
sulfonyl group by a carboxylate to yield the trifluoromethanesulfonyl ester.

The covalent intermediate shown in Figure 10 has a similar
structure except it contains carbon instead of nitrogen. The
following analog substitution reactions on the intermediate are
proposed:

1. The displacement of disulfone (T) by an epoxide, which
   leads to homopolymerization.
2. The displacement of trifyl group by an epoxide leading
   to complete deactivation.
3. A nucleophilic atack on the sulfonyl group by an expoxide
   which leads to trifluoromethanesulfonate ester and a
   partial deactivation and, in the case of styrene oxide,
   to acceleration when catalyzed by methyl disulfone.
4. A nucleophilic attack on the halide by an epoxide leading
   to partial deactivation.

Conclusions

A new family of carbon acids has allowed the investigation of
structure effects and the elucidation of the very important role of
counteranions in homopolymerization of epoxides.

A mechanism for catalyst transformation during the reaction,
which can lead to deactivation (or acceleration) has been proposed.

The most efficient catalyst appears to be unable to form the
proposed covalent intermediate and consequently <u>survived</u> during the

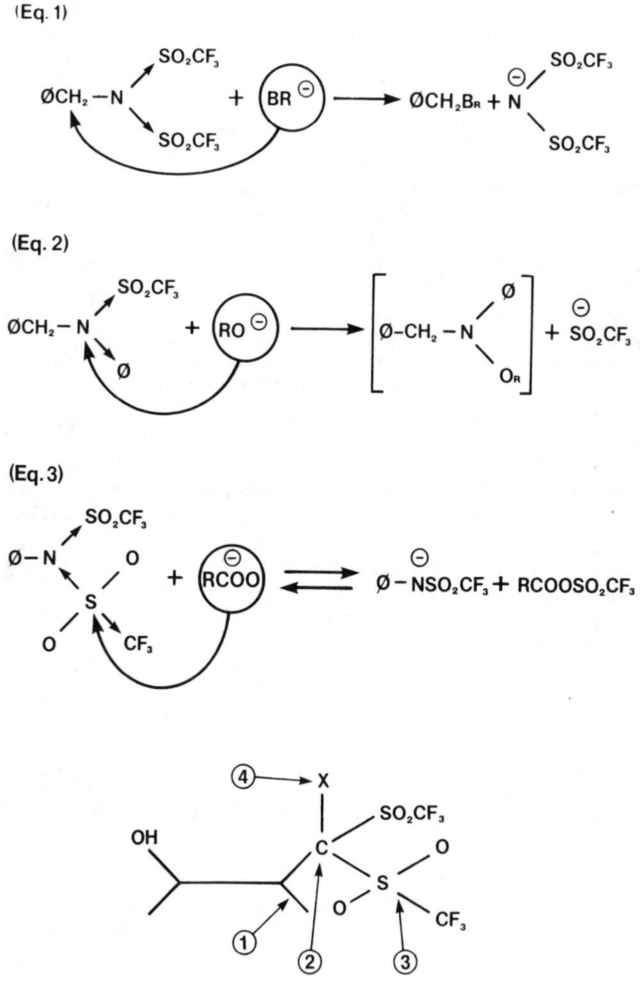

(Eq. 1)

(Eq. 2)

(Eq. 3)

SITE OF ATTACK	RESULTS	LEAVING GROUP
①	Homopolymerization	$XCT_2^{\ominus}$
②	Complete Deactivation	$CF_3SO_2^{\ominus}$
③	Partial Deactivation	$RCXT^{\ominus}$
④	— '' —	$RCT_2^{\ominus}$

Figure 10. Various modes of catalyst transformation via catalyst-epoxide intermediate by the attack of an epoxide molecule.

reaction. Tetrasulfone presents an ideal case with its internal hydrogen bonding and steric hindrance.

Literature Cited

1. Furukawa, J; Saegusa, T. Polymerization of Aldehydes and Oxides, John Wiley and Sons, 1963, p. 147.
2. Dreyfuss, P.; Dreyfuss, M. P. Polym. J., 8, 81 (1976).
3. Kawakami, Y.; Ogawa, A.; Yamashita, Y. J. Polym. Sci., Polym. Chem. Ed., 17, 3785 (1979).
4. Goethals, E. J. in Advances in Polymer Sci., 23, p. 104, Springer, NY 1977.
5. Kondo, S.; Blanchard, L. P. Polymer Letters, 7, 621 (1969).
6. Pasika, W. M. J. Polm. Sci. A3, (12), 4287 (1965).
7. Colclough, R. O.; Gee, G.; Higginson, W. C. E.; Jackson, J. B.; Litt, M. J. Polym. Sci. 34, 1971 (1959).
8. Koshar, R. J.; Mitsch, R. A. J. Org. Chem., 38, 3358 (1973).
9. Koshar, R. J. U.S. Patent No. 4,053,519, (1977).
10. Robins, J. J. Appl. Polym. Sci. 9, 821 (1965).
11. Allen, M. G. U.S. Patent No. 3,632,843, (1972).
12. Robins, J.; Kropp, J. E.; Young, C. I. U.S. Patent No. 4,115,295, (1978).
13. Hendrickson, J. B.; Bergerson, R. Tetrahedron Let. 4607 (1973).
14. Hendrickson, J. B.; Bergerson, R.; Giga Aziz; Sternback, D. J. Amer. Chem. Soc., 95, 3412 (1973).
15. Hendrickson, J. B.; Giga, A.; Wareing, J. J. Amer. Chem. Soc., 96, 2275 (1974).

RECEIVED April 9, 1985

Mechanism of Ring-Opening Polymerization of Bicycloalkenes by Metathesis Catalysts

HARRIET E. ARDILL, RUTH M. E. GREENE, JAMES G. HAMILTON, H. THOI HO, KENNETH J. IVIN, GRZEGORZ LAPIENIS, G. MALACHY McCANN, and JOHN J. ROONEY

Department of Chemistry, The Queen's University of Belfast, BT9 5AG, United Kingdom

Work on the microstructure of the ring-opened polymers of norbornene, ten of its alkyl derivatives, and some other bicycloalkenes is reviewed. The cis double bond content ranges from 0% (with $RuCl_3$) to 100% (with $ReCl_5$); the cis/trans double bond distribution ranges from statistical to blocky; the monomers without a plane of symmetry give polymers with a range of head-tail bias, from random orientation to total bias; and the ring dyad tacticity may range from atactic to fully tactic in the sense that cis and trans double bonds are associated with r and m dyads respectively. The results are discussed in terms of propagating metal carbene complexes which may or may not have a chiral reaction site and in which the stereochemistry of the previously incorporated unit, attached to the carbene ligand, may influence the next propagation step.

It is now well established that ring-opening polymerization of cycloalkenes and bicycloalkenes, initiated with olefin metathesis catalysts, is propagated by metal carbene complexes (1).

The ring-opened polymers of bicycloalkenes have a structure which is an alternating sequence of double bonds and enchained rings, as illustrated below for the polymerization of norbornene, $\underline{1}$.

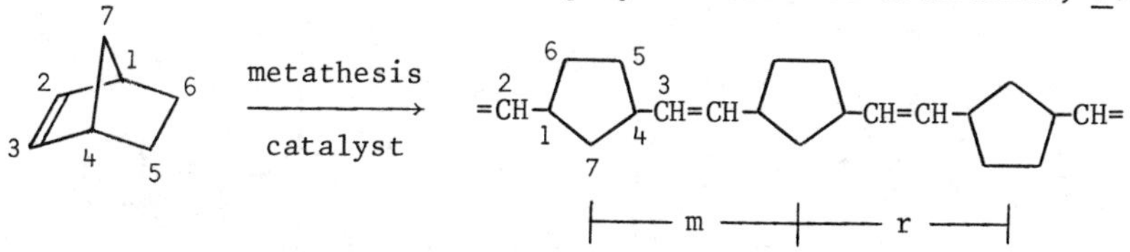

$$\underline{1} \qquad\qquad \underline{1P}$$

The structure of such a polymer is characterized by four features. First, the main-chain bonds attached to a given C_5 ring always have a cis relationship; this is a consequence of the fact that ring-opening occurs by complete rupture of the double bond.

0097–6156/85/0286–0275$06.00/0

Second, the double bonds in the polymer may be cis or trans; the
double bond pair sequences may therefore be cc, ct or tt. Third,
the rings may have one of two configurations; ring dyads may therefore be m (isotactic) or r (syndiotactic) as represented in 1P.
Fourth, when the monomer is substituted in such a way that it does
not have a plane of symmetry, there is the possibility of head-head
(HH), head-tail (HT) or tail-tail (TT) structures in the polymer.

The proportions of these various sequences provide important
information about the mechanism of the propagation reaction and the
nature of the propagating species. The cis content (σ_c), the proportions of cc, ct and tt pairs (2), and the proportions of HH, HT
and TT dyads (3) are readily determined from the ^{13}C n.m.r. spectra
of the polymers, but the ring dyad tacticity is not always so easily
found. For example, there is no fine structure in the ^{13}C n.m.r.
spectrum of 1P that can be attributed to m/r splitting. This problem was originally tackled through the use of enantiomers of
5-substituted norbornenes (4,5). In the polymer of such a monomer
an isotactic dyad is also head-tail (HT) while a syndiotactic dyad
is either head-head (HH) or tail-tail (TT). Such structures give
distinctive chemical shifts for the olefinic carbons which permits
the tacticity to be determined for dyads embracing both cis and
trans double bonds. Recently we have found that 7-substituted (6)
and certain 5,6-disubstituted norbornenes give polymers whose tacticity can be determined directly from their ^{13}C n.m.r. spectra. In
this paper we review our work in this area over the past few years
and present some new data.

Mechanistic Framework

Propagation is assumed to occur by a [2+2] reaction between the
metal carbene complex and the monomer. The approach required for
the formation of a cis double bond is illustrated in Scheme 1. The
cis double bond is formed at b following the rupture of the transition state at bonds a and c. For the formation of a trans double
bond the norbornene molecule must approach Mt=C with C^7 at the back
instead of at the front.

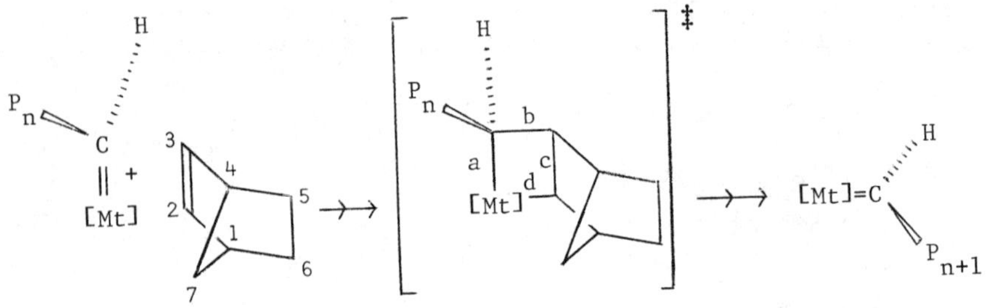

Scheme 1

The metal carbene complex [Mt]=CHP$_{n+1}$ has the formal structure
2 where n denotes the number of monomer units already added.

$$[Mt]=CH-\underset{7}{\overset{6\quad 5}{\underset{2\quad 1}{\bigtriangleup}}}\overset{3}{\underset{4}{}}-CH=CH-P_n \qquad\qquad \underline{2}$$

We shall see later that the reactivity of $\underline{2}$ and the stereochemistry of the next propagation step are governed by the following factors:
(1) whether Mt is electron-poor, in which case $\underline{2}$ may have a conformation P_c in which a cis C=C bond may remain complexed to the [Mt]=C unit;
(2) whether the C=C bond is cis or trans;
(3) whether the [Mt]=C bond is polar or non-polar;
(4) whether there is a substituent at C^1 or C^4 and sometimes whether there are substituents at C^5 or C^6;
(5) the geometry of the ligands around Mt and in particular whether the partial structure [Mt]=CH- has mirror-image forms;
(6) whether C^1 has R or S configuration.

For a complete description of a propagating species we would need to define the situation with regard to many of these factors and the symbolism would become very complex. However we shall find it sufficient to use one or two suffixes or subscripts to define the particular factors under consideration and will introduce these as necessary.

In what follows we first review the cis content of the polymers formed over a range of catalyst/monomer systems, then consider the cis/trans double bond distribution, next the HT bias that sometimes occurs with asymmetric monomers, and finally we collect and discuss the evidence on tacticity in polymers formed from various monomers.

Cis Content

The cis contents of polymers made from eight monomers having a plane of symmetry are listed in Table I. Those of polymers made from six asymmetric monomers are given in Table II.

While there is a general pattern of increasing σ_c as one proceeds down the tables for each monomer and approximate constancy as one proceeds across the tables, there are some interesting exceptions. Thus with $RuCl_3$ the norbornene derivatives always give high-trans polymers ($\sigma_c = 0$-0.1) but for $\underline{8}$ and $\underline{7}$ σ_c rises to 0.25 and 0.5 respectively, presumably because of increased strain in, and therefore reactivity of, the monomer, thereby reducing the discrimination between the two [2+2] modes of addition leading to cis and trans double bond formation respectively.

Conversely with $\underline{4}$ and $\underline{9}$, using $(mes)W(CO)_3/EtAlCl_2/EPO$ as catalyst, the cis content is markedly lower than for the other polymers. This may be attributed to the increased hindrance to cis double bond formation caused by the syn-methyl group in $\underline{4}$ and the six-membered ring in $\underline{9}$. This catalyst is in fact the only one that will polymerize $\underline{4}$ and $\underline{9}$ and in the latter case the additional presence of Me_4Sn is necessary to suppress side reactions. The propagation of the polymerization of the syn compound $\underline{4}$ will clearly be much more hindered than that of the anti compound $\underline{3}$ if it is required to present its exo face to the metal carbene complex (Scheme 1). Hence it is not surprising that with most catalyst

Table I. Fraction of cis double bonds,[a] σ_c, in ring-opened polymers of some symmetrically substituted monomers (1-9)

Monomer Catalyst[b]	1	3	4	5	6	7	8	9
RuCl$_3$	< 0.05	0.0		0.1	0.1	0.5	0.25	
IrCl$_3$	0.45	0.25						
OsCl$_3$	0.5	0.15		0.15	0.55			0.25
(mes)W(CO)$_3$/EtAlCl$_2$[c]	0.45	0.45	0.15	0.5				
Ru–TFA[d]	0.55	0.55		0.4				
WCl$_6$/Ph$_4$Sn	0.55	0.6			0.55			
WCl$_6$/Me$_4$Sn	0.55	0.55		0.8	0.7			
WCl$_6$/Bu$_4$Sn	0.7	0.5		0.95				
ReCl$_5$	1.0	0.95		1.0	> 0.95			

[a] Values have been rounded to the nearest 0.05. Variations of up to ± 0.1 are observed with some catalyst systems depending on the reaction conditions.

[b] In chlorobenzene at 20°C, except for the noble metal catalysts which generally require heating to 60-90°C and addition of ethanol to increase catalyst solubility.

[c] mes denotes mesitylene; an epoxide is generally included with this catalyst system to enhance reactivity.

[d] Ru–TFA Ruthenium trifluoroacetate complex having the approximate formula Ru$_2$(TFA)$_4$.

Table II. Fraction of cis double bonds,[a] σ_c, in ring-opened polymers of some asymmetric monomers (3-5, 9-12)

Catalyst[b]	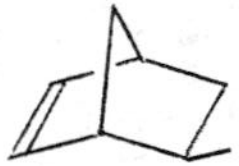 10	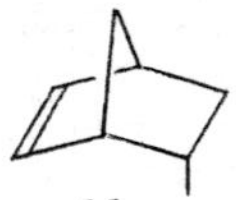 11	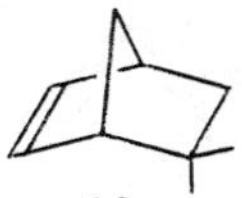 12	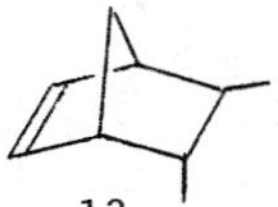 13	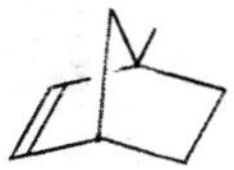 14	 15
RuCl$_3$		0.0	< 0.05	0.0	0.0	0.0
IrCl$_3$			0.25	0.1	0.2	< 0.05
OsCl$_3$			0.3		0.1	
(mes)W(CO)$_3$/EtAlCl$_2$[c]			0.25	0.3	0.3	
WCl$_6$/EtAlCl$_2$	0.5	0.5	0.6		0.5	
Ru-TFA[d]	0.6	0.5e	0.65	< 0.05	0.6	
WCl$_6$/Ph$_4$Sn			0.65		0.75	
WCl$_6$/Me$_4$Sn			0.7		0.5	
WCl$_6$/Bu$_4$Sn	0.75	0.65	0.7	0.3	0.5	0.9
ReCl$_5$	1.0	1.0	1.0	f	1.0	

a,b,c,d See corresponding footnotes to Table I.
e The endo-5-chloromethyl compound, however, gives σ_c = 0 (and no HT bias).
f No ring-opened polymer obtained.

systems 4 is totally unreactive. In the case of 9 reaction necessarily occurs at the steric equivalent of the endo face of 3 but it
must be remembered that while the electron density about the double
bond in 9 is symmetrical, in 4 it is biassed towards the exo face
(13). Hence in the one case where 4 does polymerize it is still
likely that it reacts at the exo face.

σ_c for WCl_6/R_4Sn-initiated systems shows some dependence on R
indicating that either R_4Sn or one of its fragments must be present
as a permanent ligand. The variation of σ_c with monomer in the case
of WCl_6/Bu_4Sn ($\sigma_c = 0.3$–0.95) is also to be noted. The lower σ_c for
polymers of 13 relative to those of 5 and 6 is perhaps indicative of
faster relaxation of P_c to the decoordinated form $P_{(c)}$ which then
favours formation of a trans double bond at the next addition.

In the case of $ReCl_5$ the high cis content is thought to stem
from a relatively low electron density at the [Re] centre, causing
the previously formed cis double bond to be held as a weak donor
ligand rather firmly at the Re site. This in turn has a strong
cis-directing influence on the addition of the next monomer molecule.
This concept is also used to explain the cis/trans blockiness in
high-cis polymers, as described in the next section. The higher σ_c
values generally obtained with Ru-TFA as catalyst relative to those
with $RuCl_3$ as catalyst is attributed to a reduction in electron
density at [Ru] due to the presence of hard TFA ligands. However
the low value of σ_c for the polymer of 13 indicates some other over-
riding factor, perhaps a steric effect favouring relaxation of P_c
to $P_{(c)}$.

Cis/Trans Distribution; Blockiness

In assessing data on the proportions of cc, ct and tt pairs it is
first necessary to ensure that the cis content has not been modified
by the occurrence of secondary metathesis reactions. These can
occur extremely rapidly with some polymer/catalyst systems. For
example a polymer of cycloocta-1,4-diene initially contains no tt
junctions, alternate double bonds in the polymer chain being derived
from preformed cis double bonds in the monomer; but under certain
catalyst conditions tt junctions appear within a matter of seconds
by secondary metathesis (14,15). These problems do not appear to be
serious in polymers of norbornene derivatives presumably because the
double bonds in the polymer are protected from secondary metathesis
by the adjacent C_5 rings and cannot compete when highly reactive
monomer is still present.

In polymers of norbornene it is observed (2) that the cis/trans
distribution is blocky at high cis content but statistical at low
cis content ($\sigma_c < 0.35$). At high cis content it is therefore
necessary to distinguish between a propagating species in which the
last double bond formed was cis (P_c) and one in which the last
double bond formed was trans (P_t). Four propagation steps may then
be distinguished as shown in Scheme 2. The parameters r_c and r_t
are readily obtained from the intensities of the ^{13}C peaks characteristic of cc, ct($\equiv$tc) and tt pairs: r_c = cc/ct, r_t = tt/tc. At
low cis content $r_t r_c = 1$, i.e. $k_{cc}/k_{ct} = k_{tc}/k_{tt}$, and there is no
kinetic distinction between P_c and P_t. At higher cis content $r_t r_c$
is always greater than unity, sometimes reaching values of 10 or

more. Some recent data for polymers of $\underline{5}$ and $\underline{6}$ are shown in Table III.

$$P_c + M \longrightarrow P_c \quad k_{cc}$$
$$r_c = \frac{k_{cc}}{k_{ct}}$$
$$P_c + M \longrightarrow P_t \quad k_{ct}$$

$$P_t + M \longrightarrow P_t \quad k_{tt}$$
$$r_t = \frac{k_{tt}}{k_{tc}}$$
$$P_t + M \longrightarrow P_c \quad k_{tc}$$

Scheme 2.

Values of r_t never fall significantly below unity even at high cis content, whereas values of r_c are only greater than unity when σ_c exceeds a certain value. The essential difference between systems giving a statistical cis/trans distribution and those giving blocky distributions is thought to lie in the nature of P_c. It is postulated that when the metal centre is relatively electron-rich the newly formed C=C bond rapidly detaches itself from the site of reaction regardless of whether it is cis or trans, but when the metal centre is relatively electron-poor there is a tendency for a newly formed cis double bond to remain in the vicinity of the reaction site and to influence the approach of the next monomer molecule in such a way as to favour the formation of another cis double bond at the next propagation step. This situation is represented below, $P_{(c)}$ and $P_{(t)}$ being kinetically identical, but P_c distinct.

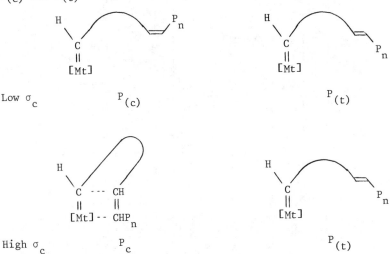

The intramolecular coordination, represented by the dashed lines, is expected to be more favorable for P_c than for P_t, for both steric and polar reasons.

The question arises as to whether P_c achieves equilibrium with $P_{(c)}$ before each propagation step. If so there will be no

Table III. r_c and r_t values[a] for polymers of 5 and 6[b]

Catalyst system	Polymers of 5				Polymers of 6			
	σ_c	r_c	r_t	$r_c r_t$	σ_c	r_c	r_t	$r_c r_t$
RuCl$_3$	0.11	c	10		0.09	c	9.4	
OsCl$_3$	0.15	c	6.8					
(mes)W(CO)$_3$/EtAlCl$_2$/epoxide	0.51	2.3	2.2	5.1	0.50	1.9	1.9	3.6
					0.51d	1.9	1.8	3.4
Ru–TFA	0.40	1.4	2.6	3.6	0.53	1.6	1.3	2.1
WCl$_6$/Me$_4$Sn	0.80	6.6	0.95	6.3	0.72	5.2	1.4	7.3
WCl$_6$/Bu$_4$Sn	0.94	23	e					

[a] Determined from C^7 peaks; polymer of 5: cc 39.71, ct≡tc 38.51, tt 37.32 ppm; polymer of 6: cc 41.74, ct≡tc 41.15, tt 40.49 ppm.

[b] See Table I for monomers and conditions of polymerization.

[c] cc peak too small to measure accurately.

[d] Monomer concentration 1.4M compared with 2.7M for the entry on the line above.

[e] tt peak too small to measure accurately.

dependence of r_c on [M]. If, however, there is a relaxation process occurring on the same time scale as the propagation reaction one should be able to observe a dependence of r_c on [M]. Some data for the polymerization of norbornene are shown in Figure 1. r_c and σ_c do fall off below [M] = 2, but at lower concentration r_t also changes (16). The effect at [M] < 0.5 is attributed to a second relaxation process, involving reorganisation of the ligand geometry for both $P_{(c)}$ and $P_{(t)}$, and represented as $P \rightarrow P'$. The geometry of P is likely to be that immediately derived from P_c or P_t, with a vacancy in place of the last-formed double bond, while P' will have the relaxed geometry normally expected of complexes with a coordination number reduced by one, e.g. penta- from hexa-.

In the polymerization of 3 using the same catalyst system the values of σ_c, r_t and r_c for the polymer prepared at [M] = 2–3M are 0.55, 1.3 and 0.9 respectively. These values are quite similar to those for the polymer of norbornene made under the same conditions. Furthermore, copolymerization studies, Table IV, show that norbornene and 3 have similar reactivities. We shall see later that polymers of 3 made under these conditions are atactic with respect to both cis and trans double bonds. It may therefore be assumed that polymers of norbornene made with this catalyst at monomer concentrations of 2–3M are also likely to be atactic. Hence the relaxation process observed at 0.6–0.1M (Figure 1) must involve either an epimerized form of P going to an achiral form or one achiral form going to another of different ligand geometry.

Table IV.   Copolymerization of norbornene (M_1) with anti-7-methylnorbornene (M_2)[a,b]

Catalyst system	Temp ($^{\circ}$C)	σ_c[c]	r_1	r_2
ReCl$_5$	20	1.0	1.25	0.75
WCl$_6$/Ph$_4$Sn[d]	20	0.5	1.2	0.8
WCl$_6$/Me$_4$Sn	20	0.55	1.0	1.0
WCl$_6$/Bu$_4$Sn	20	0.4	1.6	0.4
(mes)W(CO)$_3$/EtAlCl$_2$/EPO	20	0.5	1.7	0.6
RuCl$_3$	62	0.0	1.4	1.0

[a] Using a 1:1 mixture of syn and anti isomers. The anti isomer is copolymerized; the syn isomer acts only as a diluent.
[b] Copolymers analysed by ^{13}C n.m.r. using the upfield region to obtain F_1 and the olefinic region to obtain the proportions of M_1M_1, M_1M_2 and M_2M_2 dyads, from which single point values can be obtained (17).
[c] Value of σ_c was practically the same for M_1M_1, M_1M_2 and M_2M_2 dyads.
[d] Three experiments at different feed composition gave values of r reproducible to ± 0.2. This gives an indication of the reliability of the other single point values.

For the polymerization of 6 by (mes)W(CO)$_3$/EtAlCl$_2$/EPO (Table III) there is no detectable relaxation process in the range [M] = 2.7 to 1.4M, but the blockiness is maintained indicating propagation by P$_c$ and P$_{(t)}$. In contrast the polymerization of 14 by OsCl$_3$ gives a very blocky polymer of low cis content (σ_c = 0.13) when made at [M] = 1.5M, but an all-trans polymer when made at [M] = 0.2M, indicating both the occurrence of a relaxation process and a marked difference between r$_c$ and 1/r$_t$.

Head-tail Bias in Polymers of Asymmetric Monomers

The norbornene derivatives in Table II do not have a plane of symmetry and exist in optically active forms in which the double bonds carry a small dipole. The polymers made from racemic monomers of this kind may contain three types of dyad depending on the relative orientation of the substituents, as illustrated in 13P and 14P, and each of these has m and r forms.

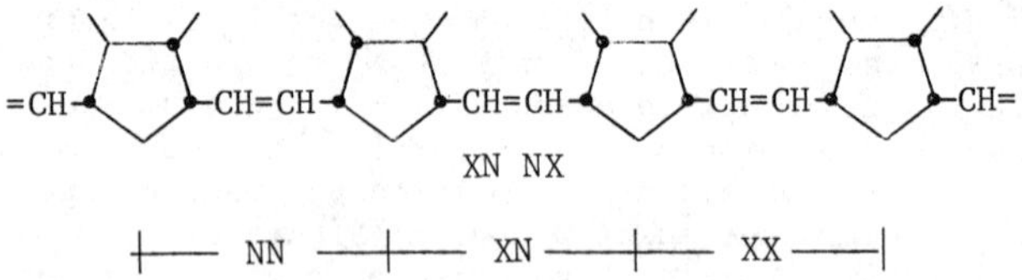

XN NX

|——— NN ———|——— XN ———|——— XX ———|

13P

TH HT

|——— HH ———|——— TH ———|——— TT ———|

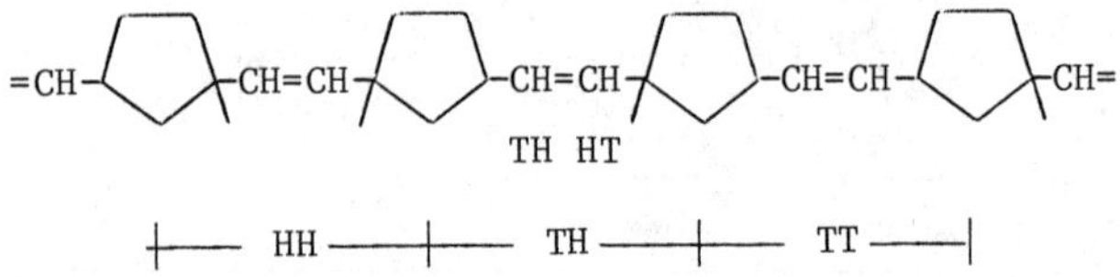

14P

Polymers of 10, 11 and 12 show relatively little HT bias but a complete range of bias has been observed in polymers of 13, 14 and 15, the effect with a given catalyst being greater for 15 than for 14 (3,8). Thus (HT + TH)/(HH + TT) ratios in the high-trans polymers made with RuCl$_3$ and IrCl$_3$ as catalysts are 2.4 and ∞ (fully biased) respectively for 15, and 1.0 (unbiased) and 3.0 respectively for 14.

Polymers of 14 containing both cis and trans double bonds can be made with other catalysts but the cis double bonds are never found within HH dyads. An example of the ^{13}C n.m.r. spectrum of the olefinic carbons in such a polymer (σ_c = 0.48) is shown in Figure 2. The trans HH signal is equal to the sum of the trans TT and cis TT signals and there is no resonance that can be ascribed to cis HH. The fine structure is due to sequences of three double bonds ccc, cct,tct etc. and the asymmetry in the HT and TH triplets is a consequence of the absence of cis HH structures. The all-cis polymer, made with ReCl$_5$ as catalyst, has no choice but to be fully biased (3,12). Cis HH structures do not form because of the severe steric

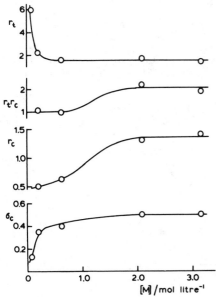

Figure 1. σ_c, r_t, r_c and $r_t r_c$ for the polymerization of norbornene at different monomer concentrations by WCl_6/Me_4Sn.

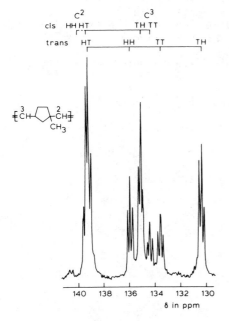

Figure 2. Olefinic region of the ^{13}C n.m.r. spectrum of a cis/trans polymer of <u>14</u> (σ_c = 0.48) made using $Mo_2(OAc)_4/EtAlCl_2$ as catalyst; (HT + TH)/(HH + TT) = 2.8. Reproduced with permission from Ref. 3. Copyright 1984, British Polymer Journal.

interaction that would be involved in their formation, resulting
from the proximity of two quaternary carbons (C^1).

The overall bias in polymers of low or intermediate cis content
may reflect to some extent the polarity of the Mt=C bond, though
steric factors may also be involved. If the carbene carbon has
cationic character the polymerization of 14 should propagate mainly
through P_H rather than P_T in accordance with the [2+2] interaction
illustrated in Equation 1.

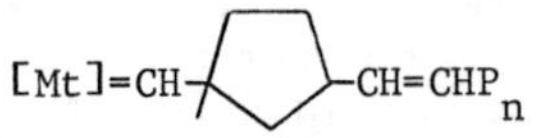

$$[Mt]=CH \diagdown CH=CHP_n \qquad (1)$$

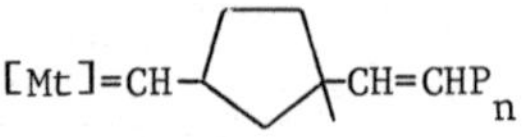

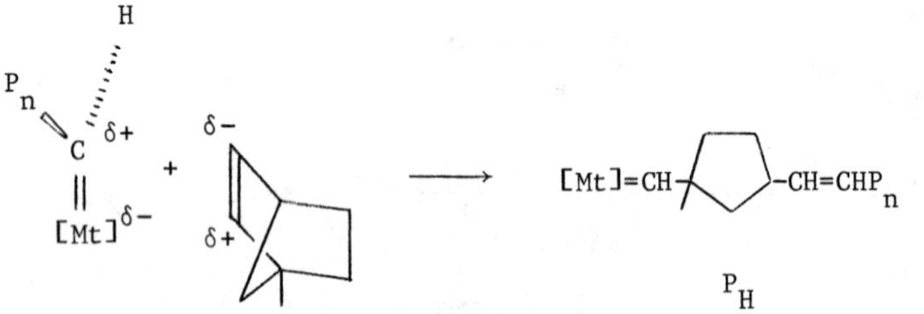

Thus the lack of bias in polymers of 14 made with $RuCl_3$ or
$W(=CHBu^t)(OCH_2Bu^t)_2Br_2/GaBr_3$ ($\sigma_c = 0.\overline{16}$) is attributed to a very
weak [Mt]=C dipole (3). In contrast the Ru-TFA complex gives a
strongly biased polymer $\left(\sigma_c = 0.5, (HT + TH)/(HH + TT) = 4.8\right)$
showing that the hard TFA ligand has not only reduced the electron
density of the ruthenium ion but has also increased the [Ru]=C
dipole.

For relatively unbiased systems the mechanism must be written
in terms of the seven steps shown in Scheme 3.

P_H + M	$\longrightarrow$	P_H	cis HT
P_H + M	$\longrightarrow$	P_H	trans HT
P_H + M	$\longrightarrow$	P_T	trans HH
P_T + M	$\longrightarrow$	P_H	cis TT
P_T + M	$\longrightarrow$	P_H	trans TT
P_T + M	$\longrightarrow$	P_T	cis HT
P_T + M	$\longrightarrow$	P_T	trans HT

Scheme 3

In this case the proportions of P_H and P_T in the reacting system should be comparable. This has been confirmed (18) by direct observation of the ^1H n.m.r. spectrum of the carbene proton in the reaction catalysed by $W(=CHBu^t)(OCH_2Bu^t)_2Br_2/GaBr_3$: P_H singlet δ 11.37, P_T doublet δ 12.11.

In more biased systems, such as that shown in Figure 2, detailed consideration of the fine structure both in the olefinic region and the upfield region, coupled with polarity considerations as in Equation 1, lead one to the conclusion that the last two steps in Scheme 3 are relatively insignificant. The polymerization may then be thought of as propagating largely through P_H (first two reactions) with an occasional reversal to form trans HH (third reaction), followed immediately by the fourth or fifth reactions so converting back to P_H.

With $IrCl_3$ as catalyst for the polymerization of 14 the bias increases from $(HT + TH)/(HH + TT) = 2$ to 3.6 as [M] is reduced from 6.2M to 0.44M, while σ_c falls from 0.28 to 0.14. This is interpreted in terms of relaxation processes $(P_c \rightarrow P \rightarrow P')$ leading to less reactive and therefore more discriminating propagating species.

Tacticity

Tacticities may be determined directly from ^{13}C n.m.r. spectra of polymers of a number of norbornene derivatives and their hydrogenated analogues. 3 is the best monomer for this purpose (6), an example of a spectrum of an all-trans atactic polymer being shown in Figure 3. In cis/trans polymers of 3 both $(\sigma_m)_t$ and, with less accuracy, $(\sigma_r)_c$ can be found; see Table V. The overall fraction of m dyads can be checked against the spectrum of the hydrogenated polymer, from which no less than four values of σ_m can be found; see Figure 4. In the spectra of polymers of 5 and 13 there is a certain amount of overlap of fine structure which prevents the determination of accurate tacticity values except in certain limiting cases (high-cis, high-trans and high tacticity).

In the earlier work (10) with monomer 12, using a more extensive range of catalysts than shown in Table V, it was found that the tacticities could be divided into four types: I, $(\sigma_r)_c = (\sigma_m)_t \sim 1$ (e.g. polymers made with $ReCl_5$ or $(mes)W(CO)_3/EtAlCl_2$ as catalysts); II, $(\sigma_r)_c \sim (\sigma_m)_t \sim 0.6 - 0.9$; III, $(\sigma_r)_c > (\sigma_m)_t$ (e.g. polymer made with $OsCl_3$ as catalyst); IV, $(\sigma_r)_c = (\sigma_m)_t \sim 0.5$ (e.g. polymers made with $RuCl_3$ or WCl_6/Ph_4Sn as catalyst). Most of the results with the other monomers fit into one or other of these categories. Occasionally we have found a fifth type: V, $(\sigma_r)_c$ and/or $(\sigma_m)_t < 0.5$; for example 3 polymerized by $IrCl_3$ gives a polymer with $\sigma_c = 0.26$, $(\sigma_m)_t = 0.64$, $(\sigma_r)_c = 0.32$; σ_c falls with increasing dilution but the tacticities are not markedly affected.

Before discussing mechanisms appropriate to some individual catalyst systems we may note that early indications (4) that there was a correlation between tacticity and cis content have not been borne out by wider studies. Thus high-cis polymers may be fully syndiotactic ($ReCl_5$ with all monomers) or nearly atactic (WCl_6/Bu_4Sn with 5), while high-trans polymers may be fully tactic (19) ($(mes)W(CO)_3/EtAlCl_2$ with 12 and 13) or nearly atactic ($RuCl_3$ with all monomers). In tactic polymers cis double bonds are always

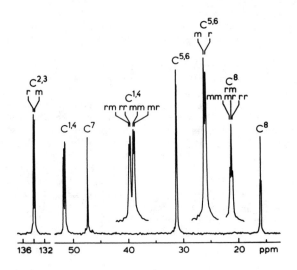

Figure 3. ^{13}C n.m.r. spectrum of an all-trans atactic polymer of <u>3</u>.

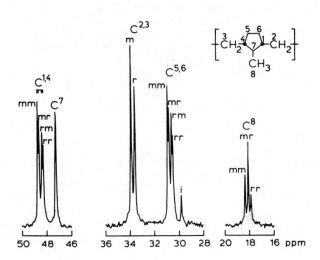

Figure 4. ^{13}C n.m.r. spectrum of a hydrogenated polymer of <u>3</u> (σ_m = 0.54). i is an impurity peak.

Table V. Dyad tacticities[a] for polymers of a series of norbornene derivatives

Catalyst	Monomer[b] 10[c]		12[c]		5[d]		13		14		3	
	$(\sigma_r)_c$	$(\sigma_m)_t$	$(\sigma_r)_c$	$(\sigma_m)_t$	$(\sigma_r)_c$	$(\sigma_m)_t$	$(\sigma_r)_c$	$(\sigma_m)_t$	$(\sigma_r)_c$	$(\sigma_m)_t$	$(\sigma_r)_c$	$(\sigma_m)_t$
RuCl$_3$			-	0.55	-	0.5	-	0.6	-	~0.5	-	0.53
OsCl$_3$			0.75	0.6		0.6				~0.5	~0.4	0.56
IrCl$_3$											0.32	0.64
(mes)W(CO)$_3$/EtAlCl$_2$		~0.5	~1	~1	~1	~1					1.00	1.00
WCl$_6$/EtAlCl$_2$	~0.5		0.6	0.6								
Ru–TFA							-	0.6			1.0	0.44
WCl$_6$/Ph$_4$Sn			0.6	0.55	0.5						0.44	0.50
WCl$_6$/Me$_4$Sn			0.7	0.6	0.6						0.5	0.50
WCl$_6$/Bu$_4$Sn	<1	<1	1.0	1.0	1.0						0.5	0.59
ReCl$_5$	1.0		1.0	-	1.0	-			1.0		1.0	~1

[a] $(\sigma_r)_c$ = fraction of cis double bonds bounded by r dyads; similarly for $(\sigma_m)_t$. Dashes indicate that the cis or trans content is too low to allow accurate measurement; cf. Tables I and II.
[b] See Tables I and II.
[c] Determined by the optical isomer method (4, 5, 10).
[d] Where only one of the pair of values is given the overlap of fine structure makes the determination of the other value too difficult.

associated with r dyads and trans double bonds with m dyads as predicted from Scheme 1. This means that in these cases propagation
occurs alternately at two adjacent ligand positions at a given chiral
site.

ReCl$_5$-initiation. All monomers give high-cis syndiotactic polymers
with this catalyst. It seems likely that the propagating species
are mainly of the type P_c, with r_c very large. The polymer of 14 is
not only syndiotactic but also all-HT. It therefore has the structure 14P' and is an alternating copolymer of the two enantiomers.
This remarkable fact is the inevitable consequence of propagation
through sites of alternating chirality, coupled with a strong preference for the formation of cis double bonds and an inability to form
cis HH dyads (12).

$$=CH-\text{[}\triangle\text{]}-CH=CH-\text{[pentagon]}-CH=CH-\text{[pentagon]}-CH=CH-\text{[pentagon]}-CH=$$

14P'

(mes)W(CO)$_3$/EtAlCl$_2$/epoxide-initiation. Polymers made with this
catalyst have σ_c = 0.15 - 0.55 but are fully tactic (cis r, trans m).
Dilution of the monomer sometimes results in a reduction of σ_c
(monomer 12) (19) and sometimes not (monomer 6). For monomer 6 the
results can be accounted for in terms of one chiral form for each of
P_c and P_t, with $r_t r_c \neq 1$. In the case of monomer 12 it is necessary
to postulate a relaxation process to account for the change of σ_c
with dilution, the relaxed form still being chiral.

RuCl$_3$-initiation. In this case the polymers are always high-trans
and nearly atactic, with $(\sigma_m)_t$ falling from 0.56 to 0.5 as [M] is
reduced from 3.3 to 0.2 mol litre^{-1} in the case of 12. This may be
accounted for in terms of either epimerization of P_t or of relaxation to an achiral form of P_t, with high r_t for both species.

IrCl$_3$ initiation. The unexpected isotactic bias on cis dyads in
polymers of 3 is interpreted in terms of control of tacticity by the
R or S configuration at the ring carbon attached to the carbene
ligand. Polymers of 1 show no cis/trans blockiness even though σ_c
falls from 0.4 to 0.2 with dilution of monomer (2). This shows that
P_c is not involved but that relaxation from P to P' is important for
both 3 and 1.

Ru-TFA-initiation. Polymerization of 3 with this initiator is a
prime example of type III: $(\sigma_r)_c$ = 1.0, $(\sigma_m)_t$ = 0.44. The polymer
is also quite blocky: r_t = 2.1, r_c = 3.2. Here it is necessary to
assume that cis double bonds are formed largely by addition of monomer to a chiral species, P_c, with high r_c, and that trans double
bonds are formed largely by addition of monomer to a relaxed achiral
species, $P_{(t)}$. The fact that $(\sigma_m)_t$ appears to be less than 0.5
suggests that in this case the configuration of the C$_5$ ring nearest
to the metal carbene centre induces a slight preference for trans r
dyads.

WCl$_6$/R$_4$Sn–initiation. The polymer of 12 is fully tactic when Bu$_4$Sn is used as cocatalyst at 20°C. However, if the cocatalyst is changed to Ph$_4$Sn (10) or the temperature raised to 100°C (10) or the monomer changed to 3 (5) the polymer formed is atactic or nearly so. Conversely the tacticity of the polymer of 5 made with WCl$_6$/Bu$_4$Sn increases as the temperature is reduced from 20° to –20°C. It is evident that in the WCl$_6$/Bu$_4$Sn systems the delicate balance between propagation and epimerization or relaxation processes is readily shifted by a change in cocatalyst, monomer or temperature.

Ph(MeO)C=W(CO)$_5$–initiation. Polymers of 13 made at 100°C have intermediate cis content (σ_c = 0.55) and tacticity $\left((\sigma_r)_c \sim (\sigma_m)_t \sim 0.75\right)$ but these values are independent of monomer concentration over the range 0.5 – 2.6 mol litre^{-1}. Hence the intermediate tacticity cannot be associated with the occurrence of an epimerization or relaxation process in competition with propagation. Rather the result must be due to a difference in probability of reaction from opposite directions in the [2+2] cycloaddition propagation step, just as in the case of type V behaviour, discussed above for IrCl$_3$ and Ru-TFA.

Conclusions

^{13}C NMR spectra of the ring–opened polymers of norbornene and its derivatives give information about their cis content, cis/trans distribution, HT or XN bias, and tacticity with respect to both cis and trans double bonds. The results indicate that the propagating metal carbene complex sometimes contains a chiral reaction site and in other cases may be epimerized or achiral with respect to the reaction site. Relaxation processes other than epimerization must also sometimes be occurring, for example a change of geometry with respect to the metal centre.

Literature Cited

1. Ivin, K. J. "Olefin Metathesis"; Academic Press, London 1983.
2. Ivin, K. J.; Laverty, D. T.; O'Donnell, J. H.; Rooney, J. J.; Stewart, C. D. Makromol. Chem. 1979, 180, 1989–2000.
3. Hamilton, J. G.; Ivin, K. J.; Rooney, J. J. British Polymer J. 1984, 16, 21–33.
4. Ivin, K. J.; Lapienis, G.; Rooney, J. J. Polymer 1980, 21, 436–443.
5. Ho, H. T.; Ivin, K. J.; Rooney, J. J. Makromol. Chem. 1982, 183, 1629–1646.
6. Hamilton, J. G.; Ivin, K. J.; Rooney, J. J. J. Mol. Catal. 1984, 17, in press.
7. Ivin, K. J.; Laverty, D. T.; Rooney, J. J. Makromol. Chem. 1977, 178, 1545–1560.
8. Hamilton, J. G.; Ivin, K. J.; Rooney, J. J. IUPAC Macro-molecular Symposium, Amherst 1982, p. 259.
9. Ivin, K. J.; Rooney, J. J.; Bencze, L.; Hamilton, J. G.; Lam, L.M.; Lapienis, G.; Reddy, B. S. R.; Ho, H. T. Pure Appl. Chem. 1982, 54, 447–460.

10. Ho, H. T.; Ivin, K. J.; Rooney, J. J. J. Mol. Catal. 1982, 15,
    245-270.
11. Ivin, K. J.; Lam, L. M.; Rooney, J. J. Makromol. Chem. 1981,
    182, 1847-1854.
12. Hamilton, J. G.; Ivin, K. J.; Rooney, J. J.; Waring, L. C.
    J. Chem. Soc., Chem. Commun. 1983, 159-161.
13. Huisgen, R.; Ooms, P. H. J.; Mingin, M.; Allinger, N. L.
    J. Am. Chem. Soc. 1980, 102, 3951-3953.
14. Ivin, K. J.; Lapienis, G.; Rooney, J. J. Polymer 1980, 21,
    367-369.
15. Syatkowsky, A. I.; Denisova, T. T.; Abramenko, E. L.;
    Khatchaturov, A. S.; Babitsky, B. D. Polymer 1981, 22,
    1554-1557.
16. Ho, H. T.; Reddy, B. S. R.; Rooney, J. J. J. Chem. Soc.,
    Faraday Trans. I 1982, 78, 3307-3317.
17. Ivin, K. J.; O'Donnell, J. H.; Rooney, J. J.; Stewart, C. D.
    Makromol. Chem. 1979, 180, 1975-1988.
18. Osborn, J. A., personal communication.
19. Devine, G. I.; Ho, H. T.; Ivin, K. J.; Mohamed, M. A.;
    Rooney, J. J. J. Chem. Soc., Chem. Commun. 1982, 1229-1231.

RECEIVED October 4, 1984

Electrophilic Ring-Opening Polymerization of New Cyclic Trivalent Phosphorus Compounds
A Novel Mechanism of Ionic Polymerization

SHIRO KOBAYASHI

Department of Synthetic Chemistry, Faculty of Engineering, Kyoto University, Kyoto 606, Japan

This paper describes the electrophilic ring-opening polymerization of seven new cyclic phosphorus(III) compounds, 1-7. The polymerization of five- and six-membered deoxophostones, 1 and 3, and of a benzoxaphosphole 2, produced poly(phosphine oxide)s 11, 35, and 23 via Arbuzov type reactions. The polymerization of five and six-membered deoxothiolphostones, 4 and 5, gave poly(phosphine sulfide)s, 36 and 37, via a new type of the C-S bond scission. The polymerization of seven- and eight-membered cyclic phosphonites, 6 and 7, produced polyphosphinates, 42 and 46, consisting of a "normal" unit. Kinetic studies of the electrophilic polymerization of monomers 1 and 2 led to a new mechanisms of ionic polymerization. Electrophilic covalent propagating species of 1 (MeI initiator) showed even greater polymerizability than ionic ones (MeOTf initiator). The electrophilic polymerization of 2 proceeds only via the covalent propagating species like 27, 29, and 30 which are very reactive; stable phosphonium species, 24 being inactive. With MeI initiator, the overall rate of polymerization is governed by the S_{Ni} process (k_1) to produce an active alkyl iodide species 32 from a phosphonium iodide 31. Based on a proposed general mechanism, k_1 values have been determined and a new classification of ionic polymerization is given.

The present paper reports the ring-opening polymerization of new cyclic trivalent phosphorus monomers. Kinetic and mechanistic studies indicate that these monomers polymerize by a novel mechanism that may be called electrophilic ring-opening polymerization. The monomers include a five-membered deoxophostone (1), a benzoxaphosphole (2), a six-membered cyclic phosphonites (6 and 7). Among these, compounds 3,4 and 5 have been prepared for the first time.

Cyclic phosphorus compounds are good starting monomers for preparing functional polymers having phosphorus groups in the main chain. They are known to undergo ring-opening polymerization via cationic, anionic, or thermal processes. Until recently, cationic (electrophilic) ring-opening polymerizations have been reported for 1,3,2-dioxaphospholanes (8) (1-10) and for 1,3,2-dioxaphosphorianes (9) (7,11,12). Generally, the polymerizations proceed via the Arbuzov reaction to produce polymers containing phosphinate or

0097–6156/85/0286–0293$06.00/0

1

2

3

4

5

6

7

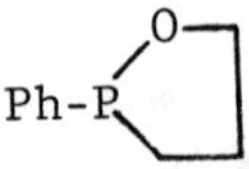

phosphonate repeating units ("normal" unit, 10a). During the polymerizations, however, a side reaction occurs to give an "isomerized" unit, 10b. The amount of the isomerized unit 10b

10a

10b

8; $X = \left(CH_2 \right)_2$

9; $X = \left(CH_2 \right)_3$

R=alkyl, aryl, alkyloxy, aryloxy

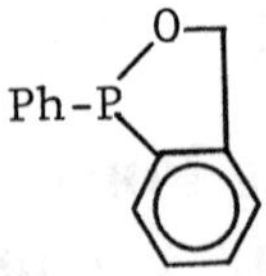

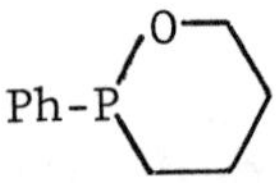

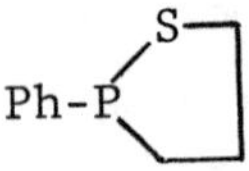

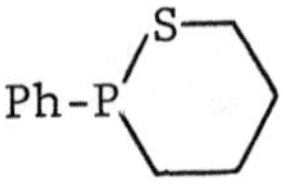

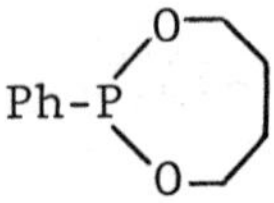

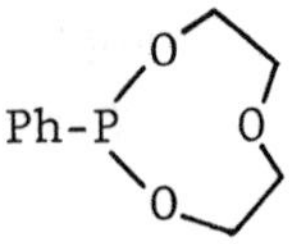

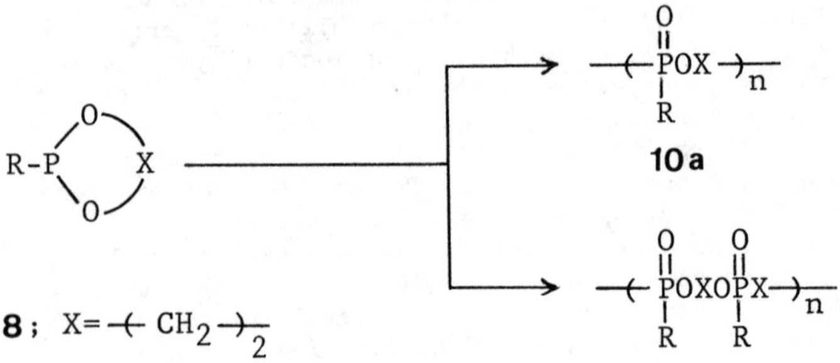

depends upon the monomer and the reaction conditions, and it sometimes exceeds that of normal unit 10a (7,9). Therefore, the polymerizaton of these cyclic phosphorus(III) monomers is complicated. Our very recent results of the electrophilic ring-opening polymerizations of monomers 1-7 have shown that they occur cleanly to produce polymers consisting exclusively of normal units without isomerization, and that they are suitable for kinetic analysis.

2-Phenyl-1,2-oxaphospholane (1)

Electrophilic Polymerization. The ring-opening polymerization of 2-phenyl-1,2-oxaphospholane, 1, a five-membered deoxophostone is induced by a cationic initiator to give a white powdery material, poly(phenyltrimethylenephosphine oxide 11 (13). Initiators like

$$\text{Ph-P} \overset{O}{\underset{}{\Big]}} \quad \overset{E^+}{\longrightarrow} \quad -\!\!\Big(\!\!\begin{array}{c} O \\ \| \\ PCH_2CH_2CH_2 \\ | \\ Ph \end{array}\!\!\Big)_{\!\!n} \quad \overset{(COCl)_2}{\underset{i\text{-}Bu_2AlH}{\longrightarrow}}$$

1   11

$$-\!\!\Big(\!\!\begin{array}{c} PCH_2CH_2CH_2 \\ | \\ Ph \end{array}\!\!\Big)_{\!\!n}$$

12

MeOSO$_2$CF$_3$(MeOTf), MeI, PhCH$_2$Br, BF$_3$·OEt$_2$, and Et$_3$O$^+$BF$_4^-$ were effective in producing polymer 11. No side reactions leading to an isomerized unit took place during the polymerization. A polymer sample having a molecular weight of 3400 was prepared. It melted at 269°C and decomposed at 465°C. This is the first example of an electrophilic ring-opening polymerization of cyclic phosphorus(III) monomer that yields a polymer with a clear-cut structure.

Reduction of 11 to polyphosphine 12 was attempted using a HSiCl$_3$/Et$_3$N reagent but resulted in the formation of a small amount of an unidentified unit (~15%) in addition to the desired phosphine units 12. A novel clean reduction method, therefore, has been developed (14). The method is an one-pot reaction, in which 11 is first treated with (COCl$_2$) and then with i-Bu$_2$AlH, giving rise to the polyphosphine 12.

Macroporous-type crosslinked chloromethylated or iodomethylated polystyrene 13 was used to initiate the polymerization of monomer 1. The product poly(styrene-g-phenyltrimethylenphosphine oxide) 14, is a white bead-like resin (n=4.1-10.5), which showed efficient chelating properties toward heavy metal ions such as UO$_2^{2+}$, Th^{4+}, Hg^{2+}, Pd^{2+}, and Cu^{2+} (15). The adsorbed ions, eg, UO$_2^{2+}$, were readily desorbed by treating the resin with 10% aqueous Na$_2$CO$_3$. This adsorption-desorption procedure could be performed repeatedly without reducing the chelating efficiency of 14 and without destroying its bead-like structure (15).

$$\text{13(X=Cl, I)} \qquad\qquad\qquad\qquad \text{14}$$

Kinetics and Mechanism. Since the ring-opening polymerizations of 1 by MeOTf and MeI are clean reactions, they are amenable to a kinetic study. The kinetic analysis was carried out by monitoring the polymerization reaction using ^{31}P NMR spectroscopy.

Meotf-Initiated System. Figure 1 shows the ^{31}P NMR spectra of the Meotf-initiated polymerization system: (a) immediately after the mixing Meotf with 1 in PhCN and (b) after 30 min at 70°C (16). The ^{31}P NMR chemical shift is relative to external 85% H_3PO_4 standard. Based on the ^{31}P NMR signals observed in Figure 1, which are assigned as cricled letters in Scheme I, as well as on ^{19}F and ^1H NMR spectra of the reaction mixture, the following scheme is proposed for the polymerization.

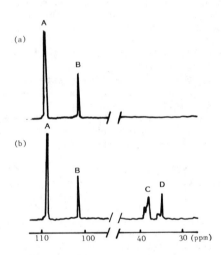

Figure 1. ^{31}P {^1H} NMR spectra of the cationic ring-opening polymerization of 1 initiated with MeOTf in PhCN([M]^o = 1.25 mol/L and [I]$_o$ = 0.125 mol/L): (a) before heating; (b) after 30 min at 70°C.   Reproduced with permission from Ref. 16.   Copyright 1984, American Chemical Society.

Scheme I

Initiation

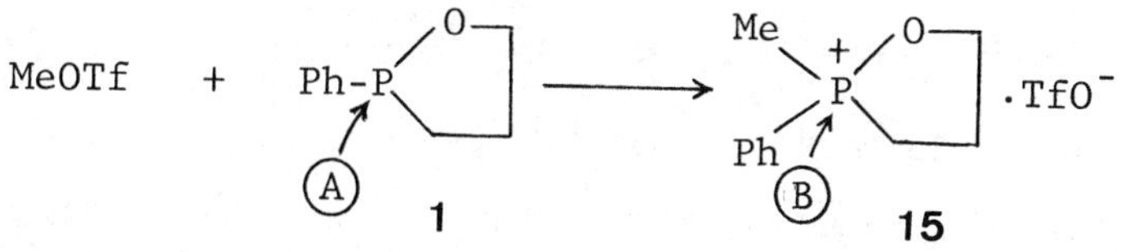

Propagation

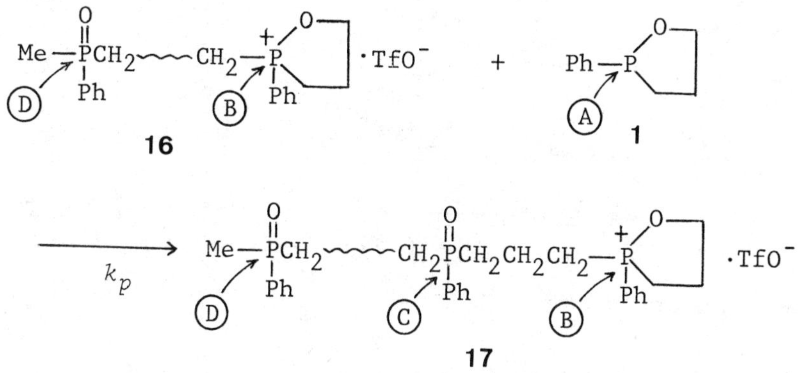

The propagating ends in this polymerization are cyclic phosphonium ions such as 15-17, which are opened by nucleophilic attack of monomer 1 to form P-phenyltrimethylenephosphine oxide units 17 via an Arbuzov-type reaction. The propagation rate constant (k_p) is obtained by the following equation, assuming as S_N2 reaction:

$$- \frac{d[M]}{dt} = k_p[P^*][M]$$

where $[P^*]$ and $[M]$ are the concentrations of propagating species and monomer. It was found that initiation is very rapid, that the concentration of propagating phosphonium species is equal to that of the initiator charged, and that this remained constant throughout the polymerization. Therefore, the integrated form of the above equation is given by

$$\ln \{[M]_1/[M]_2\} = k_p[P^*](t_2-t_1) \tag{1}$$

Plots of $\ln\{[M]_1/[M]_2\}$ versus t_2-t_1 showed a linear relationship, from which k_p values were obtained. Arrhenius plots of k_p values at four temperatures gave a straight line whose slope led to the activation parameters (Table I) (16).

Table I. Propagation Rate Constants and Activation Parameters in the Polymerization of 1

	Initiators			
	MeOTf	MeI	PhCH$_2$Br	PhCH$_2$Cl
$k_p \times 10^4$	3.52(50°C)	4.53 (50°C)	2.13(70°C)	1.88(130°C)
(L/mol.sec)	9.27(60°C)	11.5 (60°C)	4.34(80°C)	3.89(140°C)
	24.3 (70°C)	25.0 (70°C)	7.99(90°C)	5.45(150°C)
	40.3 (80°C)	43.8 (80°C)	15.0(100°C)	10.5 (160°C)
$k_p \times 10^4$(50°C) (L/mol.sec)	3.52	4.53	(0.475)[b]	(0.00538)[b]
$\Delta H^{\ddagger}$ (50°C) (kJ/mol)	73.3	66.7	63.5	74.3
$\Delta S^{\ddagger}$ (50°C) (J/K.mol)	-84.8	-103	-132	-136

a) $[M]_o$ = 1.25 mol/L and $[I]_o$ = 0.125 mol/L in PhCN solvent.

b) Calculated values.

Alkyl Halide Initiated Systems. Figure 2 shows the ^{31}P NMR spectra of the polymerization system initiated by MeI in PhCN. Peak assignments were made as circled letters in Scheme II and led to the general polymerization mechanism (Scheme II). No signal due to phosphonium species was detected under the polymerization conditions employed. This is in sharp contrast to the MeOTf-initiated system. Intermediates 18 and 21 are unstable. The stable propagating ends

Scheme II

Initiation

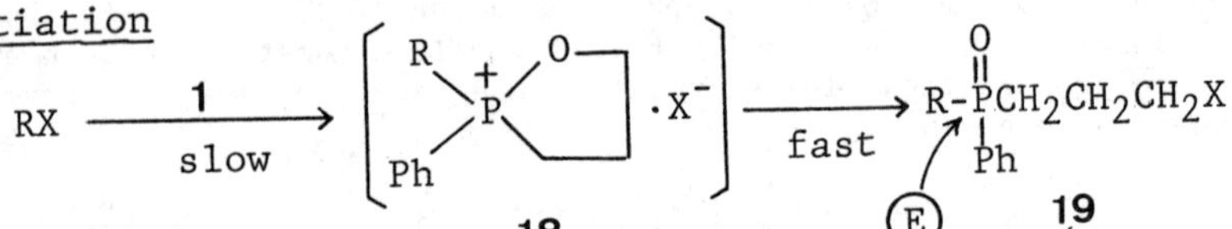

R=Me, PhCH$_2$

X=I, Br, Cl

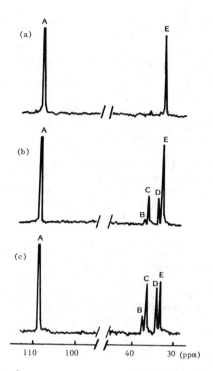

Figure 2. ^{31}P {^{1}H} NMR spectra of the cationic ring-opening
polymerization of 1 initiated with MeI($[M]_0$ = 1.25 mol/L
and $[I]_0$ = 0.125 mol/L):   (a) after 1 min at 70°C; (b)
after 10 min at 70°C; (c) after 20 min at 70°C.
Reproduced with permission from Ref. 15.   Copyright
1984, American Chemical Society.

Propagation

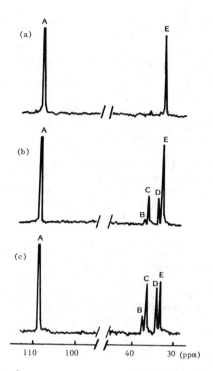

are alkyl iodide species like 19, 20, and 22.

The rare-determining steps of both initiation and propagation
are the dipole-dipole S_N2 reactions between and alkyl iodide and
monomer 1 producing transient phosphonium species such as 18 and 21,
which are converted rapidly into covalent alkyl iodide species 19

and 22 via nucleophilic attack of the iodide counteranion. From the plots of the second-order kinetics of Equation 1 k_p values were obtained. Similarly, benzyl bromide and benzyl chloride were also found to proceed via alkyl halide propagating ends as stable species. These kinetic results are given in Table I (16). The k_p values are very much dependent upon the counter-anion derived from initiator. The relative reactivities at 50°C are in the following order; MeOTf : MeI : PhCH$_2$Br : PhCH$_2$Cl = 654 : 842 :88.3 : 1.0.

The polymerization of 1 proceeds via two different mechanisms which is due to differences in the nucleophilicity of the anions TfO$^-$ and X$^-$ (X = I, Br, Cl), which affect the relative stability of the phosphonium species. The difference is reflected by $\Delta S^{\ddagger}$ values; the reduced polymerizability of the PhCH$_2$Br or PhCH$_2$Cl system is attributed to the less favorable entropy term in comparison with the MeOTf system. This can be interpreted in terms of solvation-desolvation phenomena from the initial state to the transition state (16).

In ring-opening polymerizations, cyclic onium propagating species are usually more reactive than covalently bonded ones, eg, superacid macroestertype species in cyclic ether polymerizations (17,18) and alkyl halide type species in 2-oxazoline polymerizations (19). It is to be emphasized, however, that the MeI-initiated polymerization of 1 proceeds even faster than the MeOTf-initiated system, although the difference is small. This is the first case of covalent (electrophilic) propagating species showing a higher reactivity than an ionic propagating one (16).

1-Phenyl-3H-2,1-benzoxaphosphole (2)

Ring-Opening Polymerization. Monomer 2 is a five-membered cyclic phosphinite, an analogue of dexophostone 1. 2 was prepared according to the reported procedure (20). The ring-opening polymerization of 2 produced white powdery materials of poly(phosphine oxide) 23, whose structure was determined by IR, 31$_P$, 1$_H$, and 13C NMR spectroscopy as well as by elemental analysis. The polymerization results are given in Table II (21).

2 $\longrightarrow$ $-(-\overset{\overset{\displaystyle O}{\underset{\displaystyle \|}{}}}{\underset{\displaystyle Ph}{P}}-\langle\text{benzene}\rangle-CH_2-)_n$

23

24 (R=Me, Et; X=BF$_4$, TfO)

Table II.   Ring-Opening Polymerization of $\underset{\sim}{2}^a$

Initiators (mol % for $\underset{\sim}{2}$)	Solvents	Temp. (°c)	Time (hr)	Polymer	
				Yield(%)	Mol. wt.
Et$_3$O BF$_4^-$ (17)	CH$_2$Cl$_2$	35	48	0^b	
MeOTf (19)	CDCl$_3$	35	24	0^b	
MeI (21)	CH$_3$CN	35	24	91	
MeI (18)	CDCl$_3$	35	24	100	
MeI (1.4)	CHCl$_3$	35	1440	63	4500
MeOTf (5.0)	CH$_2$Cl$_2$	50	24	0^b	
MeI (5.0)	PhCN	50	3	53	
MeI (10)	CH$_2$Cl$_2$	70	16	67	2400
none	CH$_2$Cl$_2$	70	16	0	
none	CH$_2$Cl$_2$	130	14	54	2260

a) $\underset{\sim}{2}$(0.3 g) in 1 mL of solvent in a sealed pressure under nitrogen.

b) Stable phosphonium species $\underset{\sim}{24}$ was formed.

Among the cationic initiators examined, only MeI is active for the polymerization. Et$_3$O$^+$BF$_4^-$ and MeOTf produced stable phosphonium salts $\underset{\sim}{24}$, which did not induce the polymerization of $\underset{\sim}{2}$. Anionic and radical initiators were inactive. At higher temperatures, eg, above 70°C, $\underset{\sim}{2}$ starts to show polymerizability without initiator to give polymer of the same structure as $\underset{\sim}{23}$. The mechanisms of this "thermal" polymerization is not well understood, but, at present, a zwitterion intermediate $\underset{\sim}{25}$ derived from two molecules of $\underset{\sim}{2}$ is considered to be responsible for the production of polymer $\underset{\sim}{23}$. The

25

formation of $\underset{\sim}{25}$ requires that one molecule of $\underset{\sim}{2}$ acts as a nucleophile and the other behaves as an electrophile. This "amphiphilic" nature of $\underset{\sim}{2}$ has already been confirmed in the copolymerizations, in which $\underset{\sim}{2}$ was copolymerized without initiator with an electrophilic monomer like acrylic acid and with a nucleophilic one like 2-methyl-2-oxazoline (22).

<u>Kinetics of Electrophilic Ring-Opening Polymerization of 2: A New</u>
<u>Mechanism in Ionic Polymerizations</u>. The kinetic analysis of the
polymerization initiated with MeI was carried out by ^{31}P NMR
spectroscopy at 35°C, at which the thermal polymerization does not
take place at all. Figure 3 shows a ^{31}P NMR spectrum of the
polymerization system in CH_3CN, in which a relatively large amount
of the initiator was employed for the kinetic analysis. Peak A at
119 ppm is assigned to monomer 2. Peaks due to phosphonium species
appear over a wide chemical shift range: peak A at 95 ppm is
attributed to the phosphonium iodide 26, and peaks C around 67 and
83 ppm are due to phosphonium iodides like 28, 31, and 33. Peaks D
are ascribed to various phosphine oxide units including propagating
ends such as 27 and 30, if any (Scheme III).

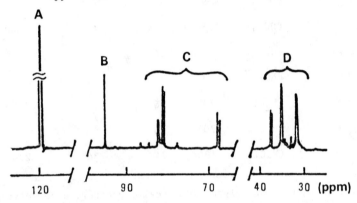

Figure 3. ^{31}P {^1H} NMR spectrum of the MeI-initiated polymerization
of 2 after 490 min at 35°C in CH_3CN. [M]$_o$ = 1.0 mol/L
and [I]$_o$ = 0.20 mol/L.

Scheme III

<u>Initiation</u>

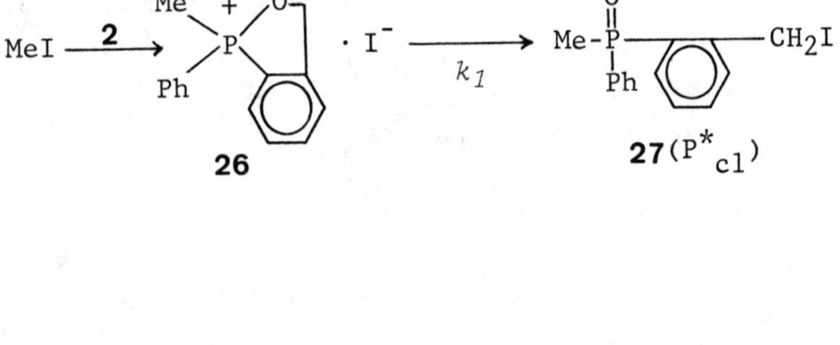

Figure 4 shows the time-conversion curves for monomer $\underset{\sim}{2}$ and the total concentrations of active species [P*]. The following

$$[P*] = [P*_i] + [P*_c]$$

relationsip holds where [P*$_i$] and [P*$_c$] are the concentrations of phosphonium species and of covalent alkyl iodide species, respectively. The initiation finished at an early stage of the reaction. Under the polymerization conditions in CH_3CN [P*$_i$] reached a constant value that was almost equal to the concentration of the initiator charged, ie, [P*] = [P*$_i$]. After [P*$_i$] became constant, the apparent rate constant of propagation ($k_{p(ap)}$) was obtained based on Equation 1. The values are very much dependent upon the solvent employed (Table III) (23). In a highly polar solvent (eg, CH_3CN) $k_{p(ap)}$ value is at least 10^2 times less than that in a less polar solvent (eg, PhCl).

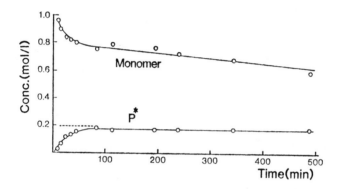

Figure 4. Time-conversion curves for monomer 2 and total active species [P*] (=[P*$_i$]) in the polymerization of 2 with MeI initiator at 35°C in CH_3CN. [M]$_0$ = 1.0 mol/L and [I]$_0$ = 0.20 mol/L.

These observations lead to the propagation mechanism shown in Scheme III. The phosphonium species $\underset{\sim}{26}$ and $\underset{\sim}{28}$ are reasonably considered as "not really active", which is supported by the fact that phosphonium salt $\underset{\sim}{24}$ did not initiate the polymerization.

Table III. Values of $k_{p(ap)}$ with MeI Initiator in Three Solvents at 35°C

Solvents	$k_{p(ap)}$ (L/mol·sec)
CH_3CN	7.3×10^{-5}
PhCN	7.6×10^{-4}
PhCl	$>1 \times 10^{-2}$

Propagation

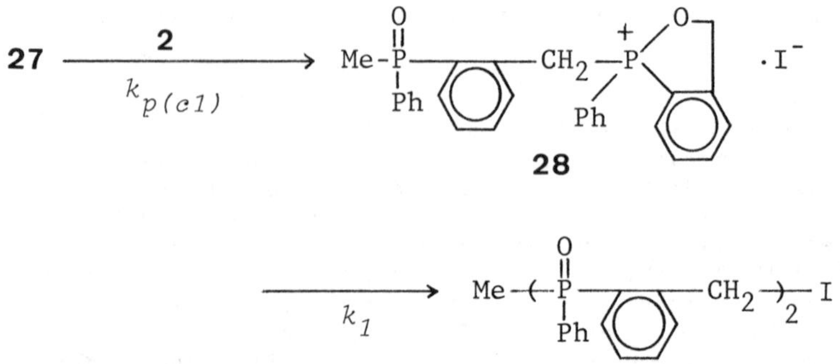

27 $\xrightarrow[k_{p(c1)}]{2}$ $Me-\overset{\overset{O}{\parallel}}{\underset{Ph}{P}}\!-\!\!\bigcirc\!\!-CH_2-\overset{+}{\underset{Ph}{P}}\!\!\diagup\!\!\overset{O-}{\diagdown}\!\!\bigcirc$ $\cdot I^-$

28

$\xrightarrow{k_1}$ $Me-\!\!\left(\!\!-\overset{\overset{O}{\parallel}}{\underset{Ph}{P}}\!-\!\!\bigcirc\!\!-CH_2\!\!-\!\!\right)_{\!2}\!\!-I$

29

Generally

$\overset{\overset{O}{\parallel}}{\underset{Ph}{\sim\!\!\sim\!P}}\!-\!\!\bigcirc\!\!-CH_2I$ $\xrightarrow[k_{p(c)}]{2}$ $\overset{\overset{O}{\parallel}}{\underset{Ph}{\sim\!\!\sim\!P}}\!-\!\!\bigcirc\!\!-CH_2-\overset{+}{\underset{Ph}{P}}\!\!\diagup\!\!\overset{O-}{\diagdown}\!\!\bigcirc$ $\cdot I^-$

30(P^*_c) **31**(P^*_i)

k_1 $2 \times k_{p(i)}$

$\overset{\overset{O}{\parallel}}{\underset{Ph}{\sim\!\!\left(\!\!-P\right.}}\!-\!\!\bigcirc\!\!-CH_2\!\!-\!\!\right)_{\!2}\!\!-I$ $\xrightarrow[k_{p(c)}]{2}$ $\overset{\overset{O}{\parallel}}{\underset{Ph}{\sim\!\!\left(\!\!-P\right.}}\!-\!\!\bigcirc\!\!-CH_2\!\!-\!\!\right)_{\!2}\!\!-\overset{+}{\underset{Ph}{P}}\!\!\diagup\!\!\overset{O-}{\diagdown}\!\!\bigcirc$ $\cdot I^-$

32(P^*_c) **33**(P^*_i)

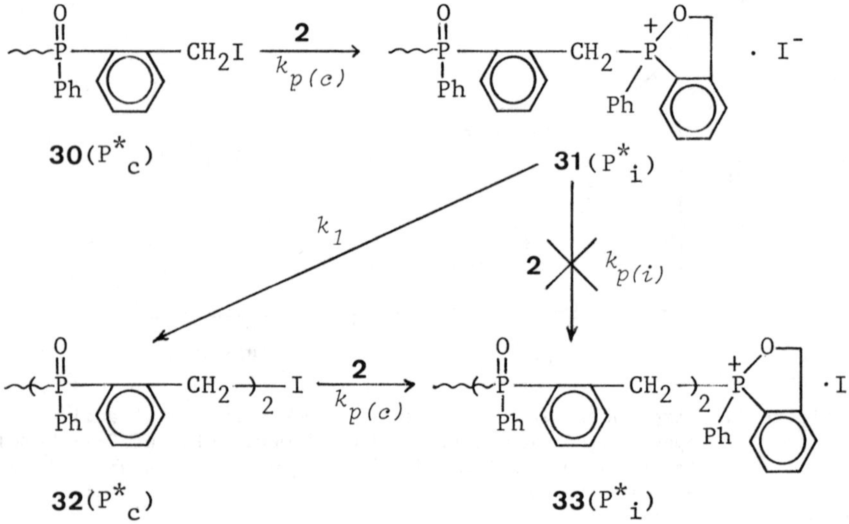

Therefore, $k_{p(i)}$ is null in the following relationship where $k_{p(c)}$

$$k_{p(ap)} = k_{p(c)} \cdot X_c + k_{p(i)} \cdot X_i$$

and $k_{p(i)}$ denote, respectively, the propagation rate constants due to covalent species like 27, 29, and 30 and to ionic species like 26, 28, and 31, and X_c and X_i are the molar fractions of covalent and ionic species, ie, $X_c + X_i = 1$, leading to

$$k_{p(ap)} = k_{p(c)} \cdot X_c \qquad\qquad (2)$$

It was not possible to determine the precise X_c value since the ^{31}P NMR signal of P^*_c overlaps with that of the product polymer unit and since $[P^*c]$ is very small especially in a highly polar solvent.

For the determination of $k_{p(c)}$ values, a model reaction was undertaken. As a model of covalent propagating species such as 30 a

1:1 addition product 27 was prepared from MeI and 2 and was isolated as crystalline material (^{31}P NMR signal of 27:+33 ppm in CDCl$_3$). 27 corresponds to the smallest propagating species P*$_{c1}$ and the bimolecular rate constant between 27 and monomer 2 denotes k$_{p(cl)}$, which can be taken as k$_{p(c)}$ (Table IV).

Table IV.   Values of $k_{p(cl)}$ in Three Solvents at 35°C

Solvents	$k_{p(cl)}$ (L/mol·sec)
CH$_3$CN	6.4×10^{-2}
PhCN	5.0×10^{-2}
PhCl	3.8×10^{-2}

Values of k$_{p(cl)}$ changed only slightly in a solvent of high or low polarity. This is probably because of the very high electrophilic reactivity of 27 which has a benzyl iodide type structure. These observations lead to a conclusion that the big difference in k$_{p(ap)}$ values is mainly due to the value of X$_c$ in Equation 2. For example, X$_c$ is approximately 0.001 in CH$_3$CN and 0.015 in PhCN. It is important to note that X$_c$ is governed by the rate of the intramolecular conversion (k$_1$) of a phosphonium species 31 to a covalent species 32. The mechanism of this type is an entirely new one which has not been known in the ionic polymerization.

It should be emphasized that P*$_c$ is the real active species, and P*$_i$ is a precursor of P*$_c$. Therefore, we propose to term P*$_i$ as the _pro-active_ species. This new concept was also confirmed by the following observation. Into a monomer solution of 2 containing an "inactive" salt 24, Bu$_4$N$^+$I$^-$ was added, and then the polymerization of 2 started to give polymer 23. This result suggests the exchange of the counteranion of TfO$^-$ with I$^-$, giving rise to a cyclic phosphonium iodide 26 which is "pro-active". Then, 26 converts into 27, which is the "real active" species responsible for the production of polymer 23. The conversion of 26 to 27 is an Arbuzov type S$_N$i (_s_ubstitution, _n_ucleophilic, _i_ntramolecular) reaction.

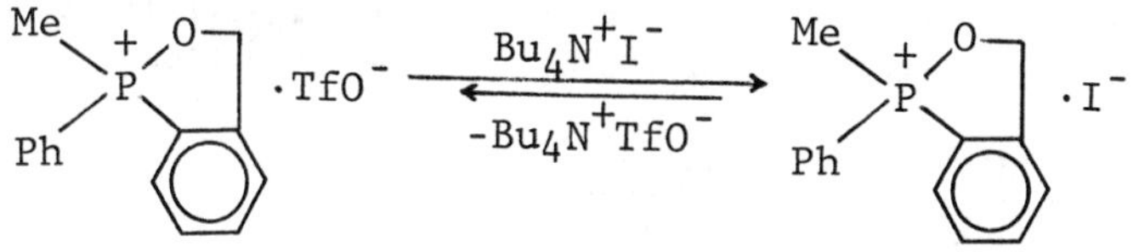

24(inactive)   **26**("pro-active")

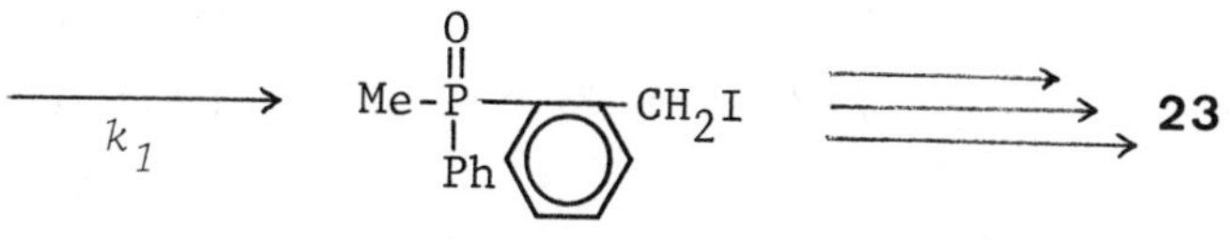

27(really active)

General Scheme of the New Electrophilic Ring-Opening Polymerization.
The new polymerization is given as the following general scheme
(Scheme IV).

Scheme IV

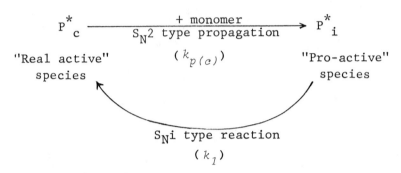

In order to evaluate k_1 for the $S_N i$ reaction, an attempt was
made to perform kinetic analyses based on Scheme IV. The rate
equations for the monomer consumption and the production of $[P^*_c]$
are given by Equations 3 and 4.

$$- \frac{d[M]}{dt} = k_{p(c)}[P^*_c][M] \qquad (3)$$

$$\frac{d[P^*_c]}{dt} = k_1[P^*_i] - k_{p(c)}[P^*_c][M] \qquad (4)$$

In a highly polar solvent the relationship (5) will be valid after
the initiator is completely consumed

$$[P^*_c] + [P^*_i] = [P^*] = [I]_0 \qquad (5)$$

where $[I]_0$ denotes the concentration of the initiator charged. It
was impossible to determine $[P^*_c]$ precisely. It is observed in
Figure 4 that the polymerization stage where $[P^*_i]$ = constant is
present. Therefore, an assumption that $[P^*_c]$ is constant is made.
Then, $d[P^*_c]/dt = 0$, leading to $[P^*_c] = k_1[I]_0/(k_{p(c)}[M]+k_1)$ which
is valid at a limited stage of polymerization. Equation 3 is then
expressed as

$$- \frac{d[M]}{dt} = \frac{k_1 k_{p(c)}[I]_0[M]}{k_{p(c)}[M] + k_1}$$

The integrated form of this equation is given as

$$\ln \frac{[M]_{t2}}{[M]_{t1}} + k_{p(c)}[I]_0(t_2-t_1) \quad k_1 = k_{p(c)} \; ([M]_{t1} - [M]_{t2}) \qquad (6)$$

By employing a $k_{p(c1)}$ value in place of a $k_{p(c)}$ value plots of
Equation 6 gave a linear relationship whose slope led to a kg value
(Table V).

On the other hand, a model reaction for the transformation of 31 to 32 by the $S_{N}i$ reaction was independently undertaken using 26 as a model of 31 and then k_1 values were successfully obtained (Table V). Values of k_1 in CH_3CN and PhCN solvents obtained by two different methods are strikingly close to each other. The value of k_1 is very small in comparsion with that of $k_{p(c)}$, indicating that the $S_{N}i$ process determines the overall rate of polymerization although the process does not consume monomer 2. These results are taken to establish the proposed new mechanism of the electrophilic ring-opening polymerization.

Table V.   Values of $k_1 (sec^{-1})$ at 35°C

Solvents	From Equation 6	From a model reaction of 26 $\xrightarrow{k_1}$ 27
CH_3CN	4.6×10^{-5}	3.8×10^{-5}
PhCN	2.6×10^{-4}	2.7×10^{-4}

New Classification of Ionic Polymerization. The polymerization of 2 with cationic initiator provides a novel system of ionic polymerization in which ionic species do not propagate but covalent electrophilic species do propagate! Several cyclic monomers are known whose cationic polymerization proceeds via covalent propagating species, but these monomers propagate also via cationic species with changing the initiator. The present monomer 2 is unique in that an electrophilic covalent species is the only species which is able to propagate.

The type of ionic polymerization is normally defined based on the nature of the propagating species. If so, there is no problem with polymerization systems involving ionic propagating species. Then, how we should regard polymerization systems involving covalent propagating species? Many examples have been accumulated until now, in which stable covalent propagating species are involved. For the systematic understanding of the systems, a classifictaion of ionic polymerization is made by taking various systems as examples (24).

Ionic Polymerization

1) Electrophilic polymerization
   a) Cationic propagating species:

16,   ~~O~~ TfO^- ,   ~~N~~ TsO^- ,   ~~$CH_2CH^+MX_m^-$~~
                                                        |
                                                        Ph

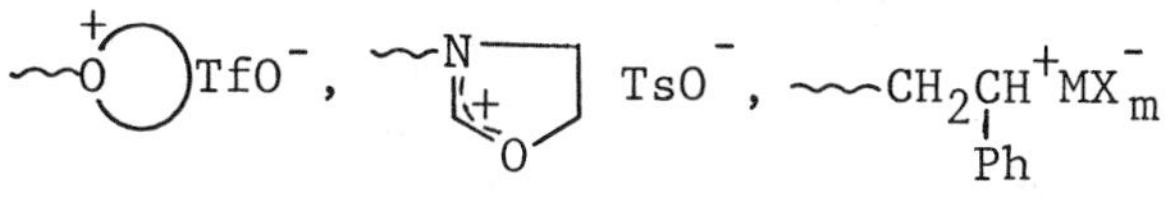

   b) Electrophilic covalent propagating species:

20,   **30**,   ~~CH_2OTf,   ~~NCH_2CH_2I,   ~~$CH_2CHOClO_3$(?)
                              | |
                             CHO   Ph

2) Nucleophilic polymerization
   a) Anionic propagating species:

$$\sim\!\!\sim\!\!CH_2S^- \ K^+, \qquad\qquad \sim\!\!\sim\!\!CH_2\underset{\underset{Ph}{|}}{CH}^- Na^+$$

b) Nucleophilic covalent propagating species:

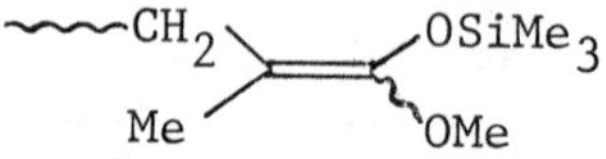

34

The species 34 is considered as a stable propagating species of
silyl ketene acetal type in the methyl methacrylate polymerization
by a "group-transfer polymerization" (25).

2-Phenyl-1,2-oxaphosphorinane (3)

A six-membered deoxophostone 3 has been obtained for the first time

(^{31}P NMR signal at +111 ppm) (26) and its ring-opening
polymerization has been achieved by cationic initiator to produce
poly(phosphine oxide) 35. The polymer is a powdery material whose
molecular weight is up to ~3000.

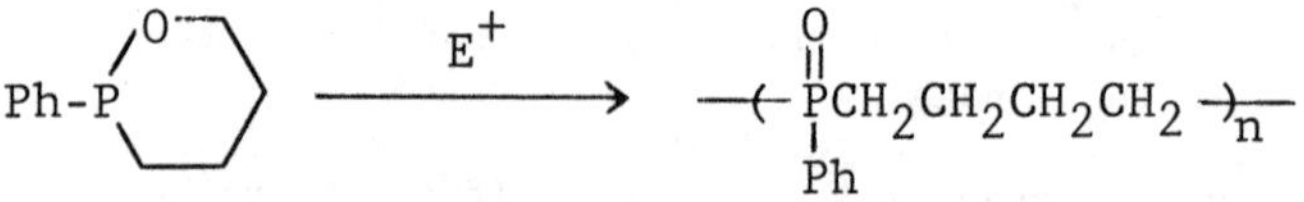

2-Phenyl-1,2-thiaphospholane(4) and 2-Phenyl-1,2-thiaphospho-rinane(5)

Five- and six-membered deoxothiolphostone (4 and 5) have been
obtained as new compounds: Then ^{31}P NMR signals are observed at +22
ppm (4) and at +5 ppm (5) (26). The ring-opening polymerizations of
4 and 5 when induced by a cationic initiator give rise to white
powdery materials of poly(phosphine sulfide)s 36 and 37 (Table VI)
(27). The structures of 36 and 37 have been demonstrated by

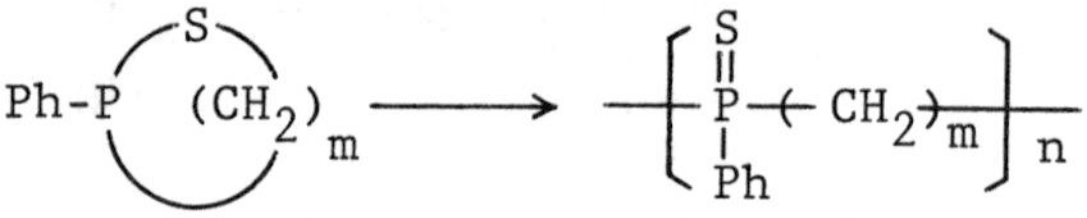

4 (m=3)   36 (m=3)

5 (m=4)   37 (m=4)

Table VI. Bulk Polymerization of 4 and 5

Monomers	Initiators (mol % for monomer)	Temp. (°C)	Time (hr)	Polymers Yield(%)	Mol. wt.
4	MeI (2.0)	80	40	62	3000
4	PhCH$_2$Br(2.8)	80	40	53	2300
4	none	80	40	0	----
4	none	150	60	81	4700
5	MeI(3.2)	100	60	68	1800
5	PhCH$_2$Br(2.8)	100	60	56	4200
5	none	200	60	a	----

a) Monomer 5 was converted to 37 in 17% checked by ^{31}P NMR.

^{1}H, ^{13}C, and ^{31}P NMR (^{31}P NMR signals; 36 at +45 ppm and 37 at +46 ppm). The polymerization of 4 took place at high temperatures, eg, 150°C, and that of 5 occurred slightly at 200°C. These behaviors are to be compared with those of the corresponding O-analogues 1 and 3 which did not thermally polymerize. Radical and anionic initiators did not cause the polymerization of 4 and 5.

The electrophilic polymerization of 4, for example, involves stable phosphonium species 38 which reacts with monomer 4 via an Arbuzov type reaction to induce a C-S bond scission leading to a phosphine sulfide unit.

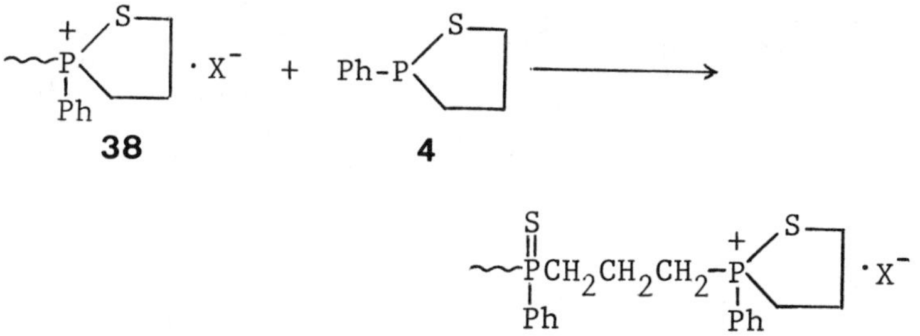

The polymerizations of 4 and 5 are the first examples of the formation of polymers with a P=S group involving an Arbuzov type reaction with C-S bond cleavage.

As to the nature of the propagating species involved in the polymerizations of 4 and 5, the following model reactions revealed an equilibrium between 39 and 40 at 80°C in CDCl$_3$ (R=Me or PhCH$_2$) (27).

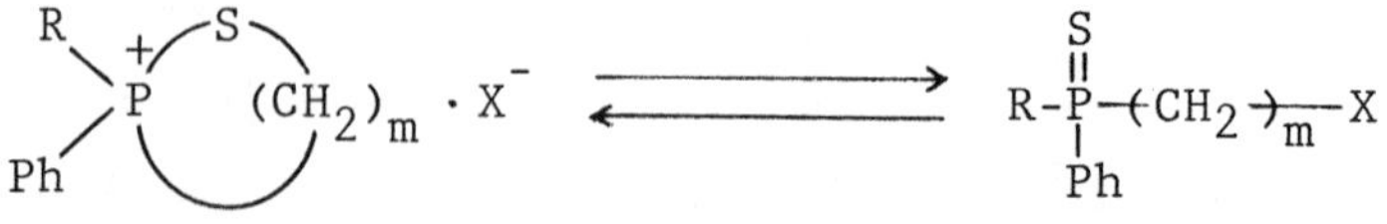

39   **40**

$$m=2, \quad X=Br$$

$$m=3, \quad X=Br, \ I$$

The course of the thermal polymerization of 4 and 5 may be considered to be similar to that of 2 involving a zwitterion 41 which corresponds to 25. Therefore, 4 and 5 have amphiphilic nature (27).

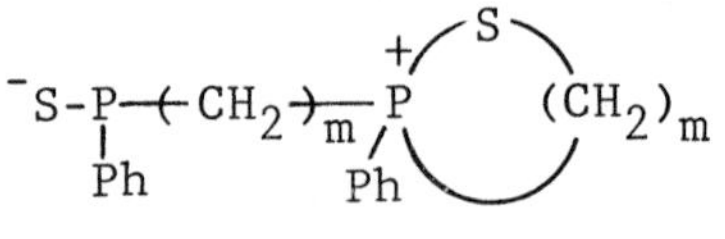

41

2-Phenyl-1,3,2-dioxaphosphepane(6)

A seven-membered cyclic phosphonite (a dioxaphosphepane, 6) has been polymerized using a cationic initiator (28). A bulk polymerization of 6 using MeI as an initiator (2.0 mol % for 6) at 100°C gave in 20 hr a polyphosphinate 42 (molecular weight = 12000). This is the

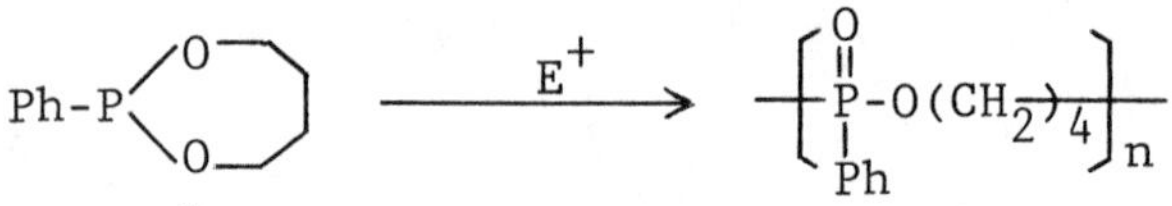

6   **42**

first instance in the electrophilic polymerization of cyclic phosphonites that produced polymer consisting exclusively of "normal" units without any "Isomerized" units.

Since the MeI-initiated polymerization of 6 occurs cleanly, the kinetics of the reaction was investigated. The polymerization proceeded via an alkyl iodide propagating species 43. At 95°C in $CHCl_3$, a k_p value of 2.48×10^{-3} L/mol·sec was obtained.

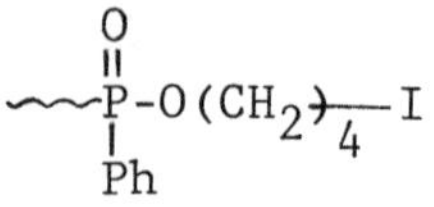

43 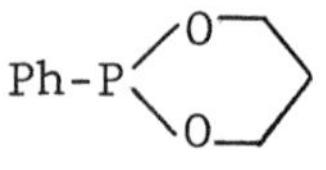   **44** 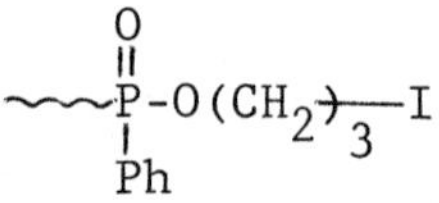   **45**

For comparing the polymerizability of the seven-membered monomer 6 with that of the corresponding six-membered one (2-phenyl-

1,3,2-dioxaphosphorinane, 44), the polymerization of 44 was examined. It has been found that the MeI-initiated polymerization of 44 produced polymer having a "normal" polyphosphinate structure containing no "isomerized" units and that a stable propagating species is an alkyl iodide 45. A k_p value obtained is 1.34×10^{-3} L/mol·sec at 95°C in CHCl$_3$. These results indicate that a seven-membered cyclic phosphonite 6 is 1.9 times more reactive in polymerizability than its six-membered analogue 44 (28).

2-Phenyl-1,3,6,2-trioxaphosphocane(7)

An eight-membered cyclic phosphonite 7 has been shown to polymerize via ring-opening with a cationic initiator such as PhCH$_2$Cl and MeOTf (29). The PhCH$_2$Cl-initiated polymerization produced a polyphosphinate 46 (molecular weight = 2800), containing mainly "normal" units, but the MeOTf-initiation gave 46 containing 34% "isomerized" units.

$$Ph-P\overbrace{}^{O}_{O} \xrightarrow{E^+} -\!\!\left(\!\!\underset{\underset{Ph}{|}}{\overset{\overset{O}{\|}}{P}}OCH_2CH_2OCH_2CH_2\!\!\right)\!\!_n\!\!-$$

7   **46**

With the PhCh$_2$Cl-initiated system, propagating species are of alkyl chloride type 47. Kinetic analyses for the PhCH$_2$Cl-initiated reaction at 150°C in PhCN give a k_p value of 2.8×10^{-4} L/mol·sec. The reduced polymerizability is due to the structure of an electron-withdrawing 2-alkoxyethyl chloride type of 47 (29).

$$\sim\!\!\sim\!\!\underset{\underset{Ph}{|}}{\overset{\overset{O}{\|}}{P}}OCH_2CH_2OCH_2CH_2Cl$$

47

Literature Cited

1. McManimie, R. J. U.S. Patent 2 893 961, 1959.
2. Petrov, K. A.; Nifantev, E. E.; Khorkhoyanu, L. V.; Merkulova, M. I.; Voblikov, V. F. Vysokomol. Soedin. 1962, 4, 246.
3. Mukaiyama, T.; Fujisawa, T.; Tamura, y.; Yokota, Y. J. Org. Chem. 1964, 29, 2572.
4. Shimidzu, T.; Hakozaki, T.; Kagiya, T.; Fukui, K. J. Polym. Sci. Part B. 1965, 3, 871.
5. Shimidzu, T.; Hakozaki, T.; Kagiya, T.; Fukui, K. Bull. Chem. Soc. Jpn. 1966, 39, 562.
6. Fujisawa, T.; Yokota, Y.; Mukaiyama, T. Bull. Chem. Soc. Jpn. 1967, 40, 147.
7. Harwood, H. J.; Patel, N. K. Macromolecules 1968, 1, 233.
8. Yamashita, Y. J. Polym. Sci. Polym. Symp. 1976, 56, 447.
9. Vogt, W.; Ahmad, N. U. Makromol. Chem. 1977, 178, 1711.
10. Kawakami, Y.; Miyata, K.; Yamashita, Y. Polym. J. 1979, 11, 175.

11. Petrov, K. A.; Nifantev, E. E.; Sopikova, I. I. Vysokomol.
     Soedin. 1960, 2, 685.
12. Singh, G. J. Org. Chem. 1979, 44, 1060.
13. Kobayashi, S.; Suzuki, M.; Saegusa, T. Polym. Bull. 1981, 4,
     315.
14. Kobayashi, S.; Suzuki, M.; Saegusa, T. Polym. Bull. 1982, 8,
     417.
15. Kobayashi, S.; Suzuki, M.; Saegusa, T. Macromolecules 1983,
     16, 1010.
16. Kobayashi, S.; Suzuki, M.; Saegusa, T. Macromolecules 1984,
     17,107.
17. Kobayashi, S.; Morikawa, K.; Saegusa, T. Macromolecules, 1975,
     8, 386.
18. Kobayashi, S.; Tsuchida, N.; Morikawa, K.; Saegusa, T.
     Macromolecules 1975, 8, 942.
19. Kobayashi, S.; Morikawa, K.; Shimizu, N.; Saegusa, T. Polym.
     Bull. 1984, 11, 253.
20. Dahl, B. M.; Dahl, O. J. Chem. Soc. Perkin I 1981, 2239.
21. Kobayashi, S.; Mizutani, T.; Tanabe, T.; Shimoyama, T.;
     Saegusa, T. Polym. Prepr. Jpn. 1983, 32, 1475.
22. Kobayashi, S.; Mizutani, T.; Tanabe, T.; Saegusa, T. Polym.
     Prepr. Jpn. 1983, 32, 230.
23. Kobayashi, S.; Shimoyama, T.; Saegusa, T. Polym. Prepr. Jpn.
     1984, 33, 197.
24. Kobayashi, S.; Kobunshi 1984, 33, 228.
25. Webster, O. W.; Hertler, W. R.; Sogah, D. Y.; Farnham, W. B.;
     RajanBabu, T. V. J. Am. Chem. Soc. 1983, 105, 5706.
26. Kobayashi, S.; Suzuki, M.; Saegusa, T. to be reported.
27. Kobayashi, S.; Suzuki, M.; Saegusa, T. Polym. Prepr. Jpn.
     1983, 32, 1399.
28. Kobayashi, S.; Tokunoh, M.; Saegusa, T. Polym. Bull. to be
     published.
29. Kobayashi, S.; Huang, M. Y.; Saegusa, T. Polym. Bull. 1981,
     4, 185.

RECEIVED February 1, 1985

Synthesis and Polymerization of Atom-Bridged Bicyclic Acetals and Ortho Esters

A Dioxacarbenium Ion Mechanism for Ortho Ester Polymerization

ANNE BUYLE PADIAS, RYSZARD SZYMANSKI[1], and H. K. HALL JR.

Department of Chemistry, University of Arizona, Tucson, AZ 85721

A review of the synthesis and polymerization of bicyclic acetals and orthoesters is presented, and the relationship between ring structure and the ability to polymerize is discussed. The highly labile bicyclic acetals and orthoesters were synthesized at high temperature in dioctyl phthalate under vacuum in order to remove the monomers as soon as they form. The ability of the bicyclic monomers to polymerize falls in the same sequence as the ring strain: [2.2.1] > [2.2.2] > [3.2.1] > [3.3.1]. The structure of the polymers is determined by two factors: first, which bond is preferentially cleaved, and second, if the reaction is stereospecific or not. The first factor is determined by Deslongchamps' theory and by the nucleophilicity of the ring oxygens. For the least strained bicyclic acetals, the S_N2 mechanism leads to stereospecific polymer. The [2.2.1] acetals and the orthoesters, except [2.2.2] and [3.3.1], yield random polymer. An A_C2 mechanism, bimolecular addition on a carbenium ion, is supported in which the carbenium ion, and not the bicyclic oxonium ion, is the growing center. The stereospecificity observed in a few cases is explained by dipole-dipole interaction of chain oxygens with the growing carbenium ion.

This review summarizes all the data we obtained on the synthesis and cationic ring-opening polymerization of bicyclic acetals and orthoesters, and discusses the relationship between ring-strain and polymerizability. This ties in with earlier work on the polymerizability of monocyclic and bicyclic lactams (1-2). A new mechanism for the propagation step in the polymerization of bicyclic orthoesters is supported.

[1]Current address: Center of Molecular and Macromolecular Studies, Polish Academy of Sciences, 90-362 Łódź, Poland

0097–6156/85/0286–0313$06.25/0

Bicyclic Acetals: Synthesis

The polymerization of unsubstituted bicyclic acetals promised to be
a very good field to study the relationship between ring strain and
polymerizability. In 1973, when we published the first paper in
this series, Schuerch had already described the polymerization of
substituted 1,6-anhydro-aldohexoses to high molecular weight
crystalline polymers (3).

Hall and Steuck described the synthesis of one of the least
strained members of the acetal series, namely, 6,8-dioxabicyclo-
[3.2.1]octane (4). Kops in Denmark (5) and Sumitomo in Japan (6)
independently described their work on this monomer at the same time,
and discussed both the synthesis and the polymerization. This
bicyclic acetal was synthesized by conventional methods of acidic
ring-closure of the cyclic alcohol.

2,6-Dioxabicyclo[2.2.2]octane is a more strained system and a
slightly unusual synthesis was used. The diol acetal was treated
under acidic conditions to obtain ring closure (7).

The bicyclo[2.2.1]heptane acetals are much more strained
systems. More vigorous conditions are required to obtain ring closure. At the same time these acetals are much more labile and
easily undergo premature oligomerization. To circumvent this thermal polymerization, the dioctyl phthalate synthesis method was
devised. These syntheses in DOP occur in two steps. The precursor
is dissolved in dioctyl phthalate and heated in the presence of a
trace of acid catalyst. At atmospheric pressure the first equivalent of alcohol formed is distilled off, resulting in a monocyclic
structure. Higher temperatures are required in the second step,
which results in the labile bicyclic compound. A vacuum is applied
to the system in this second step, and the desired bicyclic acetal
will distill out of the system as it is formed and can be collected
in a cold trap.

2,6-Dioxabicyclo[2.2.1]heptane is prepared from the diol acetal
using the DOP method (8-9).

2,7-Dioxabicyclo[2.2.1]heptane is also synthesized using the DOP method, starting from the appropriate diol acetal, or more effectively, starting from acrolein and methyl vinyl ether (8-9).

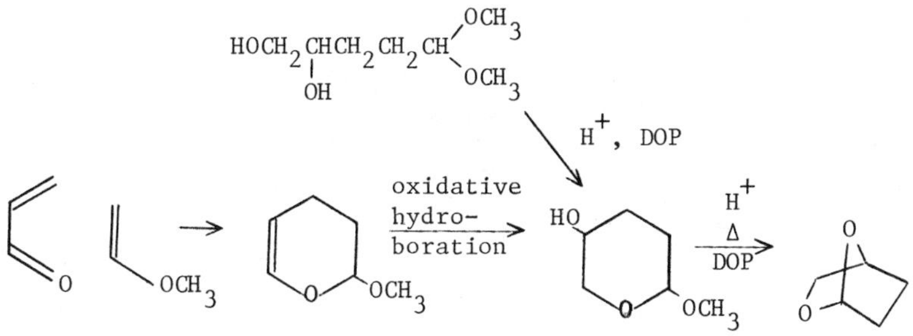

Acid Hydrolysis of Bicyclic Acetals

The reactivity of the bicyclic acetals was appraised by means of acid-catalyzed hydrolysis, and was compared to the reactivity of acyclic acetals (8-9). The studies were carried out by following the disappearance of the bridgehead acetal proton in the NMR spectrum (Table I).

Table I. Hydrolysis rates of acetals
(dichloroacetic acid as catalyst)

Acetal	k_1 (sec^{-1})	relative reactivity
dimethyl acetal		1
[3.2.1]	6×10^{-5}	7.7
[2.2.2]	1.8×10^{-4}	2.5×10^3
2,7-[2.2.1]	1.8×10^{-3}	2.5×10^4
2,6-[2.2.1]	5.3×10^{-3}	6.9×10^5

The difference in hydrolysis rates is remarkable and can be broadly correlated with the ring strain: [2.2.1] > [2.2.2] > [3.2.1].

Another striking feature is the demonstration of marked general acid catalysis in these systems. Such catalysis is rare for acetals, however, it may be encountered where steric strain is relieved in the rate-determining step, as in these cases (10).

Polymerization of Bicyclic Acetals

Reactivity. Qualitatively, the order of reactivity is the same as found for the hydrolysis rates: [3.2.1]octane < [2.2.2]octane < 2,7-[2.2.1]heptane < 2,6-[2.2.1]heptane. This order was determined by comparing the strength of the initiator needed, temperature, time and yield, etc.

6,8-dioxabicyclo[3.2.1]octane: only PF$_5$ in dichloromethane at -78°C is an effective initiator (4).

2,6-dioxabicyclo[2.2.2]octane: PF$_5$ and fluorosulfonic acid at -78°C are the most effective initiators; protonic acids such as methanesulfonic acid and trifluoroacetic acid are also effective at room temperature (7).

2,7-dioxabicyclo[2.2.1]heptane: PF$_5$ at -78°C is the best but only low molecular weight polymer is obtained; also methanesulfonic acid and methyl triflate at room temperature (9).

2,6-dioxabicyclo[2.2.1]heptane: Lewis acids are too vigorous, even at -78°C, partial gel formation occurs; even trichloroacetic acid results in quantitative yield at room temperature (9).

Polymer Structure. The structure of the acetal polymers has been determined by NMR spectroscopy. The conformational equilibrium for both cis and trans isomers of the dioxabicyclo[3.2.1]- and [2.2.2] octane polymers was calculated by the interplay of two factors: 1) the familiar preference of alkyl substitutents to be in the equatorial position, and 2) the preference of the alkoxy groups to be in the axial position, due to the anomeric effect.

For the polymer of 6,8-dioxabicyclo[3.2.1]octane, only one acetal proton is observed (it is equatorial), indicating pure trans-1,3-tetrahydropyranoside polymer. Runs at higher temperature show some absorption for the cis isomer.

For the polymer of 2,6-dioxabicyclo[2.2.2]octane, the situation is not so clear, because in this system neither the trans- nor the cis-1,4-units would be conformationally homogeneous. The observed ratio of axial and equatorial protons of the polymers formed at -78°C though are in very good agreement with the calculated values for the trans isomer. It can thus be concluded that these polymers are also pure trans-1,4-tetrahydropyranoside units. At higher temperatures, more and more cis-linkages are observed.

Examination by NMR of the polymers obtained from 2,7-dioxabicyclo[2.2.1]heptane at -78°C, showed that the polymer was exclusively composed of tetrahydrofuran structures. Cis and trans isomers are randomly distributed. Although the reaction is not stereospecific, very specific bond cleavage of the C_1-O_2 bond is observed.

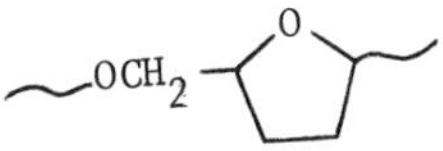

cis/trans

At higher temperatures the tetrahydrofuran link was still favored, although several initiators gave significant amounts of tetrahydropyran links.

The structure of the polymer obtained from ring-opening of 2,6-dioxabicyclo[2.2.1]heptane was examined by NMR, both proton and ^{13}C. A mixture of cis and trans isomers of 2,4-tetrahydrofuran links is observed.

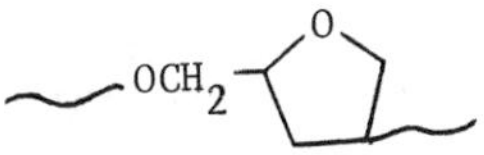

cis/trans

Mechanism. Two mechanistic aspects determine the polymer structure: first, which bond cleaves? and secondly, is the reaction stereospecific?

The polymer obtained from 6,8-dioxabicyclo[3.2.1]octane only contains tetrahydropyran rings, while the polymer obtained from 2,7-dioxabicyclo[2.2.1]heptane at low temperature only contains tetrahydrofuran rings. Very specific bond-cleavage occurs in both these cases. The other two bicyclic acetals can only result in one polymer structure, due to the symmetry of the monomers.

The stereoselective cleavage of a bond in the oxonium ion may be interpreted by the approach used by Deslongchamps (11-12). Specific cleavage of a carbon-oxygen bond occurs when two heteroatoms of the tetrahedral intermediate each have one non-bonded electron pair oriented antiperiplanar to the departing O-alkyl group. In the bicyclic acetals there is only one heteroatom present, thus the situation is not complete, but one p-orbital is nicely antiperiplanar (shaded) to the bond to be cleaved. There is another report in the literature that in order to obtain specific cleavage of a bond it is enough that one p-orbital is oriented antiperiplanar to the departing group (13). Another factor in favor of this bond cleavage is that the oxygen atoms in the larger bridges are slightly more nucleophilic.

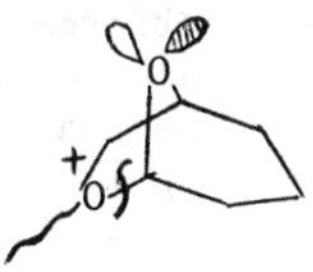

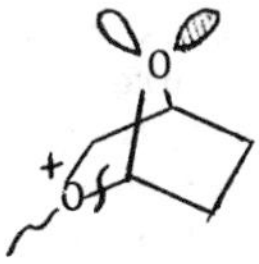

The cationic ring-opening polymerizations of 6,8-dioxabicyclo-
[3.2.1]octane and 2,6-dioxabicyclo[2.2.2]octane are highly stereoregular at low temperatures and result in pure trans linkages in the
polymer. This implies an S_N2 mechanism via bicyclic oxonium ions.

Neither [2.2.1]heptane acetal polymerizes to stereoregular
polymer. Another mechanism must be postulated in this case, and it
can be presented as follows:

The incoming monomer can attack the oxacarbenium ion from either
side, resulting in a random polymer.

A similar mechanism via an oxacarbenium ion was recently proposed in the cationic ring-opening polymerization of 1,4-anhydro-
2,3-di-O-benzyl-α-D-xylopyranose by Uryu and Matsuzaki (14). If
boron trifluoride etherate, silicon tetrafluoride or phosphorus pen-
tafluoride are used as initiator, the authors obtained stereoregular
polymer from this substituted bicyclic acetal. If the bulky initiator antimony pentachloride is used, a mixture of α- and β-linked
furanose units is obtained in the polymer chain.

This mechanism via an oxacarbenium ion is an S_N1 mechanism, if
the monomolecular cleavage of the bicyclic oxonium ion is the rate-
determining step. A variant of this mechanism will be discussed
later in the case of the polymerization of bicyclic orthoesters.

Bicyclic Orthoesters: Synthesis

There are several reports in the literature on the synthesis of
bicyclic orthoesters. Crank and Eastwood reported the synthesis of
the [3.2.1], [3.3.1] and [4.2.1] derivatives, and also noted that
these became viscous and turned to glass at room temperature (15).
Bailey reported the synthesis of more strained systems, namely,
1,4-diethyl and 4-ethyl-2,6,7-trioxabicyclo[2.2.2]octane (16).

Most bicyclic orthoesters we synthesized were obtained from the
corresponding triol and trimethyl or triethyl orthoformate. Again
it was crucial that the very sensitive monomers were removed from
the reaction mixture as formed. The reactions were run in dioctyl
phthalate in the presence of a trace of p-toluenesulfonic acid as
catalyst. The closure of the first ring was performed at
atmospheric pressure until two equivalents of alcohol were
collected. The system was then evacuated and heated while continuing very vigorous stirring. The formed bicyclic compounds are
then collected in a dry ice cooled trap or, in a modified method, as
a solid in a large sublimation apparatus.

The following unsubstituted bicyclic orthoesters were thus synthesized (17-19):

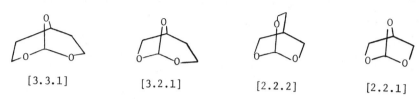

[3.3.1]   [3.2.1]   [2.2.2]   [2.2.1]

These reactions prove the generality of the DOP method, and this method can probably be extended to the synthesis of other highly strained systems. The same method was also used for the synthesis of a whole series of substituted bicyclic orthoesters.

Numerous orthoformates and orthoacetates were synthesized with different groups in the other bridgehead position (19).

R = H or CH$_3$

R' = alkyl, CH$_2$Br, CH$_2$OH, NO$_2$, NH$_2$, NMe$_2$, NHCOCH$_3$, COOMe, CH$_2$OSO$_2$C$_7$H$_7$-p

The alkyl, bromomethyl, hydroxymethyl, nitro and dimethylamino-substituted bicyclic orthoformates and orthoacetates were synthesized by the interchange reaction of the acyclic triol and acyclic orthoester. The other orthoesters were obtained by derivatization of the former.

To obtain cationic water-soluble polyorthoesters, we also synthesized two bicyclic orthoesters containing an ammonium salt substituent. 4-Trimethylammonio-2,6,7-trioxabicyclo[2.2.2]octane trifluoromethanesulfonate was obtained by direct alkylation of the corresponding dimethylamino bicyclic orthoester with methyl trifluoromethanesulfonate (20).

2,6,7-Trioxabicyclo[2.2.2]octane-4-methylene-N-pyridinium trifluoromethanesulfonate was obtained by reaction of the 4-hydroxymethyl derivative with trifluoromethanesulfonic anhydride in pyridine solution (20).

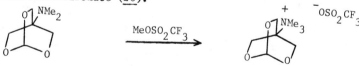

Hydrolysis of Bicyclic Orthoesters. As seen earlier, the bicyclic acetals are markedly more reactive towards acid hydrolysis than their acyclic counterparts. These rate accelerations have been attributed to partial relief of the ring strain upon ring opening hydrolysis.

The hydrolysis rates for several bicyclic orthoesters have been measured using acetic acid as catalyst, and compared to the hydrolysis rates of acyclic orthoformates (17).

Table II. Hydrolysis rates of bicyclic orthoesters

Orthoester	$10^4 \times k_1$ (sec^{-1})
CH(OCH$_3$)$_3$	19
CH(OC$_2$H$_5$)$_3$	15
[2.2.1]	10
[3.3.1]	4.3
[3.2.1]	4.1
[4.2.1]	3.2

These unexpected results were at the time attributed to a very early transition state, with very little C–O bond breaking and consequently very little strain release.

Another explanation was suggested during a more detailed study of the hydrolysis of the [2.2.1] system (21). A change in the rate-determining step may be responsible for the unusual results. Recent studies of orthoester hydrolysis in which the first step of this reaction is rate-determining, changes to one in which the third step is slower by making structural changes that accelerate the rate of the first step, or by a change in pH.

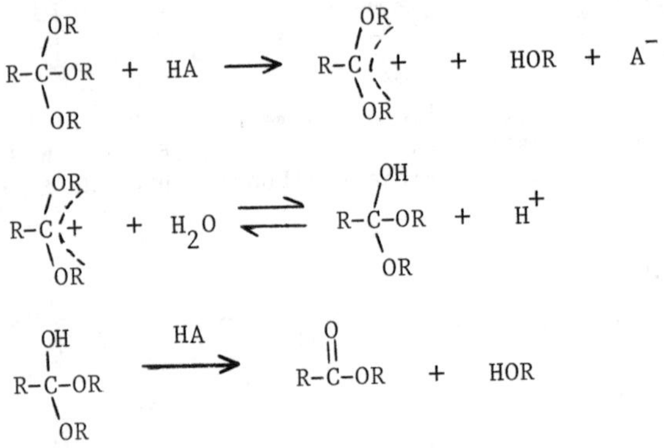

$$R-\underset{\underset{OR}{|}}{\overset{\overset{OR}{|}}{C}}-OR \; + \; HA \; \longrightarrow \; R-\underset{\underset{OR}{\backslash}}{\overset{\overset{OR}{/}}{C}}{+} \; + \; HOR \; + \; A^-$$

$$R-\underset{\underset{OR}{\backslash}}{\overset{\overset{OR}{/}}{C}}{+} \; + \; H_2O \; \rightleftharpoons \; R-\underset{\underset{OR}{|}}{\overset{\overset{OH}{|}}{C}}-OR \; + \; H^+$$

$$R-\underset{\underset{OR}{|}}{\overset{\overset{OH}{|}}{C}}-OR \; \overset{HA}{\longrightarrow} \; R-\overset{\overset{O}{||}}{C}-OR \; + \; HOR$$

Introduction of strain, as in a bicyclic system, is just such a structural change, and the measured rate constant might have been the one for the slower third step. As such, the acceleration due to the ring strain of the first step would be masked by the slower third step.

Such an effect was indeed observed, but the initial ring opening reaction proved to be not markedly faster than the corresponding reactions of monocyclic and acyclic models. Thus, the theory of an early transition state still stands.

Polymerization of Bicyclic Orthoesters

Reactivity. The reactivities of the bicyclic orthoesters can be compared by examining the conditions necessary to form polymer. Although the hydrolytic reactivities were not accelerated, these monomers were indeed very reactive in polymerization. In contrast to the behavior of the bicyclic acetals, no correlation is found between the hydrolytic reactivity and the reactivity towards cationic initiators for the bicyclic orthoesters. The following order can be proposed: [2.2.1] ≯ [2.2.2] > [3.2.1] > [3.3.1] which is the expected order from the ring strains (18–19).

Polymer structure – kinetic control. The structure of the polymer of 2,6,7-trioxabicyclo[2.2.1]heptane is dependent on the initiator at -78°C in dichloromethane (18). The polymer is mostly five-membered rings. As in the bicyclic acetals, in controlled conditions, only C_1–O_2 bond cleavage occurs.

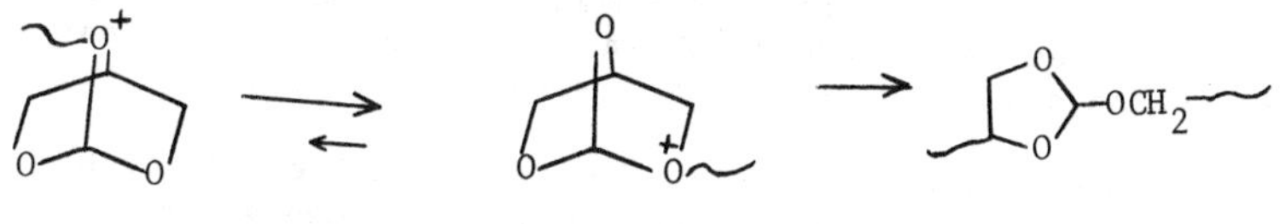

Contrary to an earlier report, the polymer obtained from the bicyclic orthoacetate is also composed of five-membered rings, as deduced from ^{13}C NMR spectra (22).

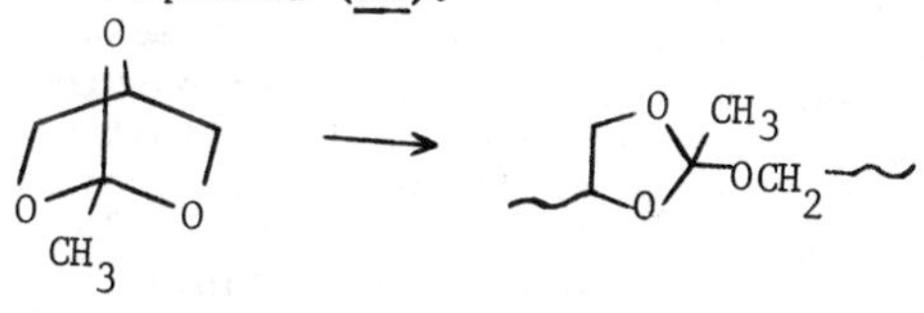

The polymer obtained from 2,6,7-trioxabicyclo[2.2.2]octane is obviously composed of six-membered rings (19). All the polymerizable substituted bicyclic [2.2.2]orthoesters also form the dioxane polymers. With the exception of 4-carbomethoxy-2,6,7-trioxabicyclo [2.2.2]octane, these are highly crystalline white powders with very high melting points with decomposition, and are not soluble in any

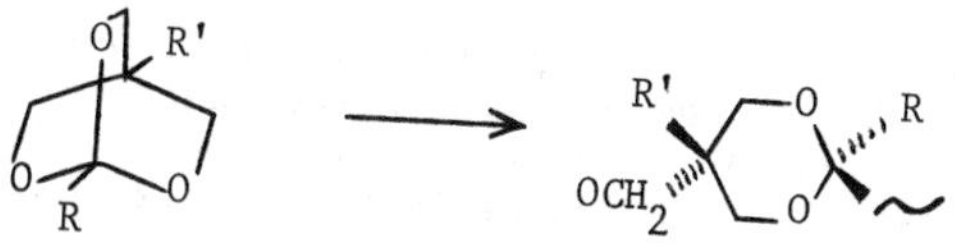

solvent without decomposition. In contrast with the polymers
obtained by Bailey, which will be discussed later, these polymers do
not show carbonyl stretching in the infrared spectrum, proving that
only one ring opens. ^{1}H and ^{13}C NMR spectra of low molecular weight
polymer of unsubstituted 2,6,7-trioxabicyclo[2.2.2]octane show only
one signal for the orthoester proton and carbon. These results
suggest that the stereoregularity of this polymer is extremely high,
probably trans.

The polymer obtained from 2,7,8-trioxabicyclo[3.2.1]octane consists mostly of five-membered rings (23). Six- and seven-membered
rings are also observed, but only for about 15%. These polymers
were analyzed by ^{1}H NMR and by comparison of these spectra with
model compounds. The ratio of cis and trans is always about .50/50.
No stereoselectivity is observed, although the presence of mainly
five-membered rings indicates a great preference to cleave the C_1-O_2
bond.

2,7,8-Trioxabicyclo[3.3.1]nonane leads to dimers and oligomers
under the influence of most cationic initiators (23). But when
trifluoromethanesulfonic acid is used as initiator in dichloromethane at -78°C, only dimer is formed. Only one orthoformate proton is observed in the NMR spectrum, which indicates that only one
conformation and configuration of monomer units occurs in the dimer.

Water-soluble polymers. Substituted water-soluble poly-orthoesters
can be regarded as polysaccharide analogs, and could have potential
medical applications as drugs or as drug carriers. Ionic substituents were introduced in the bicyclic monomers to improve the solubility of these polymers.

The two bicyclic orthoesters with the trimethylammonio- and the
pyridinium-methylene-substituents in the 4-position did not polymerize well under the influence of cationic initiators, as could be
expected. However, high molecular weight poly-orthoester could be
obtained by copolymerizing the former with 2,6,7-trioxabicyclo-
[2.2.1]heptane (20). The copolymer is soluble in water and in organic solvents.

In order to obtain an anionic poly-orthoester, 4-carbomethoxy-2,6,7-
trioxabicyclo[2.2.2]octane was polymerized using trifluoromethanesulfonic acid as initiator in dichloromethane at -78°C (20). This
is the only known poly-orthoester in the [2.2.2]-series that is

soluble in organic solvents. Also, both ^{1}H and ^{13}C NMR spectra
indicated that approximately equal amounts of two isomers are present, in contrast with all the other polymers in this series which
were stereoregular. This polymer was then hydrolyzed in aqueous
pyridine/potassium hydroxide, and the products were separated by
dialysis. Only a 50% yield of hydrolyzed product is obtained, but
this is the first report of a water-soluble poly-orthoester with
anionic substituents on the polymer chain.

Polymer structure-thermodynamic control. The polymers described
above, containing dioxolane and dioxane rings, are the kinetically
controlled products in these polymerizations. Bailey reported the
polymerization of several bicyclic [2.2.2]orthoesters and also spiro
orthoesters under much harsher conditions (24). Simultaneous
opening of both rings is achieved in this way, leading to linear
polyethers with ester branching. This is usually accompanied by
volume expansion.

We thoroughly investigated the polymerization of 2,6,7-trioxabi-
cyclo[2.2.1]heptane under different conditions (25). At low temperature (-78°C) only one ring opened, but at higher temperatures
(80°C) polyethers with formate side chains are obtained. The
opening of the second ring was also observed in dioxolane polymers,
left at room temperature for several days or heated to 80°C for a
few hours. The dioxolane-containing polymer was stable if the acid
catalyst was neutralized.

The same phenomenon was observed in the polymerization of the
[2.2.2]orthoesters: polymers with ester-branching are formed under
vigorous conditions, by isomerization of the kinetically controlled
polymers under acid catalysis (19). More recently, Matyjaszewski
also described the same phenomenon in the polymerization of a spiro
orthoester, 1,4,6-trioxaspiro[4.4]nonane (26).

Mechanism; Selective bond cleavage. Cationic polymerization of
2,6,7-trioxabicyclo[2.2.1]heptane yields a polymer consisting of

almost exclusively five-membered rings, indicating that cleavage of
the $C_1-O_{2/6}$ bond is preferred (<u>18</u>). Thermodynamically the six-
membered ring is more stable than the five-membered rings obtained
in these polymers. The polymer formation is kinetically controlled.

The selective cleavage of a specific bond in the oxonium ion is
interpreted as follows: 1) O_2 and O_6 have higher p-character than
O_7, because their bond angles are close to 100° and 90°, respectively. Accordingly, O_2 and O_6 are more nucleophilic. 2) The
stereoselective control here can again be interpreted along the
lines of Deslongchamps' theory, as discussed for the bicyclic acetals. Here though, always two p-orbitals (shaded) will be antiperiplanar to the bond to be cleaved.

The same phenomenon is observed in the polymerization of 2,7,8-
trioxabicyclo[3.2.1]octane (<u>23</u>). The five-membered rings are dominant in the polymer chain, although some six- and seven-membered
rings are also observed. The preferred C_1-O_2 bond cleavage can be
ascribed to Deslongchamps' theory, which is valid for two oxygen
atoms in the monomer, and to the higher p-character of the oxygen at
the 2-position. This oxygen is on the longest bridge and thus less
strained than the others.

<u>A dioxacarbenium ion mechanism for bicyclic orthoesters.</u> The
equilibrium between carbenium ions and oxonium ions in the cationic
polymerization of monomers containing heteroatoms has been discussed
in the literature, specifically in the case of cyclic ethers and
acetals (<u>27</u>-<u>29</u>). In analogy, the growing center in the polymerization of bicyclic orthoesters can be either a bicyclic oxonium ion or
a dioxacarbenium ion. In this reaction sequence either step can be

rate-determining. If the first step is fast, then the monomolecular
ring opening of the bicyclic oxonium ion is rate-determining and the

mechanism is a classical S_N1, which was proposed when this work was first published (18). If the attack of monomer on the dioxacarbenium ion is rate-determining, then most of the growing centers will be carbenium ions. The non-stereospecificity of the polymerization in either case is due to attack of monomer on either side of the planar dioxacarbenium ion. In the polymerization of monocyclic ethers and acetals discussed by Penczek (27), the carbenium ions are primary, with one adjacent oxygen in the latter case, and propagation occurs through the more stable oxonium ions. The obvious stabilization of the cyclic dioxacarbenium ions by the two adjacent oxygens made us reexamine this reaction.

First, kinetic studies of the polymerization of 2,6,7-trioxabicyclo[2.2.1]heptane were attempted. 1,3-Dioxolan-2-ylium salt was chosen as initiator, because of the similarity of the initiation reaction in this case to the propagation reaction.

$k_i \cong k_p$, where k_i and k_p are the initiation and propagation rate constants.

Preliminary experiments indicated that the propagation rate constant is too high to be measured by conventional methods. Even when the initiator concentration was as low as 1.1×10^{-7} M in nitromethane-d$_3$, the propagation reaction was finished before an NMR spectrum could be recorded to check the monomer/polymer ratio. Inasmuch as after 5 minutes of reaction $[M_0]/[M]>30$ for $[I_0] = 1.1 \times 10^{-7}$ M, the propagation rate constant was estimated to be higher than 1×10^5 M^{-1}s^{-1} at room temperature. This lower limit of the propagation rate constant suggests that dioxolanylium cations, and not oxonium ions, are the active species. The rate constants for combination of various carbenium ions with neutral nucleophiles are reported to be greater than 10^6 M^{-1}s^{-1} (28,30-31). For example, the rate constant of the reaction of methoxymethylium cation with methyl ether in sulfur dioxide is about 5×10^6 M^{-1}s^{-1} at 30° (28). The rate constant of the reaction of 1,3-dioxolan-2-ylium cation with oxygen nucleophiles is expected to be about one order of magnitude lower due to the stabilization provided by the two oxygen atoms.

Second, the oxonium ion-carbenium ion equilibrium in the orthoester systems was investigated. A model compound for the active center was synthesized. 4-Methoxymethyl-1,3-dioxolan-2-ylium cation is the model compound for the carbenium ion in the polymerization of 2,6,7-trioxabicyclo[2.2.1]heptane, and was obtained by hydride transfer from 4-methoxymethyl-1,3-dioxolane to triphenylmethylium cation.

The ^{1}H NMR spectra of the reaction mixture confirm that intermolecular (with unreacted acetal) as well as intermolecular
carbenium-oxonium ion equilibria are shifted far toward the carbenium cation in nitromethane-d$_3$ at room temperature. No evidence
for oxonium ion formation was found.

Unfortunately, the system could not be investigated at lower temperature because of precipitation of the salts.

5-Methoxymethyl-5-methyl-1,3-dioxane was briefly investigated as
an example for a less strained system and the precursor of the model
compound for the growing cation in the polymerization of 2,6,7-
trioxabicyclo[2.2.2]octane. Hydride transfer to triphenylmethylium
cation is so slow and the resulting 1,3-dioxan-2-ylium cation is so
unstable that a considerable amount of side products is formed
before half of the dioxane has reacted. But even in these rough
experiments the concentration of carbenium ion clearly exceeds the
concentration of oxonium ions.

Third, the reaction of 1,3-dioxolan-2-ylium cation with monocyclic orthoesters was investigated. A catalytic amount of
1,3-dioxolan-2-ylium cation brought about the exchange between
2-methoxy- and 2-ethoxy-1,3-dioxolane.

This exchange is due to combination of the 1,3-dioxolan-2-ylium cation with 2-alkoxy-1,3-dioxolane, which leads to an unstable symmetric oxonium cation.

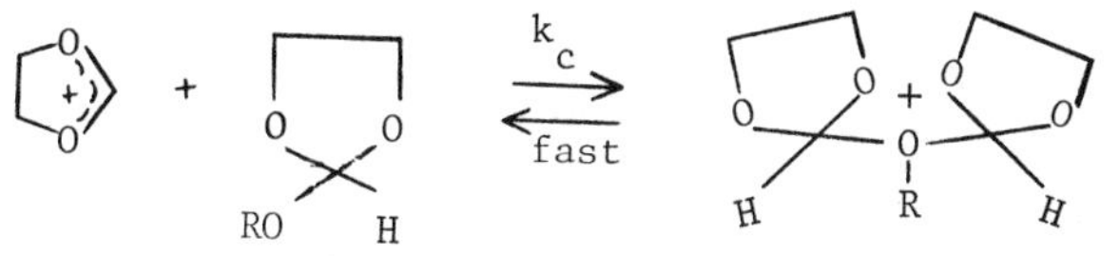

This exchange is the model reaction for the attack of monomer on the dioxacarbenium ion in the propagation step.

The combination rate constant k_c was derived from the exchange rate constant which was measured by dynamic ^{1}H NMR line shape analysis. The combination rate constant is about 3×10^4 $M^{-1}s^{-1}$ at 20° with a low enthalpy of activation, about 2 kcal/mole (see Experimental for more details). The high rate of exchange in the 2-alkoxy-1,3-dioxolane system catalyzed by 1,3-dioxolan-2-ylium cation suggests the similarity of this process to the cationic polymerization of the orthoesters, and that the active species in both processes have the 1,3-dioxolan-2-ylium structure rather than the oxonium structure. The fact that the lower limit for the propagation rate constant is higher than the rate constant for the model reaction can be explained by steric and statistical factors. The monomer has two nucleophilic centers which are less sterically hindered than the one alkoxy oxygen atom in the 2-alkoxy-1,3-dioxolanes.

Taking all these data into account, the following A_C2 mechanism (bimolecular addition on a carbenium ion) is supported for the polymerization of the [2.2.1]bicyclic orthoester. It presumably is valid for all bicyclic orthoesters, at least at room temperature or higher temperatures.

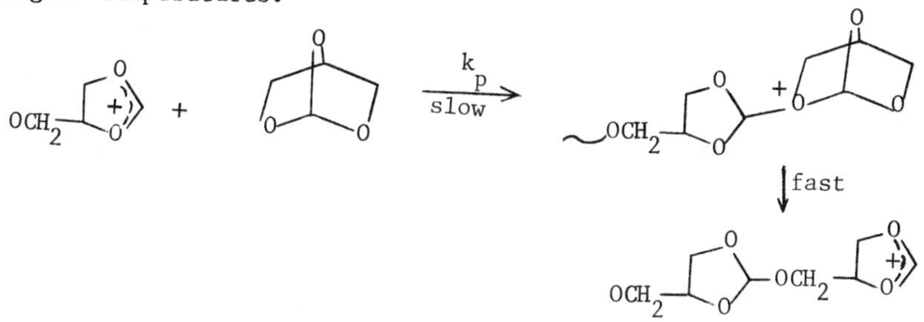

In this mechanism the rate-determining step is the bimolecular reaction of the cyclic dioxacarbenium ion with monomer. The A_C2 mechanism also predicts the non-stereospecificity of the propagation step.

The polymer obtained from 2,7,8-trioxabicyclo[3.2.1]octane is not stereospecific, and as such the A_C2 mechanism can also be proposed in this case.

High stereoselectivity is observed in the unsubstituted 2,6,7-trioxabicyclo[2.2.2]octane polymerization. In analogy to the [2.2.1] case, the A_C2 mechanism is also proposed in this case. The high stereoselectivity is due to weak dipole-dipole interactions between the dioxacarbenium ion and the first oxygen in the chain, which is in perfect position to shield one side of the dioxacarbenium ion (19).

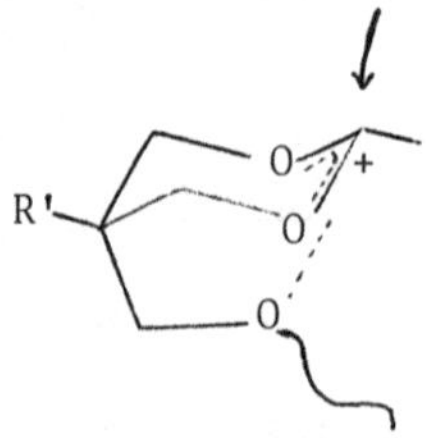

The fact that the [3.3.1] bicyclic orthoester does polymerize, is rather surprising, because it consists of two fused six-membered rings in the chair form. A small but significant ring strain is due to the repulsion between the two endo-protons at C_3 and C_7. The formation of only one isomer of the dimer is ascribed to the same weak dipole-dipole interactions between the dioxacarbenium ion and the first chain oxygen which force the incoming monomer to react in a stereoselective fashion. The formation of the dimer can then be interpreted in terms of back-biting of the growing chain end (23). From these results, the A_C2 mechanism can also be proposed in this oligomerization.

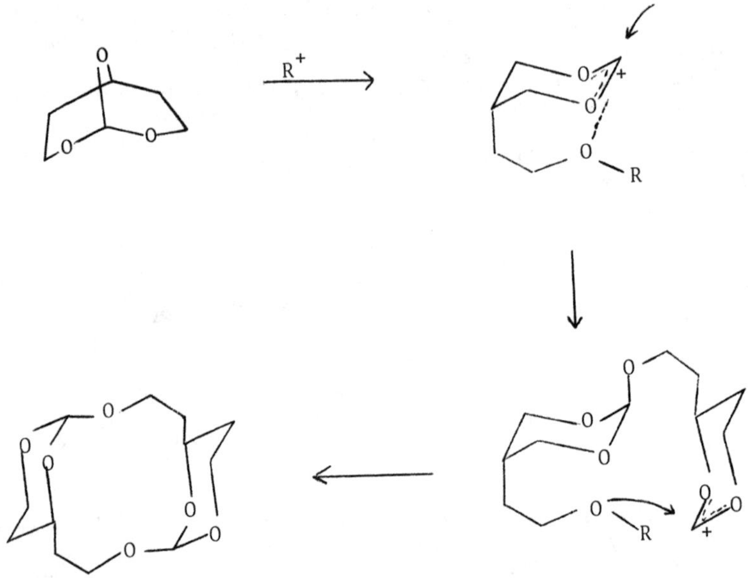

Conclusion

In conclusion, we can state that ring strain does play an important role in determining the polymerizability of bicyclic acetals and orthoesters. The sequence is as follows: [2.2.1] > [2.2.2] > [3.2.1] > [3.3.1]. The ring strain is affected by the ring size, the gauche interactions in the monomer, the conformation in the monomer, and the anomeric effect, as discussed by Yokoyama and Hall in a recent review (32). This sequence is general for all bicyclic compounds, independently of the functionalities. It is valid for the bicyclic acetals and orthoesters as described in this review, for the bicyclic lactams and lactones, bicyclic oxalactones and oxalactams (33), etc.

The preferred bond cleavage in the polymerization of these bicyclic acetals and orthoesters is determined by Deslongchamps' theory. If all oxygens are equally preferred by the former, the p-character of the oxygen atom, determined by the bond angles, will be the dominant factor.

For the least strained bicyclic acetals, the S_N2 mechanism is proposed, leading to stereospecific polymer. Polymerization of the [2.2.1] acetals leads to random polymer, and a mechanism via an oxacarbenium ion is supported.

A novel mechanism is supported for the bicyclic orthoester polymerization, namely, an A_C2 mechanism, in which the growing center is the dioxolanylium cation, and not the bicyclic oxonium ion.

Experimental of Bicyclic Orthoesters Mechanism Studies

Instrumental. ^1H NMR spectra were recorded on 60 MHz Varian EM360L NMR spectrometer and 250 MHz Brucker WM-250 FT spectrometer. ^{13}C NMR spectrum was recorded on Brucker WH-90 spectrometer.

Solvent. Nitromethane (Aldrich) was kept in a vacuum ampoule equipped with a Rotaflo stopcock over calcium hydride. It was distilled into the reaction vessels on a vacuum line.

Monomer. 2,6,7-Trioxabicyclo[2.2.1]heptane was prepared by the DOP method (18) and was kept in a vacuum ampoule (equipped with a Rotaflo stopcock) over a sodium mirror. Samples for polymerization were prepared by distillation of the monomer in a vacuum system to ampoules equipped with breakseals.

Initiator. 1,3-Dioxolan-2-ylium hexafluoroantimonate was prepared and handled as described by Stolarczyk, Kubisa and Penczek (34). Solutions of the salt were prepared in vacuum flasks equipped in Rotaflo stopcocks. Triphenylmethyl hexafluoroantimonate (Ozark-Mahoning) was purified in precipitation from dichloromethane solution by hexane.

Model Compounds. **4-Methoxymethyl-1,3-dioxolane:** 3-Methoxy-1,2-propanediol (5.3g) was mixed with 10g of dimethoxymethane and, after adding a catalytic amount of p-toluenesulfonic acid, the mixture was refluxed for 3 hr. Then the reaction mixture was distilled on a Vigreux column. Yield: 0.3g (5%), b.p. 40-50°/20mmHg, ^1H NMR (CD$_3$NO$_2$), δ: 4.87(s, 1H), 4.73(s, 1H), 4.2-3.3(m, 5H), 3.30(s, 3H) ppm. The main product was dimer: yield: 4.2g (71%), b.p. 68°/5mmHg.

   5-Hydroxymethyl-5-methyl-1,3-dioxane: A mixture of 20g of 2-hydroxymethyl-2-methyl-1,3-propanediol (Aldrich), 5g of trioxane (Matheson, Coleman and Bell) and 1.5mL of concentrated hydrochloric acid was slowly heated to 100° until the mixture became homogeneous.

Then most water was distilled off at 20mm and the residue fraction-
nated on a Vigreux column. Yield: 15g (68%), b.p. 78°/2mmHg, [1]H
NMR (acetone-d_6, δ: 4.6-4.95 (pseudo quartet, J≈6Hz, 2H), 3.3-4.0
(pseudo quartet, J≈10Hz), 3.70(s), 3.53(s) (combined 7H), 0.85(s,
3H) ppm.

5-Methoxymethyl-5-methyl-1,3-dioxane: 5-Hydroxymethyl-5-methyl-
1,3-dioxane (8g) was dissolved in 50mL of anhydrous tetrahydrofuran.
Then 1.5g of sodium and 4mL of methyl iodide was added. The reaction mixture was stirred overnight at a temperature below 40°. The
solid was filtered off and the filtrate distilled. Yield: 4.0g
(50%), b.p. 52-54°/20mmHg, ^{1}H NMR (acetone-d_6), δ: 4.6-5.0 (pseudo
quartet, J≈5Hz, 2H), 3.3-4.0 (pseudo quartet, J≈11Hz), 3.40(s),
3.33(s) (combined 9H), 0.87(s, 3H) ppm.

2-Methoxy-1,3-dioxolane and 2-ethoxy-1,3-dioxolane: Trimethyl
orthoformate, or triethyl orthoformate, was mixed with an equimolar
amount of ethylene glycol in the presence of a trace of p-toluenesulfonic acid. The mixture was heated to reflux for one hour, and
then the corresponding alcohol is distilled off. The crude products
were purified by distillation and kept in a vacuum ampoule over
calcium hydride. ^{1}H NMR (CD_3NO_2), δ: 5.7(s, 1H), 4.0(s, 4H),
3.3(s, 3H) ppm.

Polymerization kinetic studies. The polymerizations were carried in
completely sealed vacuum vessels which were equipped with two
breakseal ampoules, one with the monomer, the other with a solution
of initiator. They were also equipped with two NMR tubes. The
desired amount of solvent was distilled in on the vacuum line and
the reaction vessel was sealed off. The breakseal of the monomer
was then broken. After dissolving the monomer, a sample of the
solution was sealed off in one NMR tube to make sure that the
monomer is stable in the solution without initiator. Then the
breakseal of initiator solution was broken and the solutions mixed.
The NMR tube was filled and sealed off. Immediately ^{1}H NMR spectra
were recorded.

Vacuum line technique. In the past, the polymerizations were run in
vacuum or in a nitrogen or argon atmosphere, and septums and
syringes were used extensively. The present investigation using
high vacuum (10^{-5}mmHg) was undertaken to determine if the low molecular weights and the non-stereospecific nature of the obtained
polymers were due to the methods used or to the mechanism. The present study clearly indicates that the use of high vacuum does not
improve the stereospecificity of the polymer. The molecular weights
of the polymer were improved and closer to the expected value calculated from the monomer/initiator value. Also monomer stored over a
sodium mirror in high vacuum conditions was completely stable,
proving that the difficulties encountered before in storing the
monomer were solely due to adventitious hydrolysis.

Hydride transfer reactions. A known amount of triphenylmethyl
hexafluoroantimonate was placed in an NMR tube attached to a vacuum
line. After 30 minutes vacuum drying of the salt, the desired
amount of nitromethane-d_3 and model compound was distilled in and
the NMR tube sealed off. The reactions were carried out at room
temperature and the products were not isolated.

Orthoformate exchange studies. The rate constant for the exchange
between 2-methoxy-1,3-dioxolane and 2-ethoxy-1,3-dioxolane brought
about by a catalytic amount of 1,3-dioxolan-2-ylium cation was
measured by dynamic ^{1}H NMR line shape analysis. The following reactions occur in this system.

The dioxolane rings are marked to present the path of the exchange
between the two orthoformates.

The oxonium ion is an intermediate in the exchange reaction, and
because of its symmetry the exchange rate constant is equal to $k_c/2$
(combination rate constant). If the exchange occurs in equimolar
solution of 2-methoxy- and 2-ethoxy-1,3-dioxolane, the pseudo first
order rate constant of exchange may be expressed as

$$k_{ex} = k_c [C^+]_o/4$$

where $[C^+]_o$ is the 1,3-dioxolan-2-ylium salt concentration. This
assumption is correct if the concentration of oxonium cations is
negligible, which was proven by NMR.

The experiments were prepared in the same way as described for
the polymerization studies on a vacuum line. The dynamic NMR
spectra were recorded starting from the lowest temperature. Simulations of the spectra were made using the DNMR 3 program. The
exchange rate constant was assumed to be the value which gave the
most similar simulated spectrum. Such parameters as signal width
and maximum and saddle intensity ratios were compared. Accuracy of
determination of exchange rate constant was better than 5%.

The results of the line shape analysis study of the orthoformate
protons (2-position of the ring) are summarized in Table III.

A brief line shape analysis study was also completed on the [1]H
NMR signals of the 4,4,5,5-ring protons of 2-methoxy-1,3-dioxolane.
This is a much more complicated system. The value of the exchange
rate constant was similar to the value reported in Table III for the
orthoformate protons.

Table III. Rates of exchange and calculated rate constants of combination of 1,3-dioxolan-2-ylium hexalfuoroantimonate with 2-methoxy-1,3 dioxolane and 2-ethoxy-1,3-dioxolane in nitromethane-d3.

temperature (°C)	concentration of 1,3-dioxolan-2-ylium salt (M x 10^3)	k_{ex} (s^{-1})	k_c ($M^{-1}s^{-1}$ x10^{-4})	$\overline{k}_c$ ($M^{-1}s^{-1}$ x10^{-4})
25	1.51	5.7	1.5	
	3.59	23.5	2.6	2.4
	8.48	63.5	3.0	
0	1.51	3.5	0.93	
	3.59	10.5	1.2	1.4
	8.48	45.6	2.1	
-15	1.51	3.2	0.83	
	3.59	10.0	1.1	1.2
	8.48	34.9	1.7	
-30	1.51	2.3	0.62	
	3.59	9.5	1.1	0.90
	8.48	21.1	1.0	

[2-methoxy-1,3-dioxolane] = [2-ethoxy-1,3-dioxolane] = 0.404 M

Acknowledgments

The authors are deeply indebted to the National Institutes of Health, GM 18595, for support of this work.

Literature Cited

1. Hall, H.K., Jr. J. Amer. Chem. Soc. 1958, 80, 6404.
2. Hall, H.K., Jr. J. Amer. Chem. Soc. 1958, 80, 6412.
3. Schuerch, C. Acc. Chem. Res. 1973, 6, 184.
4. Hall, H.K., Jr.; Steuck, M.J. J. Polym. Sci., Polym. Chem. Ed. 1973, 11, 1035.
5. Kops, J. J. Polym. Sci. A-1 1972, 10, 1275.

6. Sumitomo, H.; Okada, M.; Hibino, Y. J. Polym. Sci. B 1972, 10, 871.
7. Hall, H.K., Jr.; Carr, L.J.; Kellman, R.; De Blauwe, F. J. Amer. Chem. Soc. 1974, 96, 7265.
8. Hall, H.K., Jr.; De Blauwe, F. J. Amer. Chem. Soc. 1975, 97, 655.
9. Hall, H.K., Jr.; De Blauwe, F.; Carr, L.J.; Rao, V.S.; Reddy, G.S. J. Polym. Sci., Symp. 1976, 56, 101.
10. Fife, T.H. Acc. Chem. Res. 1972, 5, 764.
11. Deslongchamps, P. Tetrahedron 1975, 31, 2463.
12. Deslongchamps, P. Heterocycles 1977, 7, 1271.
13. Kirby, A.J.; Martin, R.J.. Chem. Commun. 1978, 803; ibid. 1979, 1079.
14. Uryu, T.; Yamanouchi, J.; Hayashi, S.; Tamaki, H.; Matsuzaki, K. Macromolecules 1983, 16, 320.
15. Crank, G; Eastwood, F.W. Aust. J. Chem. 1964, 17, 1385.
16. Bailey, W.J.; Endo, T. J. Polym. Sci., Polym. Symp. 1978, 64, 17.
17. Hall, H.K., Jr.; De Blauwe, F.; Pyriadi, T. J. Amer. Chem. Soc. 1975, 97, 3854.
18. Yokoyama, Y.; Padias, A. Buyle; De Blauwe, F.; Hall, H.K., Jr. Macromolecules 1980, 13, 252.
19. Yokoyama, Y.; Padias, A. Buyle; Bratoeff, E.A.; Hall, H.K., Jr. Macromolecules 1982, 15, 11.
20. Szymanski, R.; Hall, H.K., Jr. J. Polym. Sci., Polym. Lett. Ed. 1983, 21, 177.
21. Burt, R.A.; Chiang, Y.; Hall, H.K., Jr.; Kresge, A.J. J. Amer. Chem. Soc. 1982, 104, 3687.
22. ^{13}C NMR spectrum of poly-(1-methyl-2,6,7-trioxabicyclo[2.2.1] heptane) in CDCl$_3$: δ = 121.66, 121.53(CO$_3$), 74.55(CH), 66.91(endocyclic CH$_2$), 62.60, 62.25(exocyclic CH$_2$), 21.95(CH$_3$) ppm.
23. Yokoyama, Y.; Hall, H.K., Jr. J. Polym. Sci., Polym. Chem. Ed. 1980, 18, 3133.
24. Endo, T.; Saigo, K.; Bailey, W.J. J. Polym. Sci., Polym. Lett. Ed. 1980, 18, 457, 771.
25. Hall, H.K., Jr.; Yokoyama, Y. Polym. Bull. 1980, 2, 281.
26. Matyjaszewski, K. J. Polym. Sci., Polym. Chem. Ed. 1984, 22, 29.
27. Penczek, S.; Kubisa, P.; Matyjaszewski, K. Adv. Polym. Sci. 1980, 37, 1.
28. Penczek, S.; Szymanski, R. Polymer J. 1980, 12, 617.
29. Szymanski, R.; Penczek, S. Makromol. Chem. 1982, 183, 1587.
30. Slomkowski, S.; Penczek, S. J. Chem. Soc., Perkin Trans. II 1974, 1718.
31. Penczek, S. Makromol. Chem. 1974, 175, 1217.
32. Yokoyama, Y.; Hall, H.K., Jr. Adv. Polym. Sci. 1982, 42, 107.
33. Sumitomo, H.; Okada, M. Adv. Polym. Sci. 1978, 28, 47.
34. Stolarczyk, A.; Kubisa, P.; Penczek, S. J. Macromol. Sci.-Chem. 1977, A11, 2047.

RECEIVED October 4, 1984

Radiation-Induced Cationic Polymerization of Limonene Oxide, α-Pinene Oxide, and β-Pinene Oxide

JAMES A. AIKINS and FFRANCON WILLIAMS[1]

Department of Chemistry, University of Tennessee, Knoxville, TN 37996-1600

After suitable drying, the subject monomers in the form of neat liquids undergo radiation-induced polymerization with no apparent side reactions and high conversions to precipitatable polymers of low molecular weight. A high frequency of chain (proton) transfer to monomer is indicated by the fact that the kinetic chain lengths are estimated to be several hundred times larger than the range of DP_n values (12-4). Structural characterization of the limonene oxide polymer by ^1H and ^{13}C NMR spectroscopy provides conclusive evidence that the polymerization proceeds by the opening of the epoxide ring to yield a 1,2-trans polyether. Similar NMR studies on the polymers formed from the α-pinene and β-pinene oxides show that the opening of the epoxide ring for these monomers is generally accompanied by the concomitant ring opening of the cyclobutane ring structure to yield a gem-dimethyl group in the main chain.

The radiation-induced cationic polymerization of vinyl and unsaturated monomers in the liquid state has been studied for over 25 years, and the essential features of this type of polymerization appear to be well established ([1,2]). In contrast to cationic polymerization by catalysts where the propagating species is usually described as a solvated ion pair, the distinctive characteristic of cationic polymerization induced by high energy radiation is that propagation occurs by free ions with very large rate constants, the range of k_p values for observable polymerization being from 10^4 $M^{-1}s^{-1}$ to 10^8 $M^{-1}s^{-1}$. Since the concentration of free ions is typically about $10^{-10}M$ for dose rates obtainable from kilocurie ^{60}Co gamma-radiation sources, the rates of polymerization are very sensitive to traces of impurities, including water ([3]), that can function as efficient terminating agents. Consequently, much attention has been paid to the development of stringent experimental techniques for the rigorous drying of monomers ([1-6]), since otherwise this type of polymerization may go unrecognized.

[1]Author to whom correspondence should be directed.

0097-6156/85/0286-0335$07.25/0

As compared to vinyl monomers, relatively few studies of ring-opening polymerization induced by high energy radiation have been reported in the liquid state (7). Easily the best documented example is the polymerization of 1,2-cyclohexene oxide described by Cordischi, Mele, and their co-workers (8-10). These authors found that the polymerization of this epoxide displays many of the characteristics previously observed for the radiation-induced cationic polymerization of unsaturated monomers, including the great sensitivity to water (3) and the strongly retarding effect of ammonia (11).

In view of our earlier work on the polymerization of β-pinene (12,13), it seemed of interest to attempt the ring-opening polymerization of the epoxides of limonene [1], α-pinene [2], and β-pinene [3] by irradiation in the liquid state. We were also encouraged to carry out this study by the report of Ruckel and co-workers (14) on the successful catalytic cationic polymerization of these epoxides. In particular, these authors obtained evidence for the occurrence of a molecular rearrangement during the propagation sequence for the epoxides of α- and β-pinene, the opening of the epoxide ring being followed with high probability by the concomitant opening of the cyclobutane ring structure in these monomers. This latter rearrangement is also known in the cationic polymerization of β-pinene (12-16).

Experimental

The monomers, obtained from Aldrich Chemical Co., were pre-dried for 72 hours over molecular sieves and distilled under reduced pressure immediately before use. Elaborate drying techniques similar to those used in previous studies of isobutylene (5) and vinyl ethers (6) are unsuitable for monomers of low volatility, and therefore we resorted to a simpler method using a bake-out apparatus of the type originally described by Metz and his co-workers (4). This consisted essentially of an all-glass manifold containing indicator-grade silica gel as the drying agent (13), the entire apparatus being baked out in an oven at 350°C for 2-3 days under high vacuum before being attached to a conventional vacuum line and charged with monomer. The detailed procedure for the preparation of neat monomer samples in vacuo by this technique followed closely the description given previously for studies of β-pinene (13).

The sample tubes were irradiated for the specified total doses in a Gammacell-200 (Atomic Energy of Canada Ltd.) ^{60}Co source. Allowing for ^{60}Co decay, the standard dose rate calibrated by Fricke dosimetry decreased from 0.094 Mrad hr^{-1} (August, 1982) to 0.078 Mrad hr^{-1} (January, 1984) during the period of this investigation. Although most irradiations were carried out at the ambient temperature of the Gammacell chamber (ca. 25°C), a few samples were irradiated in Dewar vessels containing ice (0°C), solid carbon dioxide (-78°C), or liquid nitrogen (-196°C).

In order to check the cationic nature of the polymerization mechanism, a few samples were doped with low concentrations (ca. 10^{-1} M) of tri-n-propylamine which was introduced directly into the vacuum line. Usually, the last sample in a given batch was treated in this way to avoid contaminating the rest of the material.

After γ irradiation, the sample tubes were opened in a glove bag providing an oxygen-free atmosphere, and the monomer-polymer mixture dissolved in methylene chloride. The polymer was precipitated as a white powder by addition of methanol and collected by filtration, after which it was assayed by drying to constant weight in a vacuum dessicator. To minimize autoxidation, which was especially severe for the polymers from α-pinene oxide and β-pinene oxide that were left standing in the atmosphere under ambient conditions, polymer samples intended for storage were reprecipitated with methanol in the presence of an antioxidant (Flectol H - polymerized 1,2-dihydro-2,2,4-trimethyl-quinoline supplied by the Monsanto Chemical Co.) so as to disperse the latter in the solid polymer.

Molecular weight measurements on the polymers were made using a Knauer vapor pressure osmometer (Utopia Instrument Co.) with pyridine as the solvent at 60°C. Additional measurements with a similar osmometer were made by Galbraith Laboratories, Inc. using chloroform as the solvent at 45°C, the osmometer calibration factor having been obtained with phenacetin.

NMR and IR spectra of the polymers were obtained using 20% solutions in chloroform ($CDCl_3$) and carbon tetrachloride, respectively. The ^{1}H and ^{13}C-NMR spectra which are presented in this paper were recorded with a 200 MHz (Nicolet NT 200) instrument while additional ^{1}H NMR spectra were taken during the course of the work with both 90 MHz (JEOL FX90 Q FT) and 60 MHz (Varian T-60) spectrometers. IR spectra were recorded with a Perkin-Elmer 727 instrument.

The thermal properties of the polymers were measured using a Perkin-Elmer differential scannning calorimeter.

Results

Kinetic Characteristics of Polymerization

(+)-Limonene Oxide. Data showing the extent of monomer conversion to polymer as a function of irradiation dose are presented in Table I. Considering the general problem associated with the great sensitivity of radiation-induced cationic polymerizations to adventitious impurities and small residual concentrations of water in the monomer (1-6), there is a reasonable degree of reproducibility in the G(-M) values obtained at 25°C from batches I and II, the average value from eight determinations being 2472 with a standard deviation (N = 8 weighting) of 601. Although the only value (G(-M) = 819) from batch III at 25°C is well below this mean, the result at 0°C from this same batch is lower by a factor of more than two. While these results at 0 and 25°C are insufficient to establish the temperature dependence of the polymerization rate in the fluid state, it is significant that the polymerization rates at -78 and -196°C are about a factor of 10 to 50 times lower. Since the monomer becomes extremely viscous at -78°C and hardens to a glassy solid at -196°C, the polymerization evidently does not proceed as readily in the solid state at low temperatures.

There is no evidence from the data in Table I that the polymerization rate, as measured by the G(-M) value, decreases by more than the monomer depletion factor with increasing percent conversion. On the contrary, the three highest conversions show above-average overall rates which may indicate that the reaction proceeds with a modest acceleration in rate as the retarders are used up or become

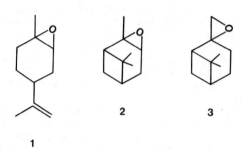

TABLE I

CONVERSION-DOSE DATA AND POLYMER MOLECULAR WEIGHTS IN THE
RADIATION-INDUCED POLYMERIZATION OF (+)-LIMONENE OXIDE

Code No.	Temp °C	Dose Mrad	Percent Conversion[a]	G(−M) monomer molecules 100 eV	$\overline{M}_n$[b]	$\overline{DP}_n$
I-1	25	1.355	79.3	3711		
I-2	25	0.927	32.7	2239	1163	7.6
I-3	25	1.823	77.5	2695		
I-4	25	0.464	13.0	1776		
II-1	25	1.840	51.0	1757		
II-2	25	1.592	67.0	2668	2302	15.1
II-3	25	0.852	28.7	2136	1443	9.5
II-4	25	1.226	54.0	2793	2838	18.7
II-5	−196	1.991	1.6	51		
III-1	25	3.715	47.5	819	1201	7.9
III-2	−78	25.50	12.0	30	1618	10.6
III-3	0	3.68	21.3	368	2333	15.3

[a] Based on recovery of precipitated polymer. [b] By vapor pressure osmometry.

less effective. It is noteworthy that very high conversions (at least 80%) to polymer can be achieved. This suggests that side reactions giving non-polymerizable products are relatively unimportant in the present case.

The expectation that the radiation-induced polymerization proceeds by a cationic mechanism was substantiated by an experiment in which 1.5 volume % of tri-$\underline{n}$-propylamine was added to a monomer sample. Even after a dose or more than 10 Mrad, no polymer was recovered by the usual precipitation technique.

An estimate of the kinetic chain length ν can be made by dividing the $\underline{G}(-M)$ value by the 100-eV yield of free ion initiators, $\underline{G}_i$. The latter quantity is generally considered to be in a narrow range between 0.1 and 0.2 ($\underline{1},\underline{2},\underline{5},\underline{6}$), at least for non-spherical molecules having low dielectric constants in the liquid phase. The polymerization of limonene oxide at 25°C is therefore characterized by ν values on the order of 10^4.

Even allowing for the scatter, the DP_n values listed in Table I are all between 7.6 and 18.7. These values are therefore about a thousand-fold smaller than the estimated kinetic chain length. Accordingly, a high frequency of chain transfer must be involved in the polymerization and this is consistent with a cationic mechanism.

α-Pinene Oxide. As seen from a comparison of the results in Table II with those in Table I, the radiation-induced polymerization of α-pinene oxide proceeds at a much slower rate than limonene oxide under the same conditions. The mean value of $\underline{G}(-M)$ for α-pinene oxide calculated from the 12 runs at 25°C is $3\overline{6}0$ with a standard deviation (N = 12 weighting) of 121. On average, therefore, the polymerization rates or $\underline{G}(-M)$ values for these two monomers differ by a factor of 6.9.

At 0°C the $\underline{G}(-M)$ values for α-pinene oxide are slightly lower than the average value at 25°C but remain well within the standard deviation for the latter set of measurements. Again, no definite trend regarding the temperature dependence of $\underline{G}(-M)$ is evident from these limited data.

Although the polymerizations were carried through to total conversions not exceeding 33.6% in this case (Table II), no significance should be attached to this feature of the results. Higher conversions would have undoubtedly been attained if the irradiation doses had been increased proportionately.

The average DP_n of 5.6 for the α-pinene oxide polymer formed at 25°C (Table II) is lower than the corresponding average of 11.8 for the limonene oxide polymer (Table I), this being also the order of the polymerization rates for these two monomers. However, the ratio (2.1) of molecular weights is somewhat less than the ratio (6.9) of $\underline{G}(-M)$ values. Despite these differences, the inequality $\nu \gg DP_n$ applies to both monomers. Thus, the kinetic chain length ($\nu \simeq 2 \times 10^3$) for α-pinene oxide polymerization is about 350 times larger than the average DP_n (5.6), again indicating the importance of chain transfer.

Although only two measurements were made, the molecular weights of the α-pinene oxide polymers prepared at 0°C (Table II) appear to be significantly higher (average DP_n of 13.4) than those prepared at 25°C. More data would be needed, however, to establish this point unequivocally.

TABLE II

CONVERSION-DOSE DATA AND POLYMER MOLECULAR WEIGHTS IN THE
RADIATION-INDUCED POLYMERIZATION OF α-PINENE OXIDE

Code No.	Temp °C	Dose Mrad	Percent Conversion[a]	$\frac{\text{G}(-\text{M})}{\text{monomer molecules}}$ 100 eV	$\overline{M}_n^b$	$\overline{DP}_n$
I-1	25	3.26	27.3	530		
I-2	25	3.27	7.6	148		
I-3	-196	4.21	5.6	85		
I-4	25	4.36	10.9	159		
II-1	25	3.09	19.3	395		
II-2	25	1.48	10.0	428		
II-3	25	3.00	14.4	304	500	3.3
II-4	25	5.11	21.7	270	1933	12.7
III-1	25	5.83	34.5	375	305	2.0
III-2	25	3.92	24.5	396	880	5.8
III-3	25	3.92	33.6	544		
III-4	25	5.83	30.0	326		
IV-2	25	3.16	22.3	447	632	4.2
IV-3	0	3.71	19.5	333	2457	16.1
IV-4	0	4.05	19.9	312	1628	10.7

[a]Based on recovery of precipitated polymer.   [b]By vapor pressure osmometry.

β-Pinene Oxide. From twelve determinations listed in Table III, the mean value of $\underline{G}(-M)$ at 25°C for this monomer is 461 with a standard deviation (N = 12 weighting) of 308. The polymerization rates are therefore quite comparable to those found for α-pinene oxide (Table II). This similarity also applies to the DP_n values, the average value of 4.0 for the β-pinene oxide polymer at 25°C being statistically indistinguishable from the results for the α-pinene oxide polymer in Table II with an average of 5.8.

It is interesting that the highest $\underline{G}(-M)$ value of 1342 was obtained in a high conversion (80.4%) experiment. This result supports the point made earlier for limonene oxide that the radiation-induced polymerization of these monomers can be carried nearly to completion, certainly without any diminution in rate and possibly even with rate enhancement as impurities are used up.

The polymerization rate was strongly retarded by the addition of 1.0 volume % tri-n-propylamine to the monomer. In contrast to a "control" sample (IV-3 in Table III) which yielded 35.0% polymer after a dose of 3.99 Mrad, the amine-doped sample gave only 1.4% polymer after exposure to the same total dose. Also, this small amount of polymer produced in the presence of the amine did not precipitate out immediately on the addition of methanol, and it was recovered only after the solution had been allowed to stand for several days. Since the $\underline{G}(-M)$ value calculated for the amine-doped sample is only 22.6 as against the "control" value of 557, it is clear that the chain character of the polymerization is seriously impaired by the amine.

Structural Characterization of Polymers

^{1}H-NMR Studies

(+)-Limonene Oxide. The ^{1}H-NMR spectra of limonene oxide and its radiation-produced polymer are shown in Figures 1(a) and (b). In each case there are strong resonances at 1.66-1.71 and 4.65-4.71 δ which can readily be assigned to the methyl and vinylidene protons, respectively, of the pendant isopropenyl (CH_3-C=CH_2) group. This comparison clearly demonstrates that the polymerization does not proceed by addition to the vinylidene double bond. The other two well-defined peaks at 1.19 and 3.5 δ in the polymer spectrum are assigned to the CH_3-C-O and $\underline{H}$-C-O protons at the C-1 and C-2 carbons of the cyclohexane ring in the repeat unit produced by opening the epoxide ring. Incidentally, the corresponding methyl protons in the monomer are responsible for the strong doublet resonance seen at 1.274-1.292 δ.

α-Pinene Oxide. Figures 2(a) and 2(b) show the ^{1}H-NMR spectra of poly(α-pinene oxide) and its monomer. In the monomer spectrum there are two well-separated peaks from methyl protons at 0.955 δ and 1.296-1.313 δ. Integration showed that the absorption from the latter group of resonances was exactly double that of the 0.955 δ peak. Since there are three non-equivalent methyl groups in α-pinene oxide, the resonances from two of these must overlap at ≈ 1.3 δ. By analogy with the assignment of the ≈ 1.28 δ peak in limonene oxide to the methyl group at the epoxide ring, one of the resonances at ≈ 1.3 δ in α-pinene can be similarly assigned. Thus the gem-dimethyl group

<u>TABLE III</u>

CONVERSION-DOSE DATA AND POLYMER MOLECULAR WEIGHTS IN THE
RADIATION-INDUCED POLYMERIZATION OF β-PINENE OXIDE

Code No.	Temp °C	Dose Mrad	Percent Conversion[a]	$\dfrac{G(-M)}{\text{monomer molecules}}$ 100 eV	$\overline{M}_n^{[b]}$	$\overline{DP}_n$
I-2	25	5.14	24.7	304		
I-3	25	4.18	25.6	388		
I-4	25	3.39	20.0	373		
II-1	25	0.88	2.7	195		
II-2	25	1.76	8.2	294		
II-3	25	2.65	8.9	213		
II-4	25	3.53	8.2	147		
III-1	25	3.80	80.4	1342		
III-2	25	4.44	46.9	670	750	4.9
III-3	25	4.46	40.3	572	460	3.0
IV-1	25	4.97	37.1	473		
IV-3	25	3.99	35.0	557		

[a]Based on recovery of precipitated polymer.   [b]By vapor pressure osmometry.

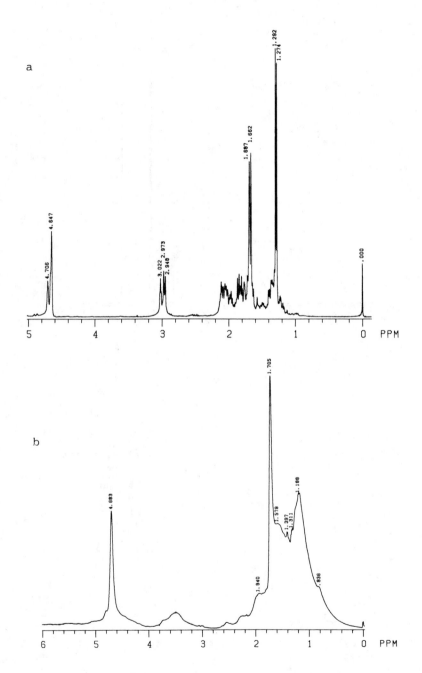

Figure 1. 200 MHz ^{1}H – NMR spectra of (a) (+)-limonene oxide and
(b) poly(limonene oxide)

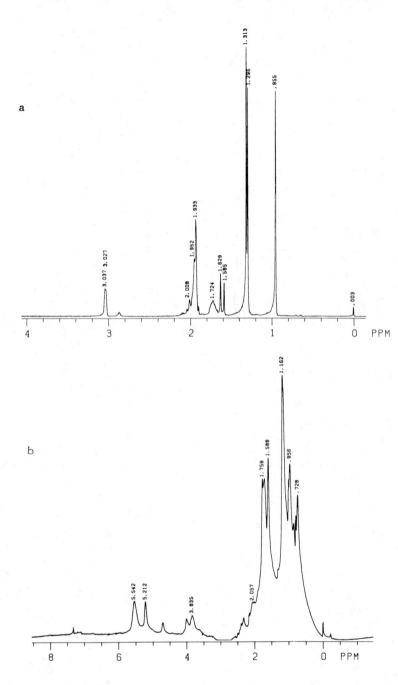

Figure 2. 200 MHz ^{1}H - NMR spectra of (a) α-pinene oxide and (b) poly(α-pinene oxide)

must absorb at 0.96 and 1.30 δ. These assignments agree with those of Ruckel et al. (14) but do not correspond with those given in the Sadtler compilation (17) which mistakenly assumes that the gem-dimethyl protons are equivalent and are responsible for the entire 1.3 δ peak.

Turning to the polymer spectrum, there are prominent peaks displayed at δ values of 0.728, 0.956, 1.16, 1.59, 1.76, 3.84, 4.7, 5.21, and 5.54. Following Ruckel et al. (14), it is helpful to consider the assignments first in terms of the rearranged repeat unit 4.

On the basis of the close correspondence with the resonance at 1.14 δ in the spectrum of poly(isobutylene oxide) (14), the very strong peak at 1.16 δ can be assigned to the protons from the gem-dimethyl group in the above structure. The protons from the remaining methyl group adjacent to the double bond would be expected to absorb at about 1.6 δ according to the ^{1}H-NMR spectrum of limonene [5] which has a resonance at about this value (1.63 δ) for a very similar methyl group (18). Therefore, we assign the 1.59 peak to this olefinic methyl in the rearranged structure 4.

Further evidence for the rearranged repeat unit 4 in the α-pinene oxide polymer comes from the observation of resonances at 5.21 and 5.54 δ which are characteristic of cyclo-olefinic hydrogens. For example, there is a similar resonance at 5.39 δ in limonene (18) (p-mentha-1(2),8(9)-diene, 5) which again serves as a suitable model compound for the cyclohexene portion of this rearranged unit. Therefore, one of the above resonances almost certainly corresponds to the cyclo-olefinic hydrogen depicted in the above structure while the other probably arises from an isomeric structure formed by a shift of the double bond, as suggested by Ruckel et al. (14). The presence of two H-C-O resonances at 3.8 – 4.0 δ is also supportive of two kinds of cyclo-olefinic repeat structures.

Hitherto, our assignments of the ^{1}H resonances in the α-pinene oxide polymer have corresponded closely to those of Ruckel et al. (14), the spectra of the two differently prepared polymers being very similar in all respects except for an additional peak at 1.30 δ in the previous work. We definitely do not find such a peak in the radiation-produced polymer, and it is curious that the previous workers also reported (14) that for polymer prepared under certain catalytic conditions, this 1.30 δ peak was either totally absent or only a trace of it was found in the spectrum. However, Ruckel et al. (14) assumed that since the peaks at 0.97 and 1.30 δ in their polymer corresponded almost exactly to the positions of the gem-dimethyl group resonances in the monomer, these peaks could be assigned to the same group in the ring-closed or unrearranged repeat unit 6. These authors went on to estimate the percentage of such ring-closed structures in the polymer from the relative intensity of the 1.30 δ absorption.

It is questionable, however, that the positions of the gem-dimethyl ^{1}H resonances remain unchanged in going from the α-pinene oxide monomer to the ring-closed polymer structure 6, one reason being that the opening of the epoxide ring results in a different placement of the gem-dimethyl group relative to the oxygen atom. Therefore, in contrast to the previous assignment (14), we propose that these methyl ^{1}H resonances in 6 can be attributed to the peaks of comparable intensity at 0.728 and 0.956 δ. This would not only account for the absence of the 1.30 δ peak in our samples, and in some of Ruckel's polymers (14,19), but also explain why the 0.73 and 0.96 δ peaks are always present together in the spectra of both types of polymers. Moreover, the present set of assignments obviates the need to explain the methyl resonance at 0.73 δ in terms of various ring-expanded repeat structures (14).

The only well-defined peaks in the poly(α-pinene oxide) spectrum which remain to be assigned are those at 1.76 and 4.70 δ. These resonance positions correspond almost precisely to those observed for methyl and olefinic protons in an isopropenyl group, as discussed earlier for limonene oxide and its polymer. Accordingly, we assign these peaks to isopropenyl end groups produced by chain (proton) transfer from a ring-opened propagating cation (see Discussion). It might be noted that Ruckel et al. (14) assigned the 1.7 δ peak to a $CH_3-C=C$ resonance in a branched mer and the 4.64 δ peak to an H-C-O resonance.

β-Pinene Oxide. The spectra of the monomer and polymer are presented together in Figures 3(a) and 3(b). In addition to the two strong peaks at 0.922 and 1.246 δ which can be assigned to the gem-dimethyl group, the monomer spectrum consists of two well-defined doublets (J ≃ 5 Hz) centered at 2.579 and 2.745 δ as well as some weaker resonances in the intermediate 1.5-2.5 δ region. The doublets are almost certainly due to the two nonequivalent exocyclic CH_2 hydrogens of the epoxide group.

The polymer spectrum is noticeably simpler than that of poly(α-pinene oxide) in Figure 2(b). In particular, only one strong resonance at 1.127 δ appears to originate from methyl hydrogens in the poly(β-pinene oxide). This is confidently assigned to the gem-dimethyl group in the ring-opened repeat unit 7 since the 1.127 δ resonance occurs at almost the same chemical shift as the 1.14 δ peak from the corresponding group in poly(isobutylene oxide) (14). There are two other prominent resonances in the polymer spectrum which are compatible with structure 7, namely the 3.729 δ peak which can be assigned to the hydrogens of the $-OCH_2-$ group, and the 5.68 δ peak which is specific for the cyclo-olefinic hydrogen. Of the remaining peaks, those at 1.729 and 4.702 δ are again characteristic of terminal isopropenyl groups produced by chain transfer from a ring-opened cation.

It should be noted that the resonances observed at 0.95 and 1.25 δ in the spectrum of the catalytically-prepared polymer (14) are virtually absent in the present case. These peaks coincide closely with the strong methyl resonances in the β-pinene oxide monomer (Figure 3(a)), and so it is conceivable that they arise from contamination by monomer. In a study of the polymerization of β-pinene (13), for instance, it was found that the polymer had to be reprecipitated from solution in order to remove all traces of monomer. However, Dr. Ruckel has informed us in a private communication that great pains were taken to remove monomer from the catalytically-prepared polymer (14).

Another difference between our results and those of Ruckel et al. (14) concerns the relative intensity of the olefinic resonance in the poly(β-pinene oxide) spectrum. They reported that it was only 30% of that expected for the ring-opened structure 7, and this led

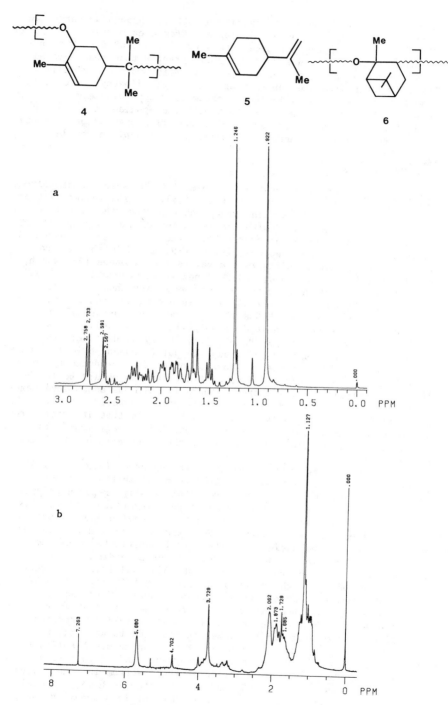

Figure 3.   200 MHz ^{1}H – NMR spectra of (a) β-pinene oxide and (b) poly(β-pinene oxide)

them to postulate a second propagation pathway through a rearranged bicyclic structure. However, it appears from a comparison of Figures 2(b) and 3(b) that the relative intensity of the 5.68 δ peak in poly-(β-pinene oxide) is equal to or slightly greater than that derived from the sum of the two olefinic peaks in poly(α-pinene oxide). Taken together with the apparent absence of gem-dimethyl absorption from bicyclic structures in the spectrum of poly(β-pinene oxide), our ^{1}H-NMR results seem to be consistent with the epoxide ring-opening being followed by a predominant cyclobutane ring-opening propagation step in the radiation-induced polymerization of β-pinene oxide.

^{13}C-NMR Studies

(+)-Limonene Oxide. The proton-decoupled ^{13}C-NMR spectrum of poly-((+)-limonene oxide) is shown in Figure 4(a). If the polymerization takes place through the epoxide group, as expected, the repeating unit will be as shown in 8 with the retention of the isopropenyl group at C-4, the numbering system being the same as in the monomer. In this case the ^{13}C resonances of C-8, C-9, and C-10 should have very similar chemical shifts to the values for limonene [5], which are located at 149.9, 108.4, and 20.7 ppm (δ_c), respectively (20). In fact, these resonances match up very well with those at 149.3, 108.5, and 20.94 ppm in Figure 4(a), reinforcing the conclusion reached earlier from the ^{1}H-NMR studies that the isopropenyl group is present in the repeating unit 8 of poly(limonene oxide).

In view of the average DP_n of 11.8, it may be difficult to detect ^{13}C resonances from end groups in poly(limonene oxide). These end groups would be expected to consist of olefins produced by chain transfer, and ^{13}C resonances from olefinic carbons should be in the low-field region of the spectrum. There is only one unassigned peak at 150.93 ppm which would fit this description, however, and its assignment must be tentative. One possibility is that it arises from an olefin end group formed between C-1 and C-7, a very low-field resonance being expected for olefinic carbons not attached to hydrogen ($\rightleftharpoons$ C=).

α-Pinene Oxide. Several peaks in the spectrum (Fig. 4(b)) of poly(α-pinene oxide) have very similar chemical shifts to those listed by Ruckel et al. (14) for the catalytically prepared polymer. Moreover, as shown in Table IV, several peaks again correspond closely to the ^{13}C resonances in sobrerol [10], a model compound possessing the polymer repeating unit 4. For example, the peaks at 132.8 and 133.9 ppm can be assigned to the methyl-substituted olefinic carbon in 4 since the resonances from the corresponding C-1 in limonene (20) and sobrerol (14) occur at 133.5 and 134.8 ppm, respectively. In addition, the peaks at 148.36, 108.5 (two), and 19.67 ppm can be assigned to isopropenyl end groups, these ^{13}C resonances having very similar chemical shifts to those described above for poly(limonene oxide). Finally, the farthest upfield ^{13}C resonance at 12.5-12.6 ppm was assigned by Ruckel et al. (14) to an angular bridgehead methyl group on the basis of similar chemical shifts for the resonance from C-10 in bornane, bornylene, and camphor (21).

β-Pinene Oxide. An examination of the spectrum in Figure 4(c) reveals resonances at 150, 108, and 21 ppm which we have previously assigned (see limonene oxide) to the carbons of isopropenyl groups.

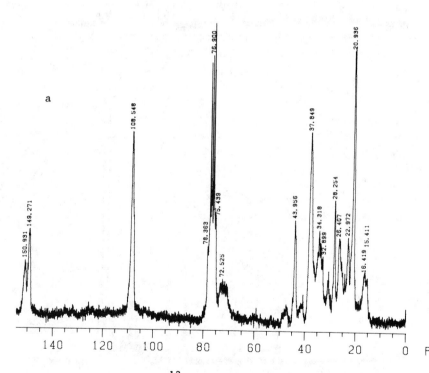

Figure 4a.   200 MHz ^{13}C-NMR spectra of poly(limonene oxide).

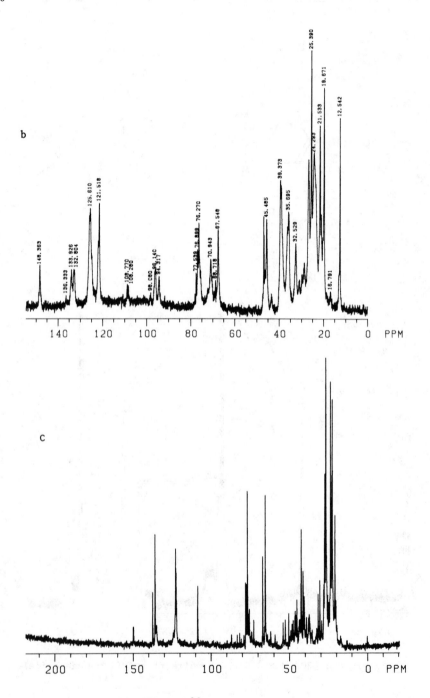

Figures 4b and 4c. 200 MHz ^{13}C-NMR spectra of poly(α-pinene oxide) and poly(β-pinene oxide).

TABLE IV

^{13}C CHEMICAL – SHIFT DATA (PPM)[†] FOR SOBREROL AND
POLY(α-PINENE OXIDE) SAMPLES

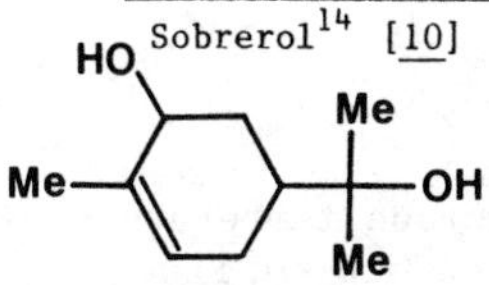

Sobrerol[14] [10]		Poly(α-pinene oxide) prepared by catalytic polymerization[14]		Poly(α-pinene oxide) prepared by radiation-induced polymerization	
CH$_3$	21.0	12.6	35.6	12.5	70.9
CH$_3$	26.0	19.8	38.9*	19.7	94.3
CH$_2$	27.3	20.1	44.3	21.5	96.1
CH$_3$	27.7	20.8	45.2	24.3	108.3
CH$_2$	32.8	21.7*	47.0	25.4	108.8
CH$_2$	38.7	25.7	68.7*	32.5	121.5
CHOH	68.0	26.6*	72.3*	35.7	125.6
COH	71.5	27.1*	121.6	39.4	132.8
=CH	124.6	27.6*	125.3*	45.5	133.9
=C<	134.8	32.8*	133.1 (?)	67.6	148.4

[†]Measured from tetramethylsilane

*Corresponds closely to a ^{13}C resonance in sobrerol[14]

It is not surprising that the expected end groups of this type are detectable in poly(β-pinene oxide), considering the relatively low DP_n ($\simeq$ 4) of this polymer. These ^{13}C results nicely confirm the earlier ^{1}H-NMR identification of the isopropenyl end groups in this polymer.

IR Spectra

The IR spectra of the three polymers taken before oxidation are shown in Figure 5(a). In addition to the characteristic absorptions at 1100 and 1350 cm^{-1} from the C-O-C and CMe$_2$ groups associated with the polymer repeat units in poly(α-pinene oxide) and poly(β-pinene oxide), there are weak bands at 890 cm^{-1} which are diagnostic of vinylidene ($\rightleftharpoons C=CH_2$) end groups in these polymers. As expected, the spectrum of poly(limonene oxide) shows the strong absorptions associated with the isopropenyl group in the repeat unit.

DSC Measurements

A typical DSC plot of the rate of heat absorption versus temperature is shown for poly(limonene oxide) in Figure 6. An endothermic transition corresponding to an enthalpy change of 0.3 cal g^{-1} was observed between 65.2 and 84.5°C, the peak being at 74.2°C. A similar plot was obtained for poly(α-pinene oxide), the peak occurring at 81.3°C with a ΔH of 0.13 cal g^{-1} measured between 78.3 and 84.3°C. Ruckel et al. (14) have reported very similar softening ranges of 67-80 and 65-80°C for the catalytically-prepared poly(α-pinene oxide) and poly(β-pinene oxide).

Discussion

The results of this work demonstrate that using neat monomers under reasonably dry conditions, the epoxides of limonene, α-pinene, and β-pinene undergo radiation-induced cationic polymerization at 0-25°C with appreciable kinetic chain lengths ν (1 x 10^4 - 2 x 10^3). Moreover, high conversions of monomer to polymer can be attained with no apparent diminution in rate, yields of 80% being obtained by suitably extending the irradiation dose. While the average $\underline{G}(-M)$ values of 2500 for limonene oxide and ca. 400 for the epoxides of α- and β-pinene are somewhat lower than those reported for the cationic polymerization of some vinyl monomers (1-6), they are only about an order of magnitude lower than the value of 10,000 often quoted as the benchmark for an economically viable radiation polymerization process (6).

In comparison with the catalytic cationic polymerization of these monomers (14), it is remarkable that the radiation-induced polymerization gives high conversions at room temperature to a precipitatable polymer. Although the catalytic polymerizations (14) were carried out in solution between -78 and -130°C under conditions which should minimize the formation of dimer oils and monomer rearrangement products, the highest reported conversion of α-pinene oxide to polymer was only 28%, and in this case the material was described as a viscous oil. Conversions to polymer in the form of a white powder varied from 1 to 20% although 76-100% of the α-pinene oxide

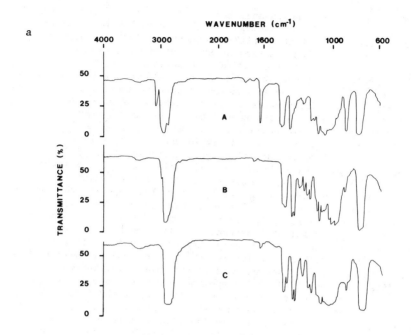

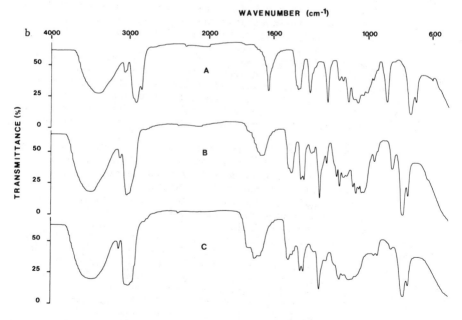

Figure 5. Infra-Red Spectra of (A) poly(limonene oxide), (B) poly(α-pinene oxide), and (C) poly(β-pinene oxide), shown before (Fig. 5a) and after autoxidation (Fig. 5b)

monomer had reacted, the balance being made up of nonprecipitatable
oligomers and isomerized monomer. Similar results were reported for
the catalytic polymerization of β-pinene oxide (14). Evidently, the
radiation-induced polymerization of these monomers is a much cleaner
polymerization reaction, even at room temperature.

   Initiation. Unfortunately, no direct method is generally available to determine the precise chemical nature of the initiation step
in the radiation-induced cationic polymerization of bulk monomers in
the liquid state. Strong inferences can, however, be drawn from
information about ion-molecule processes in the gas phase (22).
Also, the techniques of pulse radiolysis and matrix isolation can
frequently be used to characterize ionic intermediates. Indeed,
recent developments have shown that γ-irradiation of Freon solid
solids at 77 K provides a most useful method of generating solute
radical cations for ESR studies (23), and we have applied this method
to study the radical cations of several simple epoxides (24).

   The main conclusion which emerges from ESR studies of ethylene
oxide and methyl-substituted oxiranes is that the radical cation
undergoes a disrotatory C-C ring-opening of the epoxide ring to give
a planar species with an isoelectronic structure analogous to that of
allylic radicals (24). Recent work in our group (25) indicates that
the radical cations of alicyclic epoxides such as cyclopentene oxide
and cyclohexene oxide can also assume this ring-opened form. As
depicted in Scheme 1 below, we suggest that this type of species can
initiate cationic polymerization in the liquid state by proton
transfer to an epoxide molecule with the formation of an oxonium
ion. Consistent with this proposal, ESR studies have shown that the
ring-opened cation of tetramethyloxirane is a proton donor, the
neutral radical corresponding to the loss of a proton being detected
as a major decay product in certain matrices.

   Propagation and Rearrangement. Spectroscopic evidence for the
retention of the isopropenyl groups in the polymer establishes that
the radiation-induced polymerization of limonene oxide proceeds by
the opening of the epoxide ring to yield the 1,2-trans polyether
structure 9 shown below. Although the polymer is drawn as having an
isotactic structure, no information was obtained about its stereoregularity.

   As discussed by Ruckel et al., (14) the catalytic cationic polymerizations of α-pinene oxide and β-pinene oxide both involve the
concomitant opening of the epoxide and cyclobutane rings in the
propagation step. This rearrangement mechanism also clearly operates
in the radiation-induced polymerization of these monomers, and is
depicted below for α-pinene oxide (Scheme 2).

   Despite the general predominance of such ring-opened structures
in the polymers formed from both of the pinene oxides, we find ^{1}H-NMR
evidence for a significant contribution from the ring-closed structure 6 in poly(α-pinene oxide) whereas the corresponding structure
appears to be largely absent in poly(β-pinene oxide). The question
of whether other types of repeating units are present in these
polymers prepared by irradiation is more difficult to answer. As
mentioned earlier, we differ with Ruckel et al. (14) regarding the
assignment of some of the peaks in the ^{1}H-NMR spectra of poly(α-
pinene oxide) and poly(β-pinene oxide). Specifically, the assignment
of the resonances at 0.74 and 1.7 δ in poly(α-pinene oxide) to methyl

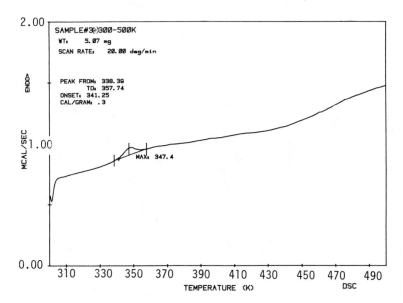

Figure 6. Differential Scanning Calorimeter (DSC) Thermogram of Poly(limonene oxide)

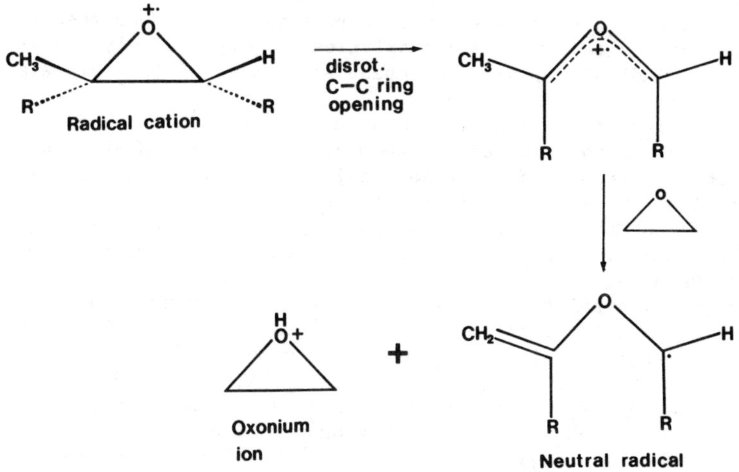

Scheme 1.

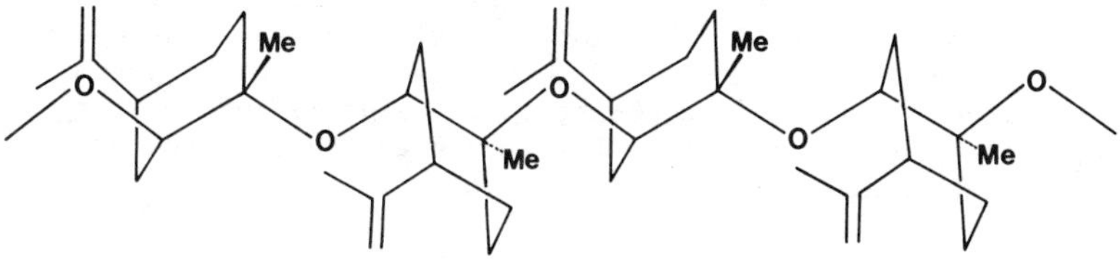

9

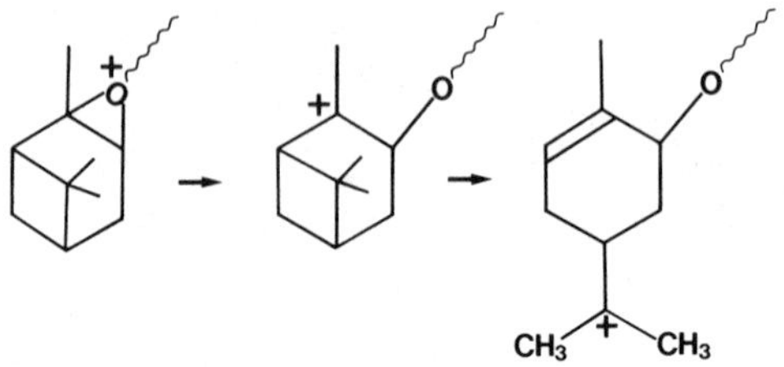

Scheme 2.

groups in ring-expanded and branched mers (14) seems questionable to us. Aside from this problem of interpretation, the spectroscopic results for the radiation and catalytically prepared polymers are comparable indicating that the repeat structures of the polymers are quite similar in the two cases.

Chain Transfer. Kinetic and structural evidence has been obtained for chain transfer in the present studies. Chain transfer is especially important in radiation-induced cationic polymerization because its frequency relative to propagation determines the molecular weight of the polymer, i.e. $DP_n = k_p/k_{tr}$. The only known exceptions to this statement are at very low temperatures which do not concern us here. In the ring-opening polymerizations of epoxides, the propagation and transfer reactions are usually written (Scheme 3) in terms of oxonium ions as the reaction intermediates (26). This is undoubtedly correct since any incipient carbonium (carbenium) ion produced by ring opening (e.g. $\sim\sim\sim O-CR_1R_2-CR_3R_4^+$) would have only a fleeting existence before becoming attached to an epoxide molecule and thereby forming another oxonium ion.

Nevertheless, it is of interest to attempt to correlate k_p/k_{tr} with the structure of this incipient carbonium ion. From the published data for cyclohexene oxide (10) ($DP_n \approx 2,500$) and the results of this work, it is clear that DP_n increases markedly with decreasing methyl substitution at the α-carbon of this cation, two methyl groups being present in the rearranged cations derived from α- and β-pinene oxides, one in the limonene oxide cation, and none in the cyclohexene oxide cation. This empirical correlation suggests that, relative to propagation, β-proton transfer is much easier from a methyl group than from other substituents at the α-carbon atom. However, other factors might well be involved.

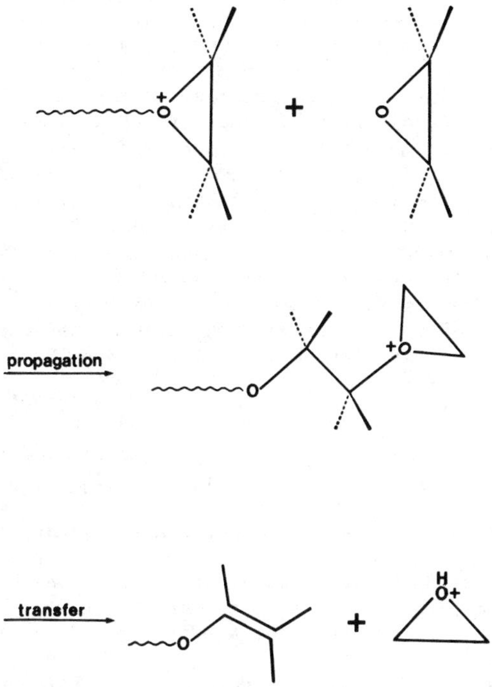

Scheme 3.

Acknowledgments

We thank Dr. Tyrone L. Vigo and Mrs. Eve Frost (USDA Textile & Clothing Laboratory, Knoxville, TN) for the DSC measurements, and Dr. Erwin R. Ruckel (Arizona Chemical Co., Stamford, CT) for his helpful comments on this work. This research was supported by the Division of Chemical Sciences, Office of Basic Energy Sciences, U.S. Department of Energy (Document No. DOE/ER/02968-151).

Literature Cited

1. F. Williams, Fundamental Processes in Radiation Chemistry, Ed. P. Ausloos, Interscience-Wiley, New York, p 515 (1968).
2. V. T. Stannett, Pure & Appl Chem., 53, 673 (1981).
3. T. H. Bates, J. V. F. Best, and F. Williams, Nature, 188, 469 (1960); Trans. Faraday Soc., 58, 192 (1962).
4. R. C. Potter, R. H. Bretton, and D. J. Metz, J. Polymer Sci. A 1, 4, 2295 (1966); C. L. Johnson and D. J. Metz, ACS Polymer Prep; 4, 440 (1963).
5. R. B. Taylor and F. Williams, J. Am. Chem. Soc., 89, 6359 (1967); 91, 3728 (1969).
6. A. M. Goineau, J. Kohler, and V. Stannett, J. Macromol. Sci.-Chem., A 11, 99 (1977).

7. (a) For a review of the radiation-induced solid state polymerization of ring compounds, see: S. Okamura, K. Hayashi, and Y.
   Kitanishi, J. Polymer Sci., 58, 925 (1962); (b) A study of the
   radiation-induced polymerization of styrene oxide in the liquid
   and solid states has been reported by Y. Tabata, J. Macromol.
   Sci. (Chem.), A1, 493 (1967). The rate of polymerization
   increased between 20 and 140°C with an apparent activation
   energy of 6.2 kcal/mol, and at 20°C the polymerization was
   inhibited by p-benzoquinone. A radical-initiated polymerization
   was suggested in which chain transfer and isomerization reactions play an important role. The nature of the propagation
   step was not elucidated, however, and both epoxide and carbonyl
   groups were reported to be present in the polymer. Also, the
   molecular weight of the polymer increased markedly with the
   irradiation time, the intrinsic viscosities being in the range
   from 0.6 to 2.0 dl/g.

8. D. Cordischi, M. Lenzi, and A. Mele, J. Polymer Sci. A 3, 3421
   (1965).

9. D. Cordischi, A. Mele, and A. Somogyi, in "Proceedings of the
   Second Symposium on Radiation Chemistry", Tihany, Hungary, p 483
   (1967).

10. D. Cordischi, A. Mele, and R. Rufo, Trans. Faraday Soc., 64,
    2794 (1968).

11. M. A. Bonin, W. R. Busler, and F. Williams, J. Am. Chem. Soc.,
    84, 4355 (1962); J. Am. Chem. Soc., 87, 199 (1965).

12. T. H. Bates, J. V. F. Best, and F. Williams, J. Chem. Soc., 1531
    (1962).

13. A. M. Adur and F. Williams, J. Polymer Sci., Polymer Chem. Ed.,
    19, 669 (1981).

14. E. R. Ruckel, R. T. Wojcik, and H. G. Arlt, Jr., J. Macromol.
    Sci. Chem., A 10, 1371 (1976); E. R. Ruckel, private communica-
    tion.

15. E. R. Ruckel, H. G. Arlt, Jr., and R. T. Wojcik, in "Adhesion
    Science and Technology", ed. L.-H. Lee, Vol. 9 A, Plenum Press,
    pp 395-412 (1975).

16. W. J. Roberts and A. R. Day, J. Am. Chem. Soc., 72, 1226 (1950).

17. Compilation of NMR Spectra, Sadtler Research Laboratories, Inc.,
    2,3-Epoxy-pinane, No. 6275 M (1969).

18. Compilation of NMR Spectra, Sadtler Research Laboratories, Inc.,
    p-Mentha-1,8-diene (limonene), No. 17189 M (1973). Since the
    methyl protons at C-10 in limonene oxide have a resonance at
    1.68 δ, we consider that the peaks at 1.63 and 1.71 δ in
    limonene should be assigned to the methyl protons at C-7 and
    C-10, respectively. These latter assignments are reversed in
    the Sadtler compilation.

19. The extreme variability of the 1.30 δ peak in the ^{1}H-NMR spectra
    of the catalytically prepared poly(α-pinene oxide) samples (14)
    could be due to contamination by monomer which has a strong
    resonance at this value. The authors of ref. 14 consider this
    explanation to be unlikely, however.

20. L. F. Johnson and W. C. Jankowski, "Carbon-13 NMR Spectra",
    Wiley-Interscience, New York, New York, p 400.

21. J. B. Stothers, "Carbon-13 NMR Spectroscopy", Academic Press,
    New York, New York, 1972, p 433.

22. S. G. Lias and P. Ausloos, "Ion-Molecule Reactions", American
     Chemical Society, Washington, D.C., 1975.
23. T. Shida and T. Kato, Chem. Phys. Letters, 68, 106 (1979).
24. L. D. Snow, J. T. Wang, and F. Williams, Chem. Phys. Letters,
     100, 193 (1979).
25. L. D. Snow and F. Williams, unpublished work.
26. G. Odian, "Principles of Polymerization", 2nd Edition, Wiley-
     Interscience, New York, New York, 1981, p 517.

RECEIVED September 14, 1984

Cationic Ring-Opening Polymerization of Epichlorohydrin in the Presence of Ethylene Glycol

YOSHIHISA OKAMOTO

BFGoodrich Research and Development Center, 9921 Brecksville Road, Brecksville, OH 44141

Cationic ring-opening polymerization of epichlorohydrin, using triethyloxonium hexafluorophosphate initiator in the presence of ethylene glycol to produce a low molecular weight polyepichlorohydrin glycol, was examined. A new polymerization mechanism was postulated that: polymerization initiates at the hydroxyl groups in the ethylene glycol, and the polymer chain propagates simultaneously at both ends through the addition of the monomer. In this polymerization system, the polymer molecular weight increases directly with the polymer conversion, and it can be controlled by the molar ratio of epichlorohydrin and ethylene glycol. The obtained polyepichlorohydrin glycols possess predominantly secondary hydroxyl groups, narrow molecular weight distribution, and a hydroxyl functionality of 2. Byproducts formed in polyepichlorohydrin glycol are identified as non-functional cyclic oligomers, mainly epichlorohydrin cyclopentamer and cyclohexamer.

It is well known that the low molecular weight polyether glycols such as poly(tetramethylene ether) glycol have been utilized in many polyurethane foams, rubbers, and castable elastomers. Recently halogen containing polyether glycols such as polyepichlorohydrin glycol attracted interest for flame resistance polyurethane application. Among the several methods to prepare such polyether glycol, a cationic ring-opening polymerization of epichlorohydrin (ECH) using water and ethylene glycol seems to be a promising method (1-2). Recently employing a similar method, poly(3,3-disubstituted trimethylene ether) glycols have been prepared using borontrifluoride initiator in the presence of alkanediols for an application in high energetic polyurethane binders for rocket propellants (3). In order to prepare polyether glycols possessing the desired molecular weight and hydroxyl functionality, research has been conducted to examine the role of using water and alkanediols in the cationic ring-opening polymerization of oxetane and tetrahydrofuran (3-5). To extend the investigation of the preparation of low molecular weight polyether

0097-6156/85/0286-0361$06.00/0
© 1985 American Chemical Society

glycols, a study on a cationic ring-opening polymerization of ECH using an oxonium type initiator in the presence of alkanediols was conducted.

Results and Discussion

The low molecular weight, 500~3000, polyepichlorohydrin glycol (PECHG) was prepared using triethyloxonium hexafluorophosphate (TEOP) initiator in the presence of various amounts of molecular weight modifiers, mostly ethylene glycol (EG).

Type of Hydroxyl Group in Polyepichlorohydrin Glycol.   First, in order to investigate the role of alkanediol in the cationic ring-opening polymerization, the type of the terminal hydroxyl groups in PECHG was examined. As shown in Figure 1, PECHG possesses predominantly secondary hydroxyl groups not only at 100% conversion, but also at as low as 40% conversion. The secondary hydroxyl group domination is also observed in cases using other glycols as molecular weight modifiers and using borontrifluoride etherate initiator (see Table I).

Table I. Effect of Molecular Weight Modifiers and Initiator on the
          Terminal Hydroxyl Groups

Glycol and Initiator		Secondary OH Group, %
EG,	TEOP	93~97
EG,	$BF_3 \cdot O(C_2H_5)_2$	98~100
2-butene-1,4-diol,	TEOP	93.0
2-butyne-1,4-diol,	TEOP	92.7
water,	TEOP	88.4~94.0

Next, small amounts of primary hydroxy groups in PECHG were examined to determine whether they belong to starting glycol or not. As shown in Table II, chemical shifts of the methylene protons adjacent to the primary hydroxy group are identical among the various types of initiators. These results clearly indicate that a small amount of primary hydroxyl groups in PECHG are not the primary hydroxyl groups belonging to the glycols used as molecular weight modifiers.

Besides ^{1}H NMR, ^{13}C NMR also indicates that PECHG prepared using EG does not contain any EG units (HO-CH$_2$-CH$_2$-O~ @ 61.5 ppm) at the polymer chain terminal position.

Disappearance of Molecular Weight Modifier.   Disappearance of the molecular weight modifier was monitored by measuring the polymer hydroxyl number at various polymerization stages. As shown in Figure 2, disappearance of EG used as a molecular weight modifier was so rapid at the beginning of the polymerization that nearly all EG was incorporated in the polymer by ~50% polymer conversion. Rapid disappearance of molecular weight modifier was also confirmed by gas

Table II. Effect of Molecular Weight Modifiers on the Terminal
Primary Hydroxyl Group

Molecular Weight Modifier	-CH$_2$-OH ^{1}H NMR Chemical Shift*
HO-CH$_2$-CH$_2$-OH	4.43 ppm
HO-CH$_2$-CH=CH-CH$_2$-OH	4.43 ppm
HO-CH$_2$-C≡C-CH$_2$-OH	4.43 ppm

* After derivatized with trichloroacetylisocyanate

chromatography and liquid chromatography, which indicated that all of
the molecular weight modifier was consumed by ~50% polymer conversion.

Polymer Molecular Weight. Polymer molecular weight was monitored by
GPC which shows that the polymer possesses narrow molecular weight
distribution (Mw/Mn = 1.1~1.2) and GPC curves shift to higher molecular weight with conversion. When the polymer molecular weight was
plotted against polymer conversion (Figure 3), the linear relationship which is a characteristic of living polymerization system
was obtained. Thus polymer molecular weight is readily adjusted by
the molar ratio of ECH/molecular weight modifier, and its molecular
weight is theoretically calculated by the following equation:

$$\overline{M}n = (92.5 \ [ECH]/[MWM]) \cdot \text{conversion}/100 + \text{mol. wt. of MWM}$$

MWM: molecular weight modifier

In a living polymerization system, Beste reports that the polymerization rate is directly related to the concentration of initiating
species (6). As shown in Figure 4, it seems that polymerization rate
of this ECH polymerization is directly related to the concentration
of initiator. No further study on the polymerization kinetics was
conducted in the present study.

Non-Functional Cyclic Oligomers. Formation of cyclic oligomers is an
inherent problem with a cationic ring-opening polymerization (7-11),
and polymerization of ECH in the presence of EG is no exception.
PECHG possessing molecular weight less than approximately 1000
contained no cyclic oligomers, while PECHG possessing molecular
weight more than approximately 1000 showed bi-modal molecular weight
distribution and contained 5~20% of oligomers. These oligomers were
isolated by preparative GPC and identified by field-desorption mass
spectrometer as non-functional cyclic oligomers, mainly ECH cyclopentamer, M$^+$·460, and cyclohexamer, M$^+$·552 (see Figure 5). No hydroxyl-
terminated linear oligomers, such as hydroxyl-terminated ECH pentamer
and hexamer, were detected.

Polymerization Temperatures. As shown in Figure 6, polymer conversion of 100% is achieved at the 30°C polymerization temperature in
8 hr. However, it seems that the polymerization rate is slower with
the higher polymerization temperature, and polymerization at 70°C

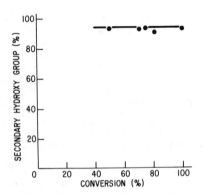

Figure 1.   Types of hydroxyl groups in polyepichlorohydrin glycol
            [ECH] = 11 mol/L, [EG] = 1.1 mol/L, [TEOP] = 2x10^{-3} mol/L

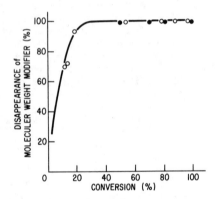

Figure 2.   Disappearance of molecular weight modifier
            [ECH] = 11 mol/L, [EG] = 1.1 mol/L, [TEOP] = 2x10^{-3} mol/L
            [O: obtained by M. P. Dreyfuss (1)]

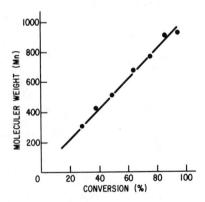

Figure 3.   Relationship between molecular weight and conversion
            [ECH] = 11 mol/L, [EG] = 1.1 mol/L, [TEOP] = 2x10^{-3} mol/L

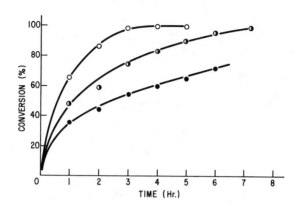

Figure 4. Effect of TEOP initiator concentration on conversion [ECH] = 11 mol/L, [EG] = 1.1 mol/L at 30°. ● = 1.1x10⁻³, ◐ = 2.2x10⁻³, and ○ = 3.7x10⁻³ mol/L

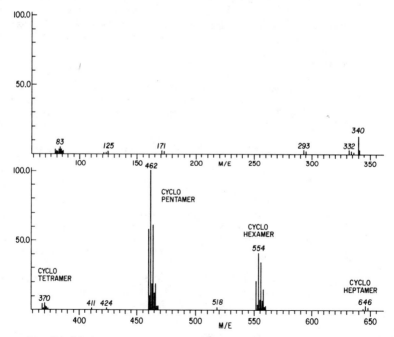

Figure 5. Field-desorption mass spectrogram of polyepichlorohydrin oligomers.

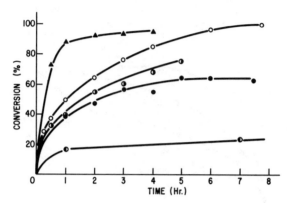

Figure 6.   Effect of polymerization temperature on conversion.
[ECH] = 11 mol/L, [EG] = 1.1 mol/L, [TEOP] = 2×10^{-3} mol/L
◑ = 0°, ◯ = 30°, ◐ = 50°, ● = 70°, and
▲ = [TEOA] = 2×10^{-3} mol/L at 70°.

reaches the plateau of approximately 60% conversion. This slower
polymerization rate with the higher polymerization temperature is
probably due to the decomposition of initiating species. When
triethyloxonium hexafluoroantimonate (TEOA), which is reported to be
more thermally stable than TEOP (12), is used as an initiator, the
polymerization reaches approximately 95% conversion in 4 hr at 70°C.
At the polymerization temperature of 0°C, the polymerization rate is
so slow that it takes about 48 hr to reach 100% polymer conversion.

Polymerization Mechanism

Based on the results mentioned above, it is doubtful that the previously postulated polymerization mechanism applies (5). In this
mechanism alkanediols were postulated to behave like chain transfer
agents:

$$HO \sim\!\!\sim\!\!\sim\!\!{}^{+}O\!\!\diagup_{CH_2Cl} \qquad + \qquad HO\text{-}R\text{-}OH$$

$$\downarrow$$

$$HO \sim\!\!\sim\!\!\sim\!\!\sim O\text{-}R\text{-}OH$$

This reaction does not seem consistent with our results which clearly
indicate that the hydroxyl groups in PECHG are not the hydroxyl
groups belonging to the alkanediols.
    Considering all the experimental results, the following polymer-
ization mechanism is postulated for ECH cationic ring-opening polymerization in the presence of alkanediols such as EG.

Initiation: Since triethyloxonium salt is a strong alkylating agent,
it probably alkylates ethylene glycol which is a stronger base than
ECH monomer (13-14). Then the proton is abstracted by ECH to form
initiating species 2 for this cationic ring-opening polymerization.

$$(C_2H_5)_3 \overset{+}{O} \cdot PF_6^- + HOCH_2CH_2OH \underset{\leftarrow}{\rightarrow} C_2H_5 \overset{+}{O}CH_2CH_2OH + (C_2H_5)_2O$$

$$\underset{\underset{1}{\sim}}{\underset{|}{H}}$$

$$\underset{\sim}{1} + \text{[epoxide]}-CH_2Cl \underset{\leftarrow}{\rightarrow} C_2H_5OCH_2CH_2OH + \text{[epoxide]}-CH_2Cl$$

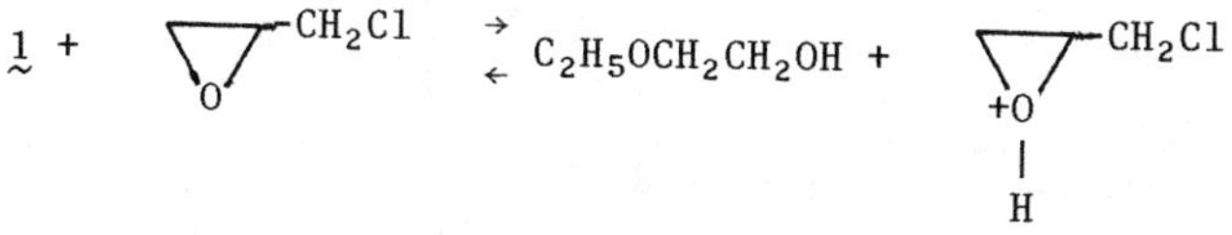

$$\underset{\underset{2}{\sim}}{\underset{|}{H}}$$

<u>Transfer</u>: The initiating species $\underset{\sim}{2}$ is then attacked by EG, instead of ECH, to produce $\underset{\sim}{3}$

$$\underset{\sim}{2} + HOCH_2CH_2OH \rightarrow HO-CH-CH_2-\overset{+}{O}-CH_2CH_2OH$$
$$\underset{\underset{3}{\sim}}{\underset{CH_2Cl \quad H}{| \quad \quad |}}$$

<u>Termination & Reinitiation</u>: The proton in $\underset{\sim}{3}$ is abstracted by ECH to form new ECH-EG adduct glycol, $\underset{\sim}{4}$, and regenerate the propagating species $\underset{\sim}{2}$.

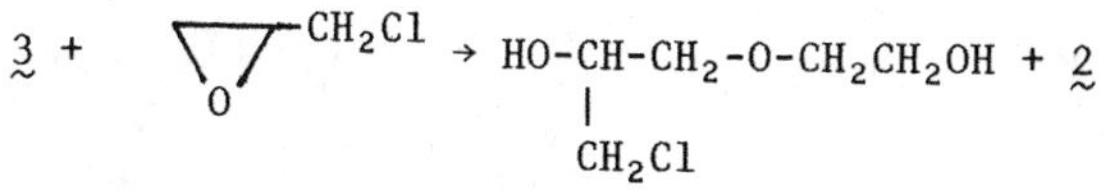

$$\underset{\sim}{3} + \text{[epoxide]}-CH_2Cl \rightarrow HO-CH-CH_2-O-CH_2CH_2OH + \underset{\sim}{2}$$
$$\underset{\underset{4}{\sim}}{\underset{CH_2Cl}{|}}$$

After all EG is consumed, the regenerated propagating species $\underset{\sim}{2}$ then reacts with the nascent glycol $\underset{\sim}{4}$ by a manner identical to that mentioned above to produce the new ECH-EG-ECH adduct glycol, $\underset{\sim}{5}$, and regenerated propagating species $\underset{\sim}{2}$.

$$\underset{\sim}{4} + \underset{\sim}{2} \rightarrow HO-CH-CH_2-O-CH_2CH_2-\overset{+}{O}-CH_2-CH-OH$$
$$\underset{CH_2Cl \quad \quad \quad H \quad \quad CH_2Cl}{| \quad \quad \quad \quad | \quad \quad |}$$

$$\text{[epoxide]}-CH_2Cl \xrightarrow{\quad\quad\quad\quad} HO-CH-CH_2-O-CH_2CH_2-O-CH_2-CH-OH \quad + \underset{\sim}{2}$$
$$\underset{\underset{5}{\sim}}{\underset{CH_2Cl \quad \quad \quad \quad \quad CH_2Cl}{| \quad \quad \quad \quad \quad \quad |}}$$

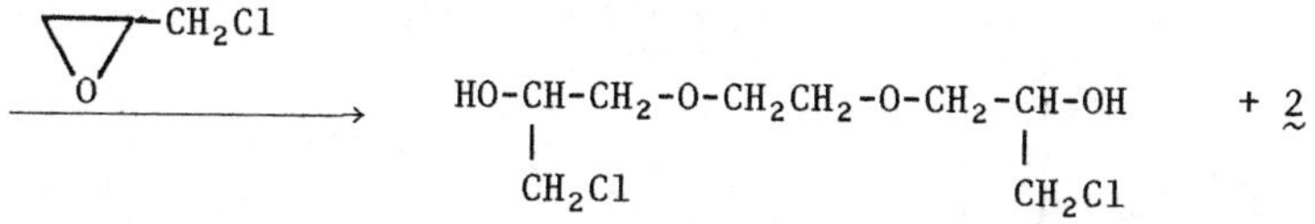

Overall polymerization proceeds by repeating the "transfer", "termination" and "reinitiation" steps schematically as shown below.

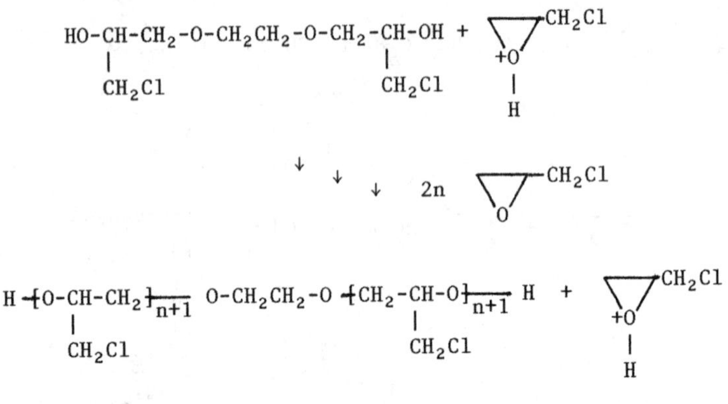

$$HO-CH-CH_2-O-CH_2CH_2-O-CH_2-CH-OH \quad + \quad \triangle^{CH_2Cl}_{+O} \atop H$$
$$| \qquad\qquad\qquad\qquad\qquad | $$
$$CH_2Cl \qquad\qquad\qquad\qquad CH_2Cl$$

$$\downarrow \quad \downarrow \quad \downarrow \quad 2n \quad \triangle^{CH_2Cl}_{O}$$

$$H \overline{\{O-CH-CH_2\}}_{n+1} \; O-CH_2CH_2-O \; \overline{\{CH_2-CH-O\}}_{n+1} \; H \quad + \quad \triangle^{CH_2Cl}_{+O} \atop H$$
$$| \qquad\qquad\qquad\qquad\qquad\qquad | $$
$$CH_2Cl \qquad\qquad\qquad\qquad CH_2Cl$$

$$\underset{\sim}{2}$$

According to this new postulated mechanism, all results obtained are readily explainable. It is reasonable that all EG was consumed at the beginning of the polymerization stage, and no EG residue is located at the terminal position of the polymer chains.

Small amounts of primary hydroxyl groups observed in PECHG are probably due to opening of the oxirane CH-O linkage in the propagating species:

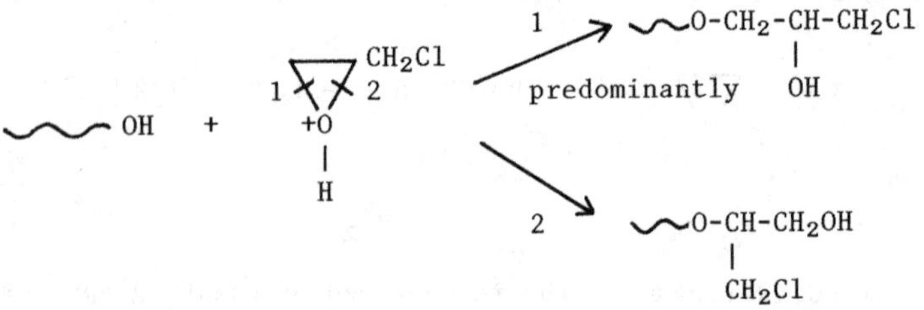

$$\sim\!OH \quad + \quad \overset{1}{\underset{}{\triangle}}\overset{CH_2Cl}{\underset{2}{}} \quad \overset{1}{\longrightarrow} \quad \sim\!\!O-CH_2-CH-CH_2Cl$$
$$\qquad\qquad\qquad +O \qquad\quad predominantly \qquad | $$
$$\qquad\qquad\qquad | \qquad\qquad\qquad\qquad\qquad OH$$
$$\qquad\qquad\qquad H$$
$$\qquad\qquad\qquad\qquad\qquad\qquad \overset{2}{\searrow} \quad \sim\!\!O-CH-CH_2OH$$
$$\qquad\qquad\qquad\qquad\qquad\qquad\qquad\qquad\qquad | $$
$$\qquad\qquad\qquad\qquad\qquad\qquad\qquad\qquad\qquad CH_2Cl$$

The oligomers are believed to be formed by so-called back-biting or tail-biting which is supposed to produce hydroxyl-terminated linear oligomers as well as non-functional cyclic oligomers. Absence of any hydroxyl-terminated linear oliogomers in PECHG can be explained as follows: since hydroxyl groups are polymer propagating sites, hydroxyl-terminated linear oligomers can participate back into the polymerization as soon as they are formed, while non-functional cyclic oligomers cannot. Thus oligomers found in PECHG are only non-functional cyclic oligomers.

These cyclic oligomers can form only when polymer molecular weight exceeds approximately 1000. According to the newly postulated mechanism, PECHG possessing the molecular weight of approximately 1000 has two polyepichlorohydrin chains each possessing ~500 molecular weight at both ends of EG residue.

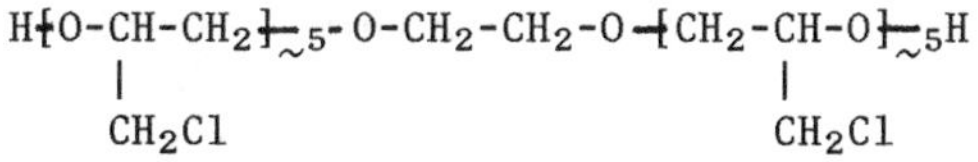

$$H\{O-CH-CH_2\}_{\sim5}-O-CH_2-CH_2-O\{CH_2-CH-O\}_{\sim5}H$$
$$\phantom{H\{O-CH-}|_{\sim5}-O-CH_2-CH_2-O\{CH_2-}|}$$
$$\phantom{H\{O-CH-}CH_2Cl\phantom{_{\sim5}-O-CH_2-CH_2-O\{CH_2-}CH_2Cl}$$

PECHG ~1000 molecular weight

These two polymer chains are too short to form cyclic oligomers such as cyclopentamer and cyclohexamer. No cyclic oligomers were, therefore, observed in PECHG possessing the molecular weight of ~1000. Cyclic oligomers formation starts taking place when molecular weight of PECHG exceeds ~1000.

The above rationalization is also supported by the other results. For example, three branched polyepichlorohydrin triol having molecular weight of 1500 does not form any cyclic oligomers.

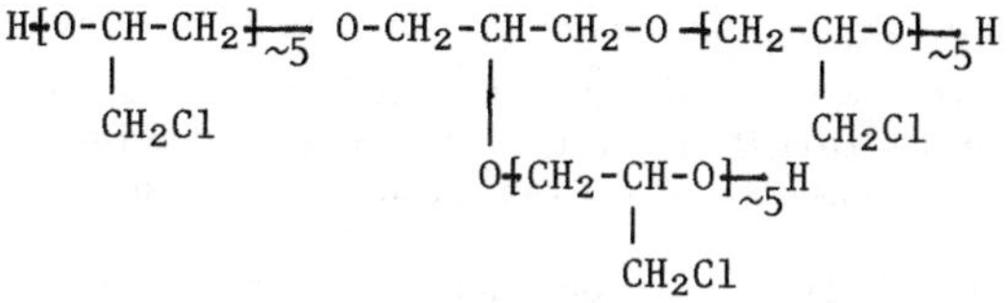

$$H\{O-CH-CH_2\}_{\sim5}\ O-CH_2-CH-CH_2-O\{CH_2-CH-O\}_{\sim5}H$$
$$\phantom{H\{O-CH-}|_{\sim5}\ O-CH_2-}|\phantom{CH_2-O\{CH_2-}|}$$
$$\phantom{H\{O-CH-}CH_2Cl\phantom{_{\sim5}\ O-CH_2-}O\{CH_2-CH-O\}_{\sim5}HCH_2Cl}$$
$$\phantom{H\{O-CH-CH_2\}_{\sim5}\ O-CH_2-CH-CH_2-}|$$
$$\phantom{H\{O-CH-CH_2\}_{\sim5}\ O-CH_2-CH-}CH_2Cl$$

polyepichlorohydrin triol ~1500 molecular weight

Polyepichlorohydrin Polyols

By utilizing the new findings mentioned above, PECHG and polyepichlorohydrin triol (PECHT), each with a molecular weight of approximately 1000, were prepared using ethylene glycol and glycerol, respectively. The results are summarized in Table III.

Table III. Polyepichlorohydrin Polyols

	PECHG	PECHT
Molecular weight ($\bar{M}n$)	954	1010
Hydroxy equivalent weight	477	335
Hydroxy functionality	2.0	3.0
Hydroxy group, primary	6.6%	6.5%
secondary	93.4%	93.5%
Dispersity ($\bar{M}w/\bar{M}n$)	1.29	1.12

Attempts to prepare polyepichlorohydrin tetraol, pentaol, and octaol using pentaerythritol, glucose, and sucrose, respectively were unsuccessful, probably due to the poor solubility of starting polyols in ECH.

Conclusion

All results mentioned above clearly support the new mechanism of cationic ring-opening polymerization of ECH in the presence of EG used as a molecular weight modifier. Polymerization initiates at the hydroxyl groups in EG and the polymer chain propagates simultaneously at both ends through the addition of the monomer. Since this polymerization possessed a characteristic of living polymerization, i.e. polymer molecular weight increased directly with polymer conversion, the polymer molecular weight was readily controlled by adjusting the ratio of ECH to EG. The obtained molecular weight PECHG possessed predominantly secondary hydroxyl groups, narrow molecular weight distribution, and a hydroxyl functionality of two.

Experimental

Reagents. Epichlorohydrin was dried over molecular sieves. Ethylene glycol, reagent grade, was used as received. 2-Butene-1,4-diol and 2-butyne-1,4-diol were freshly distilled before used. TEOP (mp 140-142°C) was reprecipitated from methylene chloride and ether, and borontrifluoride etherate was distilled before use.

Polymerization. Polymerization was carried out in a 250 mL 3-neck flask, equipped with a mechanical stirrer, a thermometer, and a rubber septum for initiator introduction. ECH, 93.8 g, and EG, 6.2 g, were charged into a flask and the flask was purged with dry nitrogen. The initiator solution, 0.045 g of TEOP dissolved in 5 mL of methylene chloride, was added to the above mixture incrementally (1 mL per every 5 min) by a hypodermic syringe while maintaining the temperature at 30°C. The polymerization was carried out at 30°C and the reaction was monitored by taking small amounts of sample during the polymerization. On completion of the polymerization, the reaction was terminated with ~300 μL of a mixture of 30% ammonium hydroxide and isopropanol (1:4 by vol), and then the polymer was dried on a rotary evaporator at 60°C in vacuo. The obtained PECHG possessed a molecular weight of 950 determined by vapor pressure osmometry (THF solvent), hydroxyl number of 118, and molecular weight distribution, Mw/Mn, of 1.23.

Determination of Hydroxyl Groups. Type of hydroxyl groups in PECHG were determined using a Bruker WH-200 superconducting NMR spectrometer employing the trichloroacetylisocyanate derivatization method (15). Typical NMR spectrum was shown in Figure 7.

Identification of Oligomers. The low molecular weight fraction was separated by a preparative GPC and identified using a Finnigan MAT 311A field desorption mass spectrometer (FD-MS).

Molecular Weight Determination.   Molecular weights were determined
using a Waters Model 200 gel permeation chromatograph equipped with a
modified Waters R4 differential refractometer detector.   The solvent
was THF; the flow was 2.0 mm³/min.   The column, 25 cm x 7.8 mm ID,
consisted of 10^6, 10^5, 10^4, 10^3, 10^2 A° waters microstyragel.

Disappearance of Molecular Weight Modifier.   Hydroxyl number was
determined using a standard procedure (16).   Disappearance of molecular weight modifier was calculated by dividing the polymer hydroxyl
number by the theoretical hydroxyl number.   It was assumed in calcu-
lating the theoretical hydroxyl number that all molecular weight
modifier was incorporated into the polymer obtained.

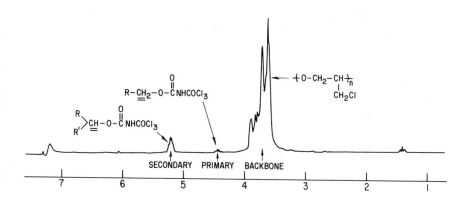

Figure 7.   ¹H NMR spectrum of polyepichlorohydrin glycol derivatized
with trichloroacetylisocyanate: CDCl₃ solvent.

Acknowledgments

The author would like to thank Dr. M. P. Dreyfuss who generously shared his expertise, Dr. D. Harmon and Mrs. B. Boose for GPC analyses, Dr. R. Lattimer for FD-MS analyses, Mr. J. Westfahl for NMR analyses, and Mrs. R. Lord for carrying out the experiments. The author wishes to express his appreciation to the BFGoodrich Chemical Group for permission to publish this work.

Literature Cited

1. Dreyfuss, P. U.S. Patent 3 850 856, 1974.
2. Young, C. I.; Barker, L. L. U.K. Patent Application 2 021 606, 1979.
3. Manser, G. E.; Guimont, J.; Ross, D. L. Presented at the 1981 JANNAF Propulsion Meeting, New Orleans, Louisiana, May 27, 1981.
4. Dickinson, L. A. J. Polymer Sci. 1962, 58, 857.
5. Hammond, J. M.; Hooper, J. F.; Robertson, W. G. P. J. Polymer Sci. A-1 1971, 9, 265.
6. Beste, L. F.; Hall, H. K., Jr. J. Phy. Chem. 1964, 68, 269.
7. Ito, K.; Usami, N.; Yamashita, Y. Polymer J. 1979, 11, 171.
8. Dreyfuss, P.; Dreyfuss, M. P. Polymer J. 1975, 8, 81.
9. Robinson, I. M.; Pruckmayr, G. Macromolecules 1979, 12, 1043.
10. Goethals, E. J. "Advances in Polymer Science", Springer-Verlag Berlin Heidelberg, New York, 1977; Vol. 23, pp 104~130.
11. Penczek, S.; Kubisa, P.; Matyjaszewski, K. "Advances in Polymer Science", Springer-Verlag Berlin Heidelberg, New York, 1980; Vol. 37.
12. Jones, F. R.; Plesch, P. H. Chem. Commun. 1969, 1231 and J. Chem. Soc., Dalton 1979, 927.
13. Meerwein, H. "Houben-Weyl Methoden der Organischen Chemie.", E. Miller Ed., Stuttgart, George Thieme Verlog, 1965; Vol VI/3, pp 359.
14. Penczek, S.; Kubisa, P.; Matyjaszewski, K. "Advance in Polymer Science Cationic Ring-Opening Polymerization", Springer-Verlag Berlin Heidelberg, New York, 1980; Vol. 37, p 6.
15. Groom, T.; Babiec, J. S., Jr.; Van Leuwen, B. G. J. of Cellular Plastics 1974, January/February, 43.
16. Sorenson, W. R.; Campbell, T. W. "Preparative Method of Polymer Chemistry", 2nd Ed., Interscience Publisher, a division of John Wiley & Sons, New York, London, Syndey, Toronto, 1968; p 155.

RECEIVED October 4, 1984

Lactone Polymerization
Pivalolactone and Related Lactones

WILLIAM H. SHARKEY[1]

Central Research and Development Department, E. I. du Pont de Nemours & Company, Wilmington, DE 19898

The polymerization of pivalolactone, α,α-dimethyl–β-propiolactone, is a remarkably easy anionic, ring-opening reaction that takes place rapidly and completely in organic media at mild temperatures. Since it appears to be similar to other β-lactones in its polymerization and the properties of its polymers have been more widely studied than those of other β-lactone polymers, the formation and properties of polypivalolactone will be emphasized in this discussion.

Early studies on pivalolactone ($\underline{1},\underline{2}$) established that its polymerization is initiated by tertiary amines and phosphines. The reaction was visualized as occurring in two steps.

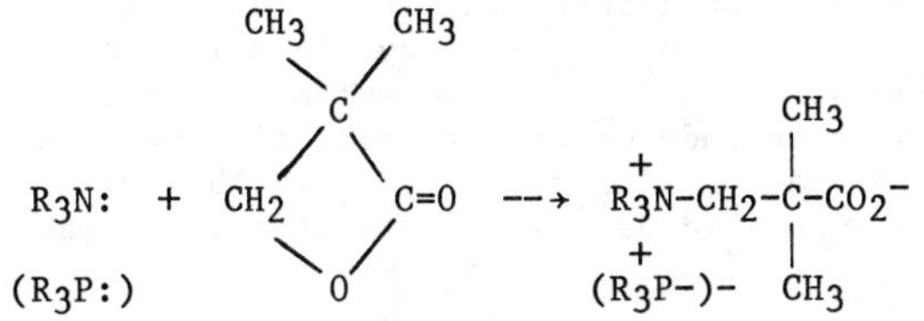

$$R_3N: \quad + \quad CH_2 \overset{CH_3 \quad CH_3}{\underset{O}{\diagup C \diagdown}} C=O \quad \longrightarrow \quad \overset{+}{R_3N}-CH_2-\overset{CH_3}{\underset{CH_3}{\overset{|}{\underset{|}{C}}}}-CO_2^- \quad \text{Initiation}$$

(R_3P:)   (R_3P-)-

$$\overset{+}{R_3N}-CH_2-\overset{CH_3}{\underset{CH_3}{\overset{|}{\underset{|}{C}}}}-CO_2^- \quad + \quad m \; CH_2 \overset{CH_3 \quad CH_3}{\underset{O}{\diagup C \diagdown}} C=O \quad \longrightarrow \quad \overset{+}{R_3N}-\left(CH_2-\overset{CH_3}{\underset{CH_3}{\overset{|}{\underset{|}{C}}}}-CO_2\right)_n-CH_2-\overset{CH_3}{\underset{CH_3}{\overset{|}{\underset{|}{C}}}}-CO_2^-$$

(R_3P-)-   (R_3P-)-   Propagation

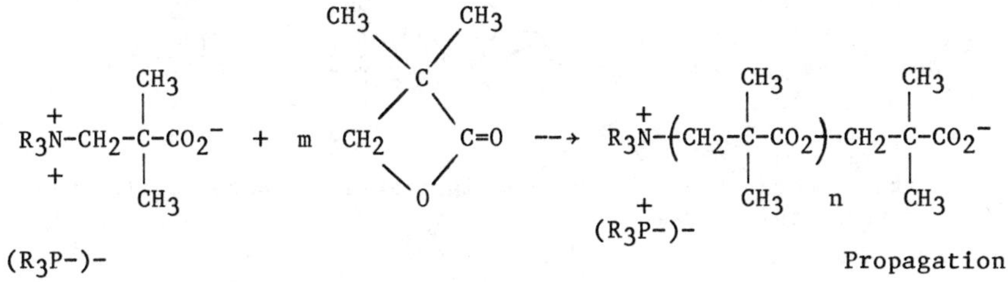

The formation of a macrozwitterion by attack of the nucleophile on the methylene group of PVL has been supported by NMR and chemical studies on propiolactone. By these means Maskevich, Pakhomeva, and Enikolopyan ($\underline{3}$) who used triethylphosphine and Mathes and Jacks ($\underline{4}$) who used triethylamine, obtained evidence for the initiation

[1]Current address: 174 S. Collier Blvd., Marco Island, FL 33937

reactions. In addition, Mathes and Jacks (5) used betaine,

$(CH_3)_3\overset{+}{N}CH_2CO_2^-$, as an initiator for propiolactone and showed by
electrophoresis that quaternary ammonium cations and carboxylate
ions are present in the same polymer chain.

The formation of polymeric zwitterions might be thought as
unlikely because of high coulombic energy of charge separation.
Mayne (6) has pointed out this need not be so because the chains may
be cyclized or paired with each other as indicated below.

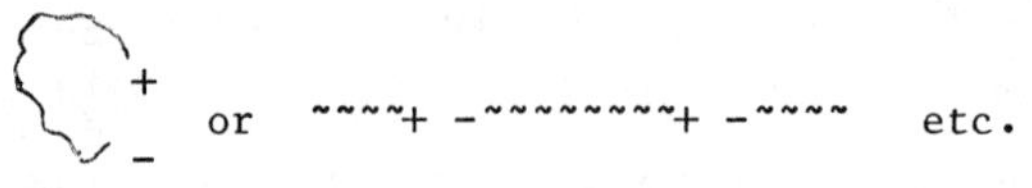

or ~~~~+ -~~~~~~~~+ -~~~~   etc.

Mayne has also shown the macrozwitterions to be living
polymers. He has used PVL oligomers of Mn 2000–7000 suspended in
refluxing hexane as initiators for the formation of high molecular
weight PPVL. He has stated that propagation by these oligomers is
faster than initiation by tributylphosphine by a factor of at least
600. This behavior is a consequence of the high reactivity of the
carboxyl anion toward the lactone.

It has been recognized by all investigators of beta-lactone
polymerization that carboxyl ions are the initiators and that they
are most effective when used as tetralkylammonium salts. Before
pursuing this point further, an interesting variation on tertiary
amine initiation should be mentioned. Thus cyclic amine initiation
was reported by Wilson and Beaman.(7) These authors showed that
cyclic tertiary amines which undergo easy thermal ring opening can
be used to synthesize amine-containing copolymers. This scheme is
illustrated below with quinuclidene. The first step is to make a
macrozwitterion.

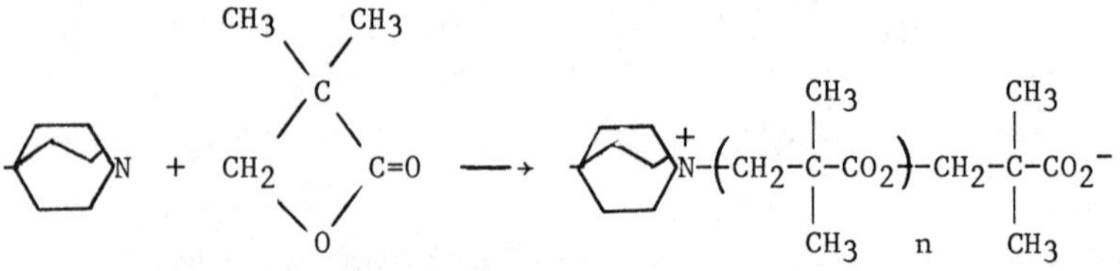

In the second step, the polymer is heated in acetonitrile under
reflux for several hours.

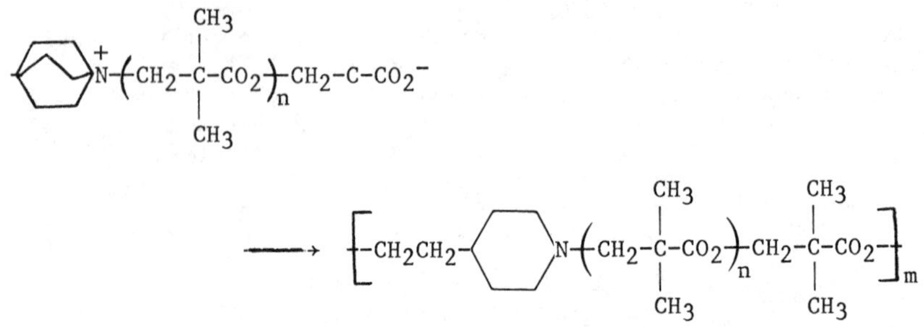

A number of strained-ring, tertiary amines were investigated including aziridines, azetidines, pyrrolidines, conidine, cycloamphidine, and isogranatanine. Copolymers were obtained containing 7 to 58 initiator units per polymer chain.

Polymerization Mechanism

As stated above, it has been established that carboxylate ions are the true initiators in β-lactone polymerization. In an early study, Hall (8) stated that in intermediate pH ranges anions or water itself attack the CH_2 group with alkyl-oxygen cleavage by SN_2 reaction that proceeds most rapidly in anhydrous acetonitrile or tetrahydrofuran. From heats of combustion he determined the heat of polymerization of pivalolactone* ΔH_{1c} to be -20.1 kcal mol^{-1}. Mayne (6) gives a value of -18.4 kcal mol^{-1} and the value for polymerization of propiolactone (9) is -18.4 kcal mol^{-1}.

Because the heat of polymerization is high, Hall was able to use a calorimetric method (10,11) to determine polymerization rates. The process can be described by initiator I attacking the β-lactone M to give the β-substituted carboxylate IM_1, which attacks additional monomer molecules to give the poly-β-lactone IM_2. In those cases when kp is only slightly less than k_i, the

$$I + M \xrightarrow{\quad k_i \quad} IM_1 \qquad \text{Initiation}$$

$$IM_1 + M \xrightarrow{\quad k_p \quad} IM_2 \qquad \text{Propagation}$$

polymerization will proceed according to approximate first-order kinetics and for them only k_i was reported. For some initiators under certain conditions k_i differed markedly form k_p and here the treatment of Beste and Hall (12) was followed. Typical results are described in **Table I** where k_i and k_p are in M^{-1} sec^{-1}. The data indicates the polymerization rate is very high and substantially higher in THF than in acetonitrile.

Bigdeli and Lenz (13) have also examined the polymerization kinetics of β-lactones. They used an IR method (14) and their results support, in broad outline, Hall's findings. Their numbers are different, but they confirm that PVL polymerizes rapidly and that THF is a better solvent than DMSO. This is shown in **Table II** where k_p is expressed in M^{-1} sec^{-1}. Note that α-methyl-α-propyl-β-propiolactone polymerizes most rapidly in THF.

Inasmuch as the polymerization of β-lactones, and pivalo-lactone in particular, is very rapid, the question arises as to how

---

*Unpublished work at Haskell Laboratory for Toxicology and Industrial medicine, E. I. du Pont de Nemours and Company, has shown that pivalolactone caused skin tumors in mice when applied as a 25% solution in acetone for most of the life span of the mice. The time required for tumor formation was greater than that for β-propiolactone.

Table I

Rate Constants for Polymerization of β-Lactones at 35°[a]

Lactone	Solvent	k_i	k_p
PVL	CH_3CN	0.36	--
$\begin{array}{c} CH_3 \quad CH_2CH_2CH_3 \\ \diagdown C \diagup \\ CH_2 \quad C{=}0 \\ \diagdown O \diagup \end{array}$	CH_3CN	0.18	--
$\begin{array}{c} CH_3 \quad CH_2CH_2CH_3 \\ \diagdown C \diagup \\ CH_2 \quad C{=}0 \\ \diagdown O \diagup \end{array}$	THF	--	0.90

[a] Rate constants are $M^{-1} sec^{-1}$.

Table II

Polymerization Rates for β-Lactones According to
Bigdeli and Lenz (13)

Lactone	Temp °C	Solvent	k_p $(M^{-1} sec^{-1})$
PVL	22	DMSO	0.17
$\begin{array}{c} CH_3 \quad CH_2CH_2CH_3 \\ \diagdown C \diagup \\ CH_2 \quad C{=}0 \\ \diagdown O \diagup \end{array}$	22	DMSO	0.20
	37	DMSO	0.34
	37	THF	0.53

it compares to other very fast anionic polymerizations. Such a comparison has been presented by Hall ($\underline{8}$), which is reproduced in Table III to which the value for α-methyl-α-propyl-β-lactone in THF has been added.

Table III

Rate Constants for Anionic Polymerizations ($\underline{8}$)

Propagation Reaction[a]	Solvent	Temp. °C	k_p M^{-1} sec^{-1}
$S^- + S$	THF	25	65,000
α-MeS$^-$ + α-MeS	THF	25	830
$I^- + I$	Heptane	20	0.65
$RCO_2^- +$			
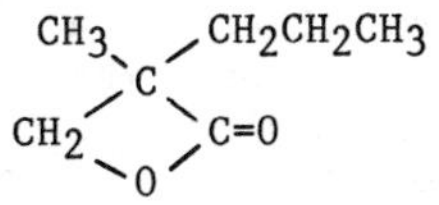 (β-lactone structure)	CH$_3$CN	35	0.40 (Ref. 8)
	THF	35	0.90 (Ref. 8)
	THF	37	0.53 (Ref.13)
$RO^- + EO$	DMSO	25	0.13
$I^-Li^+ + I$	THF	30	0.14

[a]S = styrene, α,MeS = alpha –methylstyrene, I = isoprene, EO = ethylene oxide.

Since rate constants for PVL in acetonitrile or DMSO are the same or higher than those of α-methyl-α-propylpropiolactone, the polymerization rate of PVL in THF should be just as high as that of the methyl, propyl derivative. If this is so, the above data suggest anionic polymerization of PVL proceeds as fast as anionic polymerization of isoprene in hydrocarbons to cis-1,4-polyisoprene. This is indeed quite impressive.

A consequence of rapid polymerization in which $k_i \gg k_p$ and the absence of chain transfer to monomer is the formation of polymer having a Poisson distribution of molecular weights with Mw/Mn only slightly above one.

Though there is no chain transfer to monomer, there is a high degree of chain transfer to polymer. That is, all carboxyl sites in the system are covered by PPVL chains ($\underline{15,16}$). This is the result of the rapid exchange between the tetra-alkylammonium salts and free acid groups. This effect is particularly significant with regard to

$$\sim\sim\sim\sim\sim\overset{*}{C}O_2NR_4 \quad + \quad \sim\sim\sim\sim\sim CO_2H \rightleftharpoons \sim\sim\sim\sim\sim\overset{*}{C}O_2H \quad + \quad \sim\sim\sim\sim CO_2NR_4$$

$$\sim\sim\sim\sim\sim\overset{*}{C}O_2(PVL)_nNR_4 \qquad\qquad\qquad\qquad \sim\sim\sim\sim CO_2(PVL)_nNR_4$$

synthesis of block copolymers because it allows quantitative conversion of a beta-lactone monomer to polymer without requiring quantitative conversion of carboxylic acids to salts.

In summary, pivalolactone undergoes rapid anionic, ring-opening polymerization when initiated by tetraalkylammonium carboxylates in aprotic organic solvents, of which THF is preferred. The polymer is linear, has very low polydispersity, is "living", and the living propagating terminus has exceptionally long lifetimes. Most, if not all, of these features are also found in many other substituted β-lactones.

Pivalolactone Polymers

Polypivalolactone is a highly crystalline polymer that exists in three crystalline modifications; those have been described by Oosterhoff (17) of Shell Laboratories. The main product that crystallizes from a polypivalolactone melt is the α-modification. It melts at about 245°C and has a glass transition temperature (Tg) of -10°C. In this form the polymer chains have a helical structure with two monomer units per turn. Slow cooling from the melt leads to the β-form, which has a melting point of 228°C. Annealing above 228°C converts the β-form to the α-form. The γ-form is metastable and arises from orientation. In this modification the polymer backbone has a planar zig-zag structure and is extended 1.6 times with respect to the alpha-form helix. Upon annealing it reverts to the alpha-form with concomitant shrinkage. As would be expected, this reversibility has a strong influence upon the properties of polypivalolactone.

Helical conformations with two monomer units per turn appear to be general for polymers from β-lactones.(18) Fiber repeat distances may vary, but not by a great amount. Examples are polypropiolactone and poly(D,L-α-methyl-a-n-propyl-β-propiolactone). Both are converted to a planar zig-zag form by stretching and both reverts to a helical configuration upon annealing.

Molten polypivalolactone, presumably because of its high chemical purity, has a much smaller number of nucleation sites than most other polymers. As a consequence spherulites formed during cooling may become quite large. Above 190°C they may grow to 1 mm or larger. Since such large crystals are usually undesirable, it is best to avoid them by efficient quenching or by use of nucleating agents. Normal cooling in a mold leads to 75% crystallinity and a density (20°C) of 1.19. The density of 100% crystalline material at 20°C is reported to be 1.223.(17)

The apparent melt viscosity of polypivalolactone over the range of 260-290°C at high shear rates is below that of poly(ethylene terephthalate).(17) Accordingly, poly(PVL) is amenable to melt-spinning into fibers and molding into objects. The very high rate of crystallization and spherulite growth play an important part in both processes. High linear speeds of spinning and rapid cooling lead to fibers with higher tenacities. Fibers can be easily oriented by stretching over hot plates or hot pins at temperatures of 20 to 210°C. Stretching followed by annealing and stretching again has led to fibers having very high tenacities (75-90 g./tex) and high moduli up to 1100g./tex . These fibers contain large amounts of the metastable gamma-crystalline form, which can be reconverted to the alpha-form by heat treatment at 150-200°C. The conversion, however, is accompanied by considerable shrinkage and decreases in tenacity to 45-55 g./tex (40-50 g/d; 35-44 dN/tex) and modulus to 350 g.tex (315 g/d; 275 dN/tex.).

For molded items it is best to incorporate nucleating agents to control spherulite size. Injection-molding can be accomplished at temperatures of 260-300°C with mold temperatures of room temperature to 150°C. Oosterhoff (17) has said molecular weights should be between 150,000 and 400,000. Molded polypivalolactone has good strength and modulus and good retention of properties up to 200°C. Annealed samples have exceptionally low set after breaking. This means that test samples revert almost completely to original dimensions after breaking.

Block Copolymers

The easy formation of polypivalolactone together with its rapid crystallizability and good physical properties has stimulated investigations of the attachment of polypivalolactone segments to elastomeric polymers for the purpose of developing new types of thermoplastic elastomers. In these cases the elastomeric component forms the continuous phase and the polypivalolactone blocks crystallize into discontinuous domains that are the "crosslinks". Included are low-melting β-propiolactone polymers, vinyl and acrylic polymers, polyisoprene, and poly(isobutylene).

Combinations of poly(α-methyl-α-butyl-β-propiolactone) (MBPL) and poly(α-methyl-α-propyl-β-propiolactone) (MPPL) as B blocks and polypivalolactone (poly(PVL)) as A blocks in ABA block copolymers have been studied by Lenz, Dror, Jorgensen, and Marchessault.(19) They were prepared by using the tetrabutylammonium salt of sebacic acid as a difunctional initiator to initiate the polymerization of MBPL or MPPL followed by addition of PVL.

$$Bu_4NO_2C(CH_2)_8CO_2NBu_4$$

$$\downarrow MBPL$$

$$Bu_4N(MBPL)_nO_2C(CH_2)_8CO_2(MBPL)_nNBu_4$$

$$\downarrow PVL$$

$$Bu_4N(PVL)_{m}--(MBPL)_nO_2C(CH_2)_8CO_2(MBPL)_n--(PVL)_{m}NBu_4$$

$$\downarrow H+$$

$$H(PVL)_{m}--(MBPL)_nO_2C(CH_2)_8CO_2(MBPL)_n--(PVL)_{m}H$$

The PVL blocks crystallized into domains with good phase separation from either poly(MBPL) or poly(MPPL), which constituted the principal and continuous phases. The ABA block copolymer with poly(MPPL) as the B block behaved as a thermoplastic elastomer but suffered from poor ability to snap back quickly from a considerable stretch, i.e., it was "logy".

Modification of vinyl and acrylic polymers have included attachment of poly(PVL) blocks to poly(ethylene-co-vinyl acetate-co-methacrylic acid).(20) Tetrabutylammonium salts of the copolymer in THF were reacted with PVL to lead to a copolymer modified with blocks of poly(PVL) along the chain. The highly crystalline micro-

$$-(CH_2CH_2)_n-(CH_2CH)_m-(CH_2-\underset{\underset{CO_2NBu_4}{|}}{\overset{\overset{CH_3}{|}}{C}})_o$$
$$\underset{OAc}{|}$$

$$\big|\; PVL$$
$$\downarrow$$

$$-(CH_2CH_2)_n--(CH_2-CH)_m-(CH_2-\underset{\underset{CO_2(PVL)_p}{|}}{\overset{\overset{CH_3}{|}}{C}})_o$$
$$\underset{OAc}{|}$$

domains of poly(PVL) increased minimum flow temperature, limited solubility, and reinforced the mechanical properties of the base resin.

Caywood (21) has modified poly(ethyl acrylate), poly(ethyl acrylate-co-butyl acrylate), and poly(butyl acrylate) by first saponifying some of the ester groups by reaction with tetrabutylammonium hydroxide and use of the carboxylic salts so developed to initiate the polymerization of pivalolactone. Poly([ethyl acrylate]-g-pivalolactone) was found to be easily processable on conventional rubber working equipment. It was easily processable on a two-roll mill, had excellent calendering properties, could be compression molded at 225-230°C, and could be injection molded at 225°C. Extrusion was more difficult requiring high temperatures (250°C) and slow extrusion rates. Physical properties of the graft copolymers were similar to those of the parent elastomeric polyacrylates that had been compounded with carbon block and chemically crosslinked.

Modification of poly(ethylene-co-propylene-co-1,4-hexadiene) has been described by Thamm and his associates.(22,23) This involved reactions on the side chain unsaturation of Nordel®, DuPont's hydrocarbon elastomer, a copolymer of ethylene, propylene, and 1,4-hexadiene (an EPDM). An EPDM-g-thioglycollic acid was first made by reaction of the EPDM with large excess of thioglycolic acid using a procedure described by Calhoun and Hewett.(24) This EPDM-T was then reacted with tetrabutylammonium hydroxide in THF and the product used to initiate the polymerization of pivalolactone to give EPDM-g-SCH$_2$CO$_2$(PVL)$_n$, a hydrocarbon elastomer with poly(PVL) segments distributed randomly along the polymer chain.

A second method for preparing an EPDM-g-$(PVL)_n$ was reaction of the hydrocarbon polymer with maleic anhydride in bulk at 260-300°C followed by conversion of the anhydride groups in the product to tetrabutylammonium salts and use of the salts to initiate the polymerization of pivalolactone.

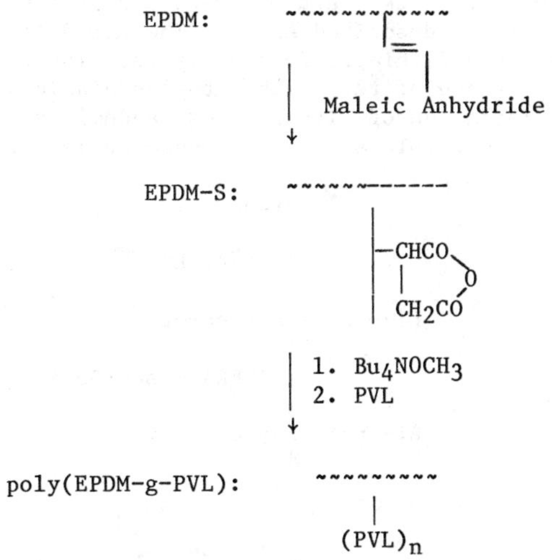

EPDM: ~~~~~~~~~~~~
|
Maleic Anhydride
↓

EPDM-S: ~~~~~~~~~~~~
|
—CHCO
| O
CH₂CO

1. Bu_4NOCH_3
2. PVL
↓

poly(EPDM-g-PVL): ~~~~~~~~~~
|
$(PVL)_n$

Tensile strengths of the above products are equal to those of their chemically crosslinked and reinforced non-grafted counterparts. Compression set values of the grafted products, especially after annealing, are unusually low. This indicates the presence of thermally and mechanically stable crosslinks and little tendency to form a second network under pressure.

Polyisoprene has also been converted to thermoplastic elastomers by attachment of segments of poly(PVL).(25,26) These are of two types, i.e., ABA triblocks or poly(pivalolactone-b-isoprene-b-pivalolactone) and block-graft copolymers or poly[(pivalolactone-b-isoprene-b-pivalolactone)-g-pivalolactone]. Both require a difunctional initiator for converting isoprene to an α-ω-dilithio-cis-1,4-polyisoprene. The initiator employed was synthesized by addition of sec-butyl lithium to 1,3-diisopropenyl-benzene followed by reaction with isoprene and modified with triethylamine.(27) It has the formula:

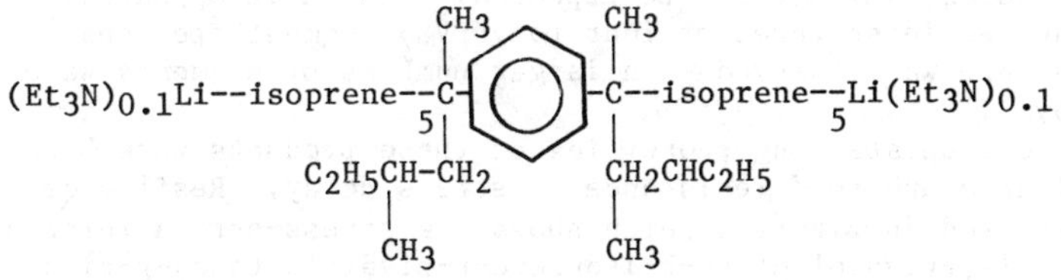

$(Et_3N)_{0.1}$Li--isoprene--C————————C--isoprene--Li$(Et_3N)_{0.1}$

Use of this initiator to polymerize isoprene in cyclohexane gave polyisoprene of high cis-1,4 content having carbanion on each end of

the polymer chains. Addition of this reaction mixture to
tetrahydrofuran saturated with carbon dioxide led to carboxylation
at the lithio sites. Conversion of the product to
tetrabutylammonium salts and use of these salts to initiate
pivalolactone gave the desired ABA triblock copolymer.

Synthesis of the block-graft polymers also started with the
difunctional initiator described above. The α,ω-dilithio-cis-1,4-
polyisoprene was then lithiated further by reaction with sec-butyl
lithium in the presence of tetramethylethylenediamine. This
resulted in the formation of lithio sites randomly spaced along the
polymer chains. The complete reaction sequence is given below.

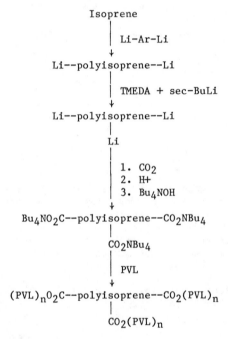

These block-graft polymers were easily melt-processable, being
readily melt spun into elastic fibers of good strength, high
elongation, and high resilience. For good properties, 35% of PVL or
higher appeared desirable. As PVL content was increased, elongation
was reduced, which would be expected. Though it appeared desirable
to have at least three or four poly(PVL) segments per chain, little
difference was observed when larger numbers of segments were
employed.

Most outstanding properties of these products were high
resilience and good resistance to stress decay. Resilience is
illustrated in Figure 1 which shows the stress-strain relationship
of a poly[(pivalolactone-b-isoprene-b-pivalolactone)-g-pivalo-
lactone] fiber as it was stretched 300% and then allowed to relax.
The shaded area is the work lost as the fiber was loaded and then
unloaded. This area amounts to 13% of the total, which shows that
work recovered was 87%. Such high resilience compares very
favorably with that of chemically-cured natural rubber.

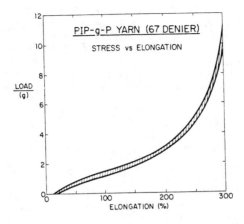

Figure 1.   Stress-strain relationship of PIP-g-P fiber.

Stress-decay is illustrated in Figure 2. Here the fiber was stretched 300% and the change is stress requird to maintain that elongation measured as a function of time.

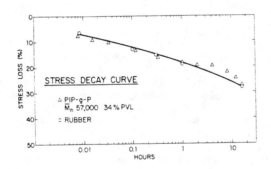

Figure 2.   Stress-decay curve of PIP-g-P fiber.

Fall off in stress as compared to a sulfur-cured gum rubber control is compared. Since both decrease at about the same rate, it is apparent the crystalline cross-links in the block-graft copolymer are as effective as the chemical cross-links in rubber.
The stress-decay data given in Figure 2 for poly[pivalolactone-b-isoprene-b-pivalolactone)-g-pivalolactone] was obtained on fibers protected with large amounts of antioxidant. If air autooxidation is an important factor in strength loss, use of a polymer backbone

for which autooxidation, minor or existent, should lead to much improved stress decay. Pursuit of this point led to work on poly(isobutylene-g-pivalolactone) compositions.(28,29)

Isobutylene was copolymerized with methylbenzenes by cationic methods to give products containing 1-2% of the comonomer. Attachment of polypivalolactone grafts was accomplished using the same chemistry as that for isoprene block-graft copolymers.

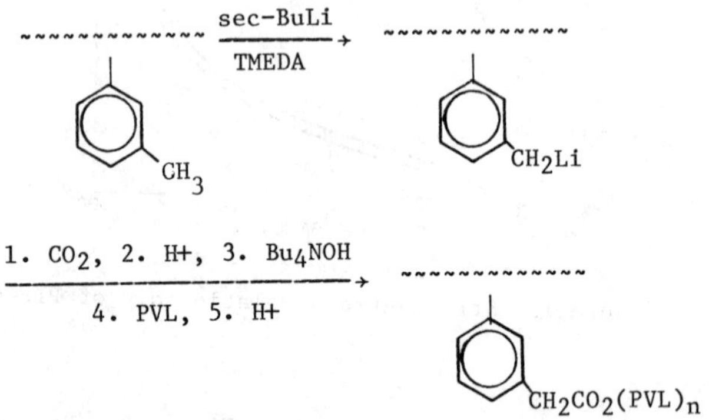

The poly(isobutylene-co-methylstyrene)-g-pivalolactone polymers were sticky semisolids when amounts of PVL were low, ranged through a rubbery-thermoplastic region as amount of PVL was increased, and were hard, non-elastic solids when PVL content was over 60%. Films obtained by melt-pressing of polymers containing 20-60% PVL were as strong or stronger than conventionally cured polyisobutylene. As is true with all other PVL containing compositions, strength increased markedly upon orientation by drawing.

The graft copolymers could be spun into fibers at 250-260°C. These fibers, after orientation by drawing, were not as strong as fibers from the isoprene compositions, but they had tenacities in the 0.4-0.5 g/d range. Stress-decay on these fibers, determined the same way as for those described in Figure 2, was only 11% for the first hour with a further loss of only 5% in the next 22 hours. Since this loss is much less than that of the isoprene block-graft fiber, which lost over 20% in 10 hours, it appears that autooxidation is a significant factor in long term stress decay.

Acknowledgments

Deep appreciation is expressed to Professor J. E. McGrath for the invitation to participate in this symposium and to R. E. Putscher of the Central Research and Development Department of the DuPont Company for library assistance.

Literature Cited

1. Fischer, N. Thesis, University of Paris, 1959.
2. Reynolds, R. J. W. and Vickers, E. J., British Patent 766.347, Jan. 23, 1967 assigned to Imperial Chemical Industries.
3. Markevich, M. A., Pakhomera, L. K., and Enikolopyan, N. S.,

Proc. Acad. Sci. USSR, Phys. Chem. Sect., 1970, 187 (1-3), 499.

4. Jaacks, V., and Mathes, N., Makromol. Chem., 1970, 131, 295.
5. Jaacks, V., and Mathes, N., Makromol. Chem., 1971, 142, 209.
6. Mayne N. R., Chem. Tech., 1972, 728.
7. Wilson, D. R., and Beaman, R. G., J. Polymer Science, 1970, A-1, 8, 2161.
8. Hall, H. K., Jr., Macromolecules, 1969, 2, 488.
9. (a) Boyesso, B., Nakase, Y., and Sunner, S., Acta, Chem. Scand., 1966, 20, 803; (b) Mansson, M., Nakase, Y., and Sunner, S., Acta. Chem. Scand., 1968, 22, 171.
10. Lueck, C. H., Beste, L. F., and Hall, H. K., Jr., J. Phys. Chem., 1963, 67, 972.
11. Hall, H. K., Jr., J. Org. Chem., 1964, 29, 3539.
12. Beste, L. F., and Hall, H. K., Jr., J. Phys. Chem., 1964, 68, 269.
13. Bigdeli, C. E., and Lenz, R. W., Macromolecules, 1978, 11, 493.
14. Eisenbach, C. D., and Lenz, R. W., Macromolecules, 1976, 9, 227.
15. Sundet, S. A., Thamm, R. C., Meyer, J. M., Buck, W. H., Caywood, S. W., Subramanian, P. M., and Anderson, B. C., Macromolecules, 1976, 9, 371.
16. Sharkey, W. H., Proceedings of China-U.S. Bilateral Symposium on Polymer Chemistry and Physics, Oct. 5-10, 1979, Beijing, China, Science Press, Beijing, China, 1981, pp. 278.
17. Oosterhoff, H. A., Polymer, 1974, 15, 49.
18. Cornibert and Marchessault, Macromolecules, 1975, 8, 296.
19. Lenz, R. W., Dror, M., Jorgensen, R., and Marchessault, R. H., Polymer Engineering and Science, 1978, 18, 937.
20. Sundet, S. A., Macromolecules, 1978, 11, 146.
21. Caywood, S. W., Rubber Chemistry and Technology, 1977, 50, 127.
22. Thamm R. C., and Buck, W. H., Polym. Prepr., Am. Chem. Soc., Div. Polym. Chem., 1976, 17(1), 205.
23. Thamm, R. C., Buck, W. H., Caywood, S. W., Meyer, J. M., and Anderson, B. C., Angew. Macromol. Chem., 1977, 58/59, 345.
24. Calhoun G. J., and Hewett, W. A., U. S. Patent 3,052,657 (1960) (to Shell Oil Company).
25. Foss, R. P., Jacobson, D.K. , Cripps, H.N., and Sharkey, W. H., Macromolecules, 1976, 9, 373.
26. Foss, R. P., Jacobson, H. W., Cripps, M. N., and Sharkey, W. H., Macromolecules, 1979, 12, 1210.
27. Foss, R. P., Jacobson, H. W., and Sharkey, W. H., Macromolecules, 1977, 10, 287.
28. Harris, J. F. Jr., and Sharkey, W. H., Macromolecules, 1977, 10, 503.
29. Foss, R. P., Harris, J. F. Jr., and Sharkey, W. H., Polymer Science and Technology, 1977, 10, 159.

RECEIVED April 18, 1985

INDEXES

Author Index

Subject Index

A

Production by Hilary Kanter
Indexing by Deborah H. Steiner
Jacket design by Pamela Lewis

Elements typeset by Hot Type Ltd., Washington, D.C.
Printed and bound by Maple Press Co., York, Pa.